CONTINUOUS GROUPS
OF
TRANSFORMATIONS

CONTINUOUS GROUPS

OF

TRANSFORMATIONS

By

LUTHER PFAHLER EISENHART

Dod Professor of Mathematics
in Princeton University

PRINCETON

PRINCETON UNIVERSITY PRESS

LONDON: HUMPHREY MILFORD

OXFORD UNIVERSITY PRESS

1933

PRINTED IN THE U. S. A.
BY THE MAPLE PRESS COMPANY, YORK, PA.

PREFACE

The study of continuous groups of transformations inaugurated by Lie resulted in the development of an extensive theory by him and under his inspiration by Engel, Killing, Scheffers and Schur. Cartan in his thesis placed many of these results on a more rigorous basis. Bianchi and Fubini developed geometrical applications of the theory. This chapter of the history closed about thirty years ago. A new chapter began about ten years ago with the extended studies of tensor analysis, Riemannian geometry and its generalizations, and the application of the theory of continuous groups to the new physical theories.

This book sets forth the general theory of Lie and his contemporaries and the results of recent investigations with the aid of the methods of the tensor calculus and concepts of the new differential geometry. The first three chapters contain in the main the results of the first period. Chapter Four is devoted to the theory of the adjoint group and the sequence of theorems basic to the characterization of semi-simple groups, as developed by Cartan and recently by Weyl and Schouten. Although geometrical ideas are used throughout the book, Chapter Five contains the geometrical applications of the theory both in the space of the transformations and in the group-space, and here are to be found particularly the concepts of the new differential geometry. The theory of contact transformations with applications to geometry and mechanics is set forth in the closing chapter.

Of the many exercises in the book some involve merely direct applications of the formulas of the text, but many of them constitute extensions of the theory which would properly be included as portions of a more extensive treatise. References to the sources of these exercises are given for the benefit of the reader. All references are to the papers listed in the Bibliography.

In the writing of this book I have had invaluable assistance and criticism by my colleagues Professors Bohnenblust, Knebelman and Robertson, and by my former student Dr. T. S. Graham particularly in connection with Chapter Four. The galley proof has been read by the students in my course this year and revised on the basis of their suggestions; Mr. F. D. Cubello has aided me considerably while seeing the book through the press.

LUTHER PFAHLER EISENHART.

May, 1933.

CONTENTS

CHAPTER I
THE FUNDAMENTAL THEOREMS

Chapter IV
THE ADJOINT GROUP

Chapter V
GEOMETRICAL PROPERTIES

Chapter VI
CONTACT TRANSFORMATIONS

65. Homogeneous contact transformations of maximum rank. 263
66. Geometrical properties of continuous groups of maximum rank.
      Waves 267
67. Application to geodesics of a Riemannian space 274
68. Application to dynamics 277
69. Function groups. 281
70. Homogeneous function groups. 287

     Bibliography 293

     Index 297

CHAPTER I

THE FUNDAMENTAL THEOREMS

1. Certain systems of partial differential equations. Mixed systems. In this section and the next we establish certain theorems concerning differential equations which are used later.

Consider a system of partial differential equations

$$(1.1) \quad \frac{\partial \theta^\alpha}{\partial x^i} = \psi_i^\alpha(\theta^1, \cdots, \theta^m; x^1, \cdots, x^n) \equiv$$

$$\psi_i^\alpha(\theta; x) \quad (\alpha = 1, \cdots, m; i = 1, \cdots, n),$$

where the ψ's are functions of the θ's and the x's; it is understood that the following treatment applies to a domain in which the ψ's are analytic in the θ's and x's. These equations are equivalent to the system of *total* differential equations

$$(1.2) \quad d\theta^\alpha = \psi_i^\alpha dx^i \quad (\alpha = 1, \cdots, m; i = 1, \cdots, n).$$

In these equations we have made use of the convention, to be used throughout, that when the same index appears twice in a term this term stands for the sum of the terms obtained by giving the index each of its values; thus the right-hand member of (1.2) stands for the sum of n terms as i takes the values 1 to n.

The conditions of integrability of (1.1) are

$$(1.3) \quad \frac{\partial \psi_i^\alpha}{\partial x^j} + \frac{\partial \psi_i^\alpha}{\partial \theta^\beta}\psi_j^\beta = \frac{\partial \psi_j^\alpha}{\partial x^i} + \frac{\partial \psi_j^\alpha}{\partial \theta^\gamma}\psi_i^\gamma \quad \begin{pmatrix} i, j = 1, \cdots, n; \\ \alpha, \beta, \gamma = 1, \cdots, m \end{pmatrix};$$

we remark that in these equations β and γ are summed from 1 to m. If these equations are satisfied identically, the system (1.1) is said to be *completely integrable*. In this case the solution is expressible in the form

$$(1.4) \quad \theta^\alpha = c^\alpha + \left(\frac{\partial \theta^\alpha}{\partial x^i}\right)_0 (x^i - x_0^i) + \frac{1}{2}\left(\frac{\partial^2 \theta^\alpha}{\partial x^i \partial x^j}\right)_0 (x^i - x_0^i)(x^j - x_0^j)$$

$$+ \cdots,$$

1

where $\left(\dfrac{\partial \theta^\alpha}{\partial x^i}\right)_0 = \psi_i^\alpha(c; x_0)$ and the other coefficients are obtained by differentiating (1.1) and evaluating by equating the x's to the x_0's and the θ's to the c's. Thus for the domain of the x's for which the series (1.4) are convergent we have a solution of (1.1) determined by m constants. We denote such a solution by

$$(1.5) \qquad \theta^\alpha = \varphi^\alpha(x^1, \cdots, x^n; c^1, \cdots, c^m).$$

If equations (1.3) are not satisfied identically, we have a set F_1 of equations, which establish conditions upon the θ's as functions of the x's. If we differentiate each of these equations with respect to the x's and substitute for $\dfrac{\partial \theta^\alpha}{\partial x^i}$ from (1.1), either the resulting equations are a consequence of the set F_1 or we get a new set F_2. Proceeding in this way we get a sequence of sets $F_1, F_2, \ldots,$ of equations, which must be compatible, if equations (1.1) are to have a solution. If one of these sets is not a consequence of the preceding sets, it introduces at least one additional condition. Consequently, if the equations (1.1) are to admit a solution, there must be a positive integer N such that the equations of the $(N+1)$th set are satisfied because of the equations of the preceding N sets; otherwise we should obtain more than m independent equations which would imply a relation between the x's. Moreover, from this argument it follows that $N \leq m$.

Conversely, suppose that there is a number N such that the equations of the sets

$$(1.6) \qquad F_1, \ldots, F_N$$

are compatible and each set introduces one or more conditions independent of the conditions imposed by the equations of the other sets, and that all of the equations of the set

$$(1.7) \qquad F_{N+1}$$

are satisfied identically because of the equations of the sets (1.6). Assume that there are $p(<m)$ independent conditions imposed by (1.6), say $G_\gamma(\theta, x) = 0$. Since the jacobian matrix $\left\|\dfrac{\partial G_\gamma}{\partial \theta^\alpha}\right\|$ is of rank p, these equations may be solved for p of the θ's in terms of the remaining θ's and the x's, and the equations are then of the form (by suitable numbering)

$$(1.8) \qquad \theta^\sigma - \varphi^\sigma(\theta^{p+1}, \cdots, \theta^m, x) = 0 \quad (\sigma = 1, \cdots, p).$$

From these equations we have by differentiation

$$\frac{\partial \theta^\sigma}{\partial x^i} - \frac{\partial \varphi^\sigma \partial \theta^\nu}{\partial \theta^\nu \partial x^i} - \frac{\partial \varphi^\sigma}{\partial x^i} = 0 \quad (\nu = p+1, \cdots, m).$$

Replacing $\dfrac{\partial \theta^\sigma}{\partial x^i}$ by means of (1.1), we have

$$(1.9) \qquad \psi_i^\sigma - \frac{\partial \varphi^\sigma}{\partial \theta^\nu}\psi_i^\nu - \frac{\partial \varphi^\sigma}{\partial x^i} = 0,$$

which are satisfied because of the sets (1.6) and (1.7), as follows from the method of obtaining the latter. Accordingly we have by subtraction

$$(1.10) \qquad \frac{\partial \theta^\sigma}{\partial x^i} - \psi_i^\sigma - \frac{\partial \varphi^\sigma}{\partial \theta^\nu}\left(\frac{\partial \theta^\nu}{\partial x^i} - \psi_i^\nu\right) = 0.$$

From these equations it follows that, if the functions $\theta^{p+1}, \ldots,$ θ^m are chosen to satisfy the equations

$$(1.11) \qquad \frac{\partial \theta^\nu}{\partial x^i} = \bar{\psi}_i^\nu(\theta^{p+1}, \cdots, \theta^m, x),$$

where $\bar{\psi}_i^\nu$ is obtained from ψ_i^ν on replacing θ^σ $(\sigma = 1, \cdots, p)$ by their expressions (1.8), then equations (1.1) for $\alpha = 1, \cdots, p$ are satisfied by the values (1.8). Since the equations of the set F_1 are satisfied identically because of (1.8), it follows that equations (1.11) are completely integrable; for, the equations arising from expressing their conditions of integrability are in the set F_1 because of (1.9). Consequently there is a solution in this case and it involves $m - p$ arbitrary constants.

When $p = m$, we have in place of (1.8) $\theta^\alpha = \varphi^\alpha(x)$ and in place of (1.10) that the functions θ^α satisfy (1.1). In this case there are no constants of integration. Hence we have:

[1.1] *In order that a system of equations* (1.1) *admit a solution, it is necessary and sufficient that there exist a positive integer* $N(\leqq m)$ *such that the equations of the sets* $F_1, \ldots, F_N$ *are compatible for all values of the* x's *in a domain, and that the equations of the set* F_{N+1} *are satisfied because of the former sets; if* p *is the number of independ-*

ent equations in the first N sets, the solution involves m − p arbitrary constants.*

It is evident from the above considerations that when an integer N exists such that the conditions of the theorem are satisfied, they are satisfied also for any integer larger than N. However, it is understood in the theorem and in the various applications of it that N is the least integer for which the conditions are satisfied.

The above theorem can be applied also to the case where there are certain p functional relations between the θ's and x's which must be satisfied in addition to the differential equations (1.1), say

$$(1.12) \qquad\qquad f^{\sigma}(\theta; x) = 0 \qquad\qquad (\sigma = 1, \cdot\cdot\cdot, p).$$

We say that (1.1) and (1.12) constitute a *mixed system*. In this case we denote by F_0 this set of conditions, and include in the set F_1 of the theorem also such conditions as arise from F_0 by differentiation and substitution from (1.1). Then the theorem proceeds as above with the understanding that the sets $F_0, F_1, \ldots, F_N$ shall be compatible, and that the set F_{N+1} shall be satisfied because of the former.

We return to the consideration of the case when (1.1) is completely integrable. Since the functions φ^{α} in (1.5) are such that

$$\varphi^{\alpha}(x_0^1, \cdot\cdot\cdot, x_0^n; c^1, \cdot\cdot\cdot, c^m) = c^{\alpha},$$

it follows that the jacobian

$$(1.13) \qquad\qquad \left|\frac{\partial\varphi}{\partial c}\right| \equiv \frac{\partial(\varphi^1, \cdot\cdot\cdot, \varphi^m)}{\partial(c^1, \cdot\cdot\cdot, c^m)}$$

is different from zero. Consequently equations (1.5) may be solved for the c's, which we indicate by

$$(1.14) \qquad\qquad f^{\alpha}(\theta; x) = c^{\alpha}, \qquad\qquad (\alpha = 1, \cdot\cdot\cdot, m).$$

Each of the functions f^{α} is an integral of (1.1) in the sense that, if $^{\alpha}$ is differentiated with respect to x^i, the result is reducible to zero by (1.1). In this sense any function of the f's is an integral.

If in any one of equations (1.14) we replace the θ's by their expressions (1.5), we obtain an identity in the x's and c's. Conse-

---

*For an historical account of this theorem see 1927, 1, p. 17. References are to the Bibliography at the end of the book.

quently each of the f's is a solution of the system of n differential equations

$$(1.15) \qquad \frac{\partial f}{\partial x^i} + \frac{\partial f}{\partial \theta^\alpha} \psi_i^\alpha = 0 \quad (i = 1, \cdots, n; \alpha = 1, \cdots, m).$$

Moreover any function of the f's is a solution of these equations. Suppose that in addition to the functions f^α in (1.14) we have another solution f^{m+1} of (1.15), and consider the matrix of $m + 1$ rows and $m + n$ columns

$$\left\| \frac{\partial f^\sigma}{\partial x^1}, \cdots, \frac{\partial f^\sigma}{\partial x^n}, \frac{\partial f^\sigma}{\partial \theta^1}, \cdots, \frac{\partial f^\sigma}{\partial \theta^m} \right\| \quad (\sigma = 1, \cdots, m + 1).$$

In consequence of (1.15) any determinant of order $m + 1$ of this matrix is zero. Hence f^{m+1} is a function of $f^1, \ldots, f^m$, which are independent of one another as previously shown.

Conversely, suppose that $f^1, \ldots, f^m$ are m independent solutions of (1.15) and that the above matrix for $\sigma = 1, \cdots, m$ is of rank m. Then the determinant $\left| \frac{\partial f^\alpha}{\partial \theta^\beta} \right|$ for $\alpha, \beta = 1, \cdots, m$ is different from zero, since in consequence of (1.15) every other determinant of order m of this matrix is a multiple of $\left| \frac{\partial f^\alpha}{\partial \theta^\beta} \right|$. Hence if we consider the corresponding system of equations (1.14), where the c's are constants, they define implicitly the θ's as functions of the x's and the c's. Accordingly we have

$$\frac{\partial f^\alpha}{\partial x^i} + \frac{\partial f^\alpha}{\partial \theta^\beta} \frac{\partial \theta^\beta}{\partial x^i} = 0 \quad (\alpha, \beta = 1, \cdots, m; i = 1, \cdots, n)$$

Subtracting these equations from (1.15), we have

$$\frac{\partial f^\alpha}{\partial \theta^\beta} \left(\frac{\partial \theta^\beta}{\partial x^i} - \psi_i^\beta \right) = 0,$$

which are equivalent to (1.1), since $\left| \frac{\partial f^\alpha}{\partial \theta^\beta} \right| \neq 0$. If we call (1.15) the system *associated* with (1.1), we have:

[1.2] *Given a system of equations* (1.1) *and the associated system* (1.15), *if m independent solutions of the latter are equated to arbitrary constants they define implicitly a solution of* (1.1); *and a solution of* (1.1) *determines m independent solutions of the associated system.*

Also it follows from the preceding discussion that:

[1.3]   *If* $F(x; \theta)$ *is any solution of* (1.15) *and the θ's are replaced by a solution of* (1.1), *then* $F(x; \theta)$ *is a constant and thus is an integral of equations* (1.1).

2. Linear operators.   Complete systems of linear partial differential equations.   Each of the r expressions

$$(2.1) \qquad X_a f \equiv \xi_a^i \frac{\partial f}{\partial x^i} \equiv \xi_a p_i \quad (a = 1, \cdots, r; i = 1, \cdots, n),$$

where the ξ's are functions of the x's, is called a *linear operator;* hereafter p_i will denote $\dfrac{\partial f}{\partial x^i}$.   By definition

$$X_a X_b f = \xi_a^i \frac{\partial}{\partial x^i}\left(\xi_b^j \frac{\partial f}{\partial x^j}\right) = \xi_a^i \frac{\partial}{\partial x^i}(\xi_b^j)\frac{\partial f}{\partial x^j} + \xi_a^i \xi_b^j \frac{\partial^2 f}{\partial x^i \partial x^j} =$$

$$\frac{\partial f}{\partial x^i} X_a \xi_b^i + \xi_a^i \xi_b^j \frac{\partial^2 f}{\partial x^i \partial x^j}.$$

Consequently

$$(2.2) \qquad (X_a, X_b)f \equiv X_a X_b f - X_b X_a f = (X_a \xi_b^i - X_b \xi_a^i)p_i.$$

The quantity $(X_a, X_b)f$ thus defined is called the *Poisson operator;* it is also called the *commutator* of the operators $X_a f$ and $X_b f$.   As an immediate consequence of (2.2) we have

$$(2.3) \qquad\qquad (X_a, X_b)f = -(X_b, X_a)f.$$

Also it can be shown by direct calculation that the following *Jacobi identity* holds

$$(2.4) \quad ((X_a, X_b), X_c)f + ((X_b, X_c), X_a)f + ((X_c, X_a), X_b)f = 0$$
$$(a, b, c = 1, \cdots, r),$$

whatever be the functions ξ_a^i, ξ_b^i, and ξ_c^i.

If we put

$$(2.5) \qquad\qquad X_a' f = \lambda_a^c X_c f,$$

where the λ's are any functions of the x's, then

$$(X_a', X_b')f = (\lambda_a^c X_c, \lambda_b^d X_d)f$$

$$(2.6) \qquad = \lambda_a^c X_c(\lambda_b^d) X_d f - \lambda_b^d X_d(\lambda_a^c) X_c f + \lambda_a^c \lambda_b^d (X_c, X_d)f$$

$$= [\lambda_a^c X_c(\lambda_b^d) - \lambda_b^c X_c(\lambda_a^d)] X_d f + \lambda_a^c \lambda_b^d (X_c, X_d)f.$$

In order to obtain this last expression we replaced the repeated index d by c and the repeated index c by d; evidently this is possible since the letter indicating a repeated index, sometimes called a *dummy* index, is of no significance.

Consider now the system of homogeneous linear partial differential equations

$$(2.7) \qquad X_a f = \xi_a^i p_i = 0 \quad (a = 1, \cdots, r; i = 1, \cdots, n),$$

for which the rank of the matrix

$$(2.8) \qquad M = \|\xi_a^i\|$$

is r, that is, the equations (2.7) are independent. If $r = n$, the only solution of (2.7) is evidently $f = $ const. If $r < n$, there is a possibility of solutions other than the trivial one $f = $ const. From (2.2) it is evident that any solution of (2.7) satisfies

$$(2.9) \qquad (X_a, X_b)f = 0 \qquad (a, b = 1, \cdots, r).$$

If

$$(2.10) \qquad (X_a, X_b)f = \gamma_{ab}^c X_c f \quad (a, b, c, = 1, \cdots, r),$$

where the γ's are functions of the x's, then the system (2.7) and (2.9) is in fact equivalent to (2.7). We adjoin to (2.7) all the equations obtained by equating to zero those commutators which are not expressible in the form (2.10), and in this way we have $s(\geqslant r)$ independent equations. If $s > r$, we repeat this process and obtain a system of $t(\geqslant s)$ equations. If $t > s$, we continue the process and so on. Finally we have either n independent equations in which case the only solution of (2.7) is $f = $ const., or we get a number u less than n for which a, b, c in (2.10) take the values $1, \ldots, u$. In this case we say that the system is *complete* of order u. Hence the solution of a system of the form (2.7) is that of the solution of complete systems in which (2.10) hold.

If (2.7) is a complete system, so also is

$$(2.11) \qquad X_b' f = \lambda_b^a X_a f = 0,$$

where the λ's are any functions of the x's such that the determinant

$|\lambda^a_b|$ is not zero. For, from (2.10) it follows that the right-hand member of (2.6) is linear in $X_1 f, \ldots, X_r f$, and since the latter are expressible from (2.11) linearly in the $X' f$'s, we have from (2.6) expressions of the form (2.10), which proves the statement.

Since the rank of M (2.8) is r, there is no loss in generality in assuming that $|\xi^b_a|$, for $a, b = 1, \cdots, r$ is different from zero. Hence we may solve equations (2.7) for $p_1, \ldots, p_r$ and write the result in the form

$$(2.12) \qquad X'_a f = p_a + \psi^t_a p_t = 0$$

$$(a = 1, \cdots, r; t = r + 1, \cdots, n).$$

Evidently these are of the form (2.11) and hence form a system equivalent to (2.7); when a complete system is expressed in this form, it is called *Jacobian*. If we put $X'_a f = \xi'^i_a p_i$, we have that

$$(2.13) \qquad \xi'^b_a = \delta^b_a, \quad \xi'^p_a = \psi^p_a$$

$$(a, b = 1, \cdots, r; p = r + 1, \cdots, n),$$

where δ^b_a, called the Kronecker deltas, are defined by

$$(2.14) \qquad \delta^b_a = 1 \text{ or } 0, \text{ as } a = b \text{ or } a \neq b;$$

the Kronecker deltas are used frequently throughout this book with different indices but they always have the values (2.14) for the indices involved.

Analogously to (2.10) we have

$$(X'_a, X'_b) f = \gamma'^c_{ab} X'_c f.$$

From (2.2) and (2.13) we have

$$(X'_a, X'_b) f = (X'_a \xi'^c_b - X'_b \xi'^c_a) p_c + (X'_a \psi^t_b - X'_b \psi^t_a) p_t$$
$$= (X'_a \psi^t_b - X'_b \psi^t_a) p_t$$

$$(c = 1, \cdots, r; t = r + 1, \cdots, n).$$

Since $p_1, \ldots, p_r$ do not appear in this last expression, it follows that $\gamma'^c_{ab} = 0$, that is, for a complete system in the Jacobian form

$$(X'_a, X'_b) f = 0.$$

Hence

$$X'_a \psi^t_b - X'_b \psi^t_a = 0, \text{ that is,}$$

$$\frac{\partial \psi^p_b}{\partial x^a} + \psi^q_a \frac{\partial \psi^p_b}{\partial x^q} = \frac{\partial \psi^p_a}{\partial x^b} + \psi^q_b \frac{\partial \psi^p_a}{\partial x^q} \quad \left(\begin{array}{c} a, b = 1, \cdots, r; \\ p, q = r + 1, \cdots, n \end{array} \right).$$

Comparing these equations with (1.3), we see that the equations

$$(2.15) \qquad \frac{\partial x^p}{\partial x^a} = \psi_a^p \quad (a = 1, \cdots, r; p = r + 1, \cdots, n)$$

are completely integrable. Moreover equations (2.12) are the associated system of (2.15). Hence from the results of §1 we have that equations (2.12), and consequently (2.7), admit $n - r$ independent solutions, and that they do not admit more than $n - r$ independent solutions. Hence we have:

[2.1] *A complete system of r homogeneous linear partial differential equations of the first order in $n(>r)$ variables admits exactly $n - r$ independent solutions.*

3. Essential parameters in a set of functions. Consider a set of n functions f^i of n variables $x^1, \ldots, x^n$ and r parameters $a^1, \ldots, a^r$, say

$$(3.1) \quad f^i(x^1, \cdots, x^n; a^1, \cdots, a^r) \equiv f^i(x; a) \quad (i = 1, \cdots, n),$$

which are continuous in the x's and a's, as are also their derivatives with respect to these to as high an order as enters in the subsequent discussion. This requirement will be understood throughout the book. The parameters a^α are said to be *essential* unless it is possible to choose $r - 1$ functions of them, say $A^1, \ldots, A^{r-1}$, such that we have identically

$$f^i(x^1, \cdots, x^n; a^1, \cdots, a^r) = F^i(x^1, \cdots, x^n; A^1, \cdots, A^{r-1}).$$

Since the A's are functions of the a's, the jacobian matrix $\left\Vert \dfrac{\partial A}{\partial a} \right\Vert$ is at most of rank $r - 1$, and consequently a set of functions φ^α of the a's, not all identically zero, exist which satisfy the equations

$$\varphi^\alpha(a)\frac{\partial A^\sigma}{\partial a^\alpha} = 0 \quad (\alpha = 1, \cdots, r; \sigma = 1, \cdots, r - 1).$$

Hence the A's and any function of them satisfy the equation

$$(3.2) \qquad \varphi^\alpha(a)\frac{\partial f}{\partial a^\alpha} = 0.$$

Consequently the functions F^i satisfy (3.2), since the x's do not appear in the φ's, and hence the functions f^i satisfy (3.2). Conversely, suppose that the functions f^i satisfy an equation of the form (3.2). Evidently this equation admits $r - 1$ independent solutions which are functions of the a's alone, say $A^1, \ldots, A^{r-1}$, and any solution of (3.2) is a function of these A's. Consequently each of the functions f^i is a function of the x's and the A's, and therefore the a's are not essential.   Hence we have:

[3.1]   *A necessary and sufficient condition that the r parameters a^α in (3.1) be essential is that the functions f^i do not satisfy an equation of the form (3.2).*

By the same reasoning it follows that, if the functions f^i satisfy a complete system of $s(<r)$ equations

$$(3.3) \qquad \varphi_\sigma^\alpha(a)\frac{\partial f}{\partial a^\alpha} = 0 \quad (\alpha = 1, \cdots, r; \sigma = 1, \cdots, s),$$

then f^i are functions of the x's and of $r - s$ independent solutions of this system, which are functions of the a's, and consequently are expressible in terms of $r - s$ essential parameters.

It is our purpose now to obtain a criterion for the number of essential parameters.   From the foregoing it is evident that the parameters are essential, if the rank of the matrix

$$M_0 = \left\| \frac{\partial f^i}{\partial a^\alpha} \right\|$$

is r.   Suppose that its rank μ_0 is less than r.   Differentiating the first set of the following equations with respect to the x's, we have

$$(3.4) \qquad \varphi^\alpha\frac{\partial f^i}{\partial a^\alpha} = 0, \quad \varphi^\alpha\frac{\partial f^i,_j}{\partial a^\alpha} = 0, \quad f^i,_j \equiv \frac{\partial f^i}{\partial x^j}.$$

We denote by μ_1 the rank of the matrix

$$M_1 = \left\| \frac{\partial f^i}{\partial a^\alpha}, \frac{\partial f^i,_j}{\partial a^\alpha} \right\|.$$

Evidently $\mu_1 \geq \mu_0$.   If $\mu_1 = r$, equations (3.4) admit only the solutions $\varphi^\alpha = 0$ and consequently the parameters are essential. If we put

$$f^i,_{i_1 \cdots i_s} \equiv \frac{\partial^s f^i}{\partial x^{j_1} \cdots \partial x^{j_s}},$$

we have from (3.4) by repeated differentiation

$$(3.5) \qquad \varphi^\alpha\frac{\partial f^i,_{i_1 \cdots i_s}}{\partial a^\alpha} = 0,$$

and we denote by μ_s the rank of the matrix

$$(3.6) \qquad \left\| \frac{\partial f^i}{\partial a^\alpha}, \frac{\partial f^i, \, i_1}{\partial a^\alpha}, \ldots, \frac{\partial f^i, \, i_1 \cdots i_s}{\partial a^\alpha} \right\|.$$

We get thus a sequence of ranks

$$(3.7) \qquad \mu_0 \leq \mu_1 \leq \mu_2 \cdots \leq \mu_s \cdots \leq r.$$

If any μ is equal to r, the φ's are zero and the r parameters are essential.

Suppose that $\mu_{s-1} < r$ and that $\mu_s = \mu_{s-1}$, then we shall show that $\mu_{s+1} = \mu_s$. Since $\mu_s = \mu_{s-1}$, it follows that

$$\frac{\partial f^i, \, i_1 \cdots i_s}{\partial a^\alpha} = \sum \lambda \frac{\partial f^i, \, i_1 \cdots i_\nu}{\partial a^\alpha} \qquad (\nu < s)$$

the sum on the right involving terms of the matrix M_{s-1}. Differentiating with respect to x^k, we have that the terms $\dfrac{\partial f^i, \, i_1 \cdots i_s k}{\partial a^\alpha}$ are expressible linearly in terms of the members of the matrix M_{s-1}, which proves the statement. Hence the terms of the sequence (3.7) continually increase and attain the maximum r, in which case the r parameters are essential, or they cease increasing after a certain one $\mu_s (<r)$, in which case they are not essential. We shall show that in the latter case μ_s of the parameters are essential.

If $\mu_s < r$, the equations (3.4) and the sequence up to and including (3.5) being of rank μ_s, $r - \mu_s$ of the φ's may be chosen arbitrarily and the others are then determined. Let $\varphi^1, \ldots, \varphi^{r-\mu_s}$ be these functions and take them as functions of the a's but not the x's. Then

$$(3.8) \qquad \varphi^\sigma = \lambda_\rho^\sigma \varphi^\rho \quad \left(\begin{matrix} \rho = 1, \cdots, r - \mu_s; \\ \sigma = r - \mu_s + 1, \cdots, r \end{matrix} \right),$$

where the λ's are at most functions of the a's and x's.

If we differentiate with respect to x^k, the sequence of equations (3.4), . . . , (3.5), and remark that

$$\varphi^\alpha \frac{\partial f^i, \, i_1 \cdots i_s k}{\partial a^\alpha} = 0,$$

we obtain equations which result from (3.4) and (3.5) when

each φ^α is replaced by $\dfrac{\partial \varphi^\alpha}{\partial x^k}$, and consequently we have

$$\frac{\partial \varphi^\sigma}{\partial x^k} = \lambda_\rho^\sigma \frac{\partial \varphi^\rho}{\partial x^k},$$

from which it follows that φ^σ $(\sigma = r - \mu_s + 1, \cdots, r)$ do not involve the x's, since φ^ρ for $(\rho = 1, \cdots, r - \mu_s)$ were chosen as functions of the a's alone. There are accordingly $r - \mu_s$ independent equations (3.2). These may be obtained by taking $\varphi^1 = 1$, $\varphi^2 = \cdots = \varphi^{r-\mu_s} = 0$; $\varphi^2 = 1$, $\varphi^1 = \varphi^3 = \cdots = \varphi^{r-\mu_s} = 0$ and so on. These independent equations form a complete system. For, the commutator (§2) of any two equated to zero admits f^i as solutions and hence it must be a linear combination of the given equations. Hence the functions are expressible in terms of μ_s essential parameters, and we have:

[3.2] *The number of essential parameters in terms of which the functions $f^i(x; a)$ are expressible is equal to the maximum number attained in the sequence (3.7).*

Exercises

1. Find the mixed system of equations (1.1) for which $u = ax + by$, $v = axy$, $w = bx^2 + ay^2$ is the general solution, a and b being arbitrary constants, and determine the system F_0 for this case.

2. When the functions ψ_i^α in (1.1) are linear and homogeneous in the θ's, the equations of the sets $F_1, \ldots, F_N$ have this property, and in order that (1.1) admit a solution it is necessary and sufficient that there exist a positive integer $N(\leq m)$ such that the rank of the matrix of the sets $F_1, \ldots, F_N$ is $m - q(q \gtrless 1)$ and that this be the rank also of the sets $F_1, \ldots, F_{N+1}$.

3. A complete system admitting p independent functions $\varphi^1, \ldots, \varphi^p$ as solutions is obtained by choosing for ξ_a^i for $a = 1, \cdots, n - p$ any $n - p$ independent solutions of the equations

$$\xi^i \frac{\partial \varphi^\delta}{\partial x^i} = 0 \qquad\qquad (\delta = 1, \cdots, p);$$

all complete systems admitting the same solution are in the relation (2.11).

4. Find a complete system admitting the solution

$$x^1 x^2 + x^3, \quad x^3 + x^4.$$

5. Find the most general complete system in 3 variables admitting the solution $(x^1)^2 + (x^2)^2 + (x^3)^2$.

6. If for a complete system of r equations in n variables a number $s(<n - r)$ of independent solutions are known ,say $\varphi^1, \ldots, \varphi^s$, and the transformation of coordinates

$$x'^a = \varphi^a, \ x'^t = x^t \quad (a = 1, \cdots, s; \ t = s + 1, \cdots, n)$$

is effected, the new system involves only $x'^{s+1}, \ldots, x'^n$ as independent variables and $x'^1, \ldots, x'^s$ as parameters.

7. If the r equations (2.7) form a complete system, the first $r - 1$ of these equations admits at most $n - r + 1$ independent solutions, and there are $n - r$ independent functions of the latter which are solutions of the rth equation.

8. Show that the following two systems are complete and apply the result of Ex. 7 to determine the solution of each system:

(1) $\qquad\qquad\qquad\qquad p_1 = 0, \quad x^i p_i = 0 \qquad\qquad\qquad (i = 1, 2, 3);$

(2) $\qquad\qquad x\dfrac{\partial f}{\partial x} + y\dfrac{\partial f}{\partial y} = 0, \quad y\dfrac{\partial f}{\partial y} + z\dfrac{\partial f}{\partial z} = 0.$

4. Groups and groups of transformations. Any n independent variables x^i, for $i = 1, \cdots, n$, may be thought of as the coordinates of an n-dimensional space V_n in the sense that each set of values of the variables defines a point of V_n, that is, an ordered set of numbers defines a point.

Let $f^i(x^1, \ldots, x^n; a^1, \ldots, a^r)$ be n independent functions of the x's and r essential parameters $a^1, \ldots, a^r$, where r is finite. A necessary and sufficient condition that the functions be independent is that the jacobian of the f's with respect to the x's be not identically zero; the condition that the a's be essential is given in §3. The equations

$$(4.1) \quad x'^i = f^i(x^1, \cdots, x^n; a^1, \cdots, a^r) \equiv f^i(x; a) \quad (i = 1, \cdots, n)$$

for each set of values of the a's define a transformation of a point $P(x)$ in V_n into a point $P'(x')$. If the values of the x's and a's are such that the jacobian of the f's with respect to the x's is not zero, that is,

$$(4.2) \qquad\qquad\qquad \left|\frac{\partial f}{\partial x}\right| \neq 0,$$

in accordance with the theory of implicit functions* equations (4.1) may be solved uniquely for the x's, at least in the neighborhood of a point for which (4.2) holds, so that we have

$$(4.3) \qquad\qquad\qquad x^i = \bar{f}^i(x'; a) \qquad\qquad (i = 1, \cdots, n).$$

These equations define the inverse transformation of P' into P.

---

* *Fine*, 1927, 4, p. 253.

It is understood that the functions f^i are continuous in the x's and in the a's, and that the same is true of their derivatives of such orders as arise in the subsequent discussion. From this it follows that if $P'(x')$ is the transform of a point $P(x)$ for particular values of the a's, then for small changes in the values of the a's we get points in the neighborhood of P'. Accordingly we say that the transformations are *continuous*.

We indicate the transformations (4.1) and (4.3), respectively for a set of values a_1^α of the a's in the symbolic form

$$(4.4) \qquad T_{a_1} x = x', \qquad T_{a_1}^{-1} x' = x,$$

that is, T_{a_1} is the operation which sends a point $P(x)$ into a point $P'(x')$, the latter depending upon P and the values a_1^α. If we apply a transformation T_{a_1} and then another T_{a_2} we obtain a transformation which we indicate by $T_{a_2} T_{a_1}$, which is called the *product* of the two.* It does not follow necessarily that the product of two transformations is a member of the set (4.1), that is, that there are values of the a's, say a_3^α, such that $T_{a_2} T_{a_1}$ and T_{a_3} have the same effect. From the definition of $T_{a_1}^{-1}$ it follows that

$$(4.5) \qquad T_{a_1}^{-1} T_{a_1} x = x, \qquad T_{a_1} T_{a_1}^{-1} x' = x',$$

that is, either product $T_{a_1}^{-1} T_{a_1}$ or $T_{a_1} T_{a_1}^{-1}$ leaves each point unaltered, which transformation is called the *identity*. However, it does not follow that there are values of the a's in (4.1) yielding either the inverse transformation or the identity.

Before proceeding further with a discussion of the particular operators T_a defined by equations such as (4.1), we consider a set of operators T_a abstractly. They constitute the elements of a class of objects. What in the above discussion we called the product of two operators constitutes an operation between two members of the class; we indicate the result of the operation on two elements T_a and T_b in this order by $T_b T_a$. Such a class, however its elements and the operation are defined, are said to form a *group* with respect to the operation under the following conditions:

1° If T_a and T_b are elements of the class, then $T_b T_a$ is uniquely defined and is an element of the class;

2° The associative law holds for the operation, thus

---

* Earlier writers indicated this product by $T_{a_1} T_{a_2}$, but many recent writers use the notation proposed.

$$T_c(T_b T_a) = (T_c T_b)T_a,$$

that is, the element given by the operation upon the element $T_b T_a$ and T_c is the same as when the operation is applied to T_a and $T_c T_b$;

3° There is in the class an element I called the identity such that $T_a I = I T_a = T_a$ for every element T_a;

4° For each element T_a of the class there is an element T_a^{-1} such that

$$T_a^{-1}T_a = T_a T_a^{-1} = I.$$

The question to what extent these conditions are independent has been investigated by various writers,* and for a discussion of this question the reader is referred to the articles indicated.

If only the conditions 1° and 2° of the above set are satisfied, the elements are said to form a *semi-group* for the operation.

As an example of a group we consider the set of all non-singular analytic transformations

$$x'^i = f^i(x^1, \cdots, x^n),$$

that is, those for which the jacobian $\left|\dfrac{\partial f}{\partial x}\right|$ is of rank n for the domain of the x's considered. Then from the theory of implicit functions† it follows that the above equations admit an inverse

$$x^i = \bar{f}^i(x'^1, \cdots, x'^n).$$

If T_1 and T_2 are given by $x' = f_1(x)$, $x' = f_2(x)$, then $T_2 T_1$ is given by $x' = f_2(f_1(x))$, and thus the conditions 1° is satisfied for the domain in which the functions so defined are regular; moreover, the jacobian of this function, being the product of the jacobians of f_1 and f_2, is not zero for those values of the x's for which these two functions satisfy this condition. Whenever the elements of a class are defined by equations, as in this case, the condition 2° is satisfied, because the process of substitution is associative. Since $x'^i = x^i$ are equations of the set, all the conditions are satisfied, and thus we have the group of analytic transformations. Since the functions f^i may be thought of as power series in the x's (including polynomials), the transformations involve an infinite number of arbitrary constants and consequently it is an *infinite continuous* group.

---

* *Cf. Dickson*, 1905, 2; *Huntington*, 1905, 3; *Moore*, 1905, 4.
† *Fine*, 1927, 4, p. 253.

Equations such as (4.1) involve a finite number of arbitrary constants, and when they define a group for continuous values of the a's, the group is said to be *finite* and *continuous*. Before we proceed to the determination of the conditions which (4.1) must satisfy in order that they define a group, we wish to point out that, when these conditions are realized, they may obtain only for a limited domain about the point to which any transformation of the group is applied. For such a point the functions f^i and their derivatives of required order must be continuous in the x's and a's. Furthermore, the functions must be such that for a value of the a's, say a_0^α, the equations (4.1) are $x'^i = x^i$, so that (4.2) must hold for these values of the a's, and consequently for values of the a's sufficiently near a_0^α, since we are dealing with continuous functions. Accordingly when we speak of a group of transformations in what follows it is understood that the statements hold for those values of the x's and the a's for which the conditions 1°, 2°, 3° and 4° hold.

We return to the consideration of equations (4.1) and observe that in order that these equations satisfy the condition 1° of a group it is necessary that for each set of values a_1^α and a_2^α there exist a set of values a_3^α, functions of the former, say

$$(4.6) \qquad a_3^\alpha = \varphi^\alpha(a_1; a_2) \qquad (\alpha = 1, \cdots, r),$$

such that the equations

$$(4.7) \qquad f^i[f(x; a_1); a_2] = f^i(x; a_3)$$

are identities in the x's, a_1's and a_2's, and thus for arbitrary values of the a_1's and a_2's we have

$$(4.8) \qquad T_{a_2} T_{a_1} = T_{a_3}.$$

When these conditions are satisfied, the inverse of any transformation is not necessarily a member of the set, that is, it is not necessarily true that to each set of values a^α there corresponds a set $\bar{a}^\alpha$, such that

$$(4.9) \qquad x^i = \bar{f}^i(x'; a) = f^i(x'; \bar{a}),$$

where the functions $\bar{f}^i$ are defined by (4.3), or symbolically

$$(4.10) \qquad T_a^{-1} x' = T_{\bar{a}} x' = x.$$

If this condition is satisfied, we have

$$(4.11) \qquad T_a T_{\bar{a}} x' = x', \qquad T_{\bar{a}} T_a x = x,$$

that is, when the inverse of each transformation is a member of the set, so also is the identity. Hence there is a set of values a_0^α, such that

$$a_0^\alpha = \varphi^\alpha(a; \bar{a}),$$

and

(4.12) $$f^i(x; a_0) = x^i.$$

Since the parameters in (4.1) are essential, there is only one set of values a_0^α yielding the identity. Thus T_{a_0} denotes the identity I, and from (4.11) we have

$$T_{a_0} = T_a T_{\bar{a}} = T_{\bar{a}} T_a,$$

so that in addition to (4.12) we have

(4.13) $$a_0^\alpha = \varphi^\alpha(a; \bar{a}) = \varphi^\alpha(\bar{a}; a)$$

and also

(4.14) $$a^\alpha = \varphi^\alpha(a; a_0) = \varphi^\alpha(a_0; a).$$

Hence when (4.6) are satisfied and the inverse of each transformation is a member of the set, the transformations form a finite *continuous group* G_r. If only (4.6) are satisfied, the transformations form a *finite, continuous semi-group*.

If T_{a_1} and T_{a_2} are two transformations of a semi-group, and $T_{a_1}^{-1}$ and $T_{a_2}^{-1}$ are their respective inverses, then $T_{a_1}^{-1} T_{a_2}^{-1} T_{a_2} T_{a_1}$ is the identity. From this result and (4.8) we have that

(4.15) $$T_{a_1}^{-1} T_{a_2}^{-1} = T_{a_3}^{-1},$$

and consequently

[4.1] *The inverses of the transformations of a semi-group constitute a semi-group.*

These two semi-groups are the same, when and only when they are groups.

If we denote by S the transformation defined by

(4.16) $$y^i = \psi^i(x^1, \cdots, x^n),$$

for which the jacobian $\left|\dfrac{\partial \psi}{\partial x}\right|$ is not identically zero, and we denote by S^{-1} its inverse, then the transformation

(4.17) $$\bar{T}_a = S T_a S^{-1}$$

is called the *transform of T_a by S*. It is readily shown that if a set of transformations satisfy (4.8) then $\overline{T}_{a_2}\overline{T}_{a_1} = \overline{T}_{a_3}$. Hence we have:

[4.2]   *If a set of transformations form a group, the transforms of these transformations form a group.*   (S-G)*

From (4.17) we have

$$(4.18) \qquad\qquad T_a = S^{-1}\overline{T}_a S$$

and consequently T_a is the transform of $\overline{T}_a$ by S^{-1}.

If the inverse of (4.16) is written

$$(4.19) \qquad\qquad x^i = \overline{\psi}^i(y^1, \cdots, y^n),$$

the equations of $\overline{T}_a$ are

$$(4.20) \qquad y'^i = \psi^i(f(x; a)) = \psi^i(f(\overline{\psi}(y); a)).$$

If instead of looking upon (4.16) as defining a point transformation in V_n we consider it to be a non-singular transformation of coordinates, then equations (4.20) are the equations of the transformations T_a in the coordinates y^i. Accordingly the foregoing theorems show that:

[4.3]   *The group property of a set of transformations is invariant under a change of coordinates.*   (S-G)

This result is evident from the abstract definition of a group as given, since it does not depend upon the particular coordinates used.

5. Fundamental differential equations of a group. In this section we determine the conditions which the functions f^i in (4.1) must satisfy so that the transformations form a group G_r, that is, so that there exist functions $\varphi^\alpha(a_1; a_2)$ for which

$$(5.1) \qquad f^i[f(x; a_1); a_2] \equiv f^i(x; \varphi(a_1; a_2)).$$

In the first place we shall show that

$$(5.2) \qquad\qquad \left|\frac{\partial \varphi}{\partial a_1}\right| \neq 0, \qquad \left|\frac{\partial \varphi}{\partial a_2}\right| \neq 0.$$

---

* In what follows we shall be concerned primarily with groups as distinguished from semi-groups, but some of the theorems hold equally well for the latter more general class; we indicate this by the notation (S-G) at the end of a theorem.

From (5.1) we have

$$\frac{\partial f^i[f(x;a_1);a_2]}{\partial a_2^\alpha} = \frac{\partial f^i(x;\varphi)}{\partial \varphi^\beta}\frac{\partial \varphi^\beta}{\partial a_2^\alpha} \quad (\alpha,\beta = 1,\cdots,r).$$

If $\left|\dfrac{\partial \varphi}{\partial a_2}\right| \equiv 0$, there exist a set of functions A_2^α of the a_1's and a_2's such that

$$A_2^\alpha \frac{\partial \varphi^\beta}{\partial a_2^\alpha} = 0.$$

Then from the preceding set of equations we have, on replacing $f^i(x;a_1)$ by x'^i,

$$A_2^\alpha \frac{\partial f^i(x';a_2)}{\partial a_2^\alpha} = 0,$$

that is, the a's are not essential parameters in $f^i(x;a)$.

If we write (5.1) in the form

(5.3) $$f^i(x';a_2) \equiv f^i(x;\varphi(a_1;a_2)),$$

we have

$$\frac{\partial f^i}{\partial x'^i}\frac{\partial x'^i}{\partial a_1^\alpha} = \frac{\partial f^i}{\partial \varphi^\beta}\frac{\partial \varphi^\beta}{\partial a_1^\alpha}.$$

If $\left|\dfrac{\partial \varphi}{\partial a_1}\right| \equiv 0$, there exist functions A_1^α of the a_1's and a_2's, such that $A_1^\alpha \dfrac{\partial \varphi^\beta}{\partial a_1^\alpha} = 0$, and consequently

$$\frac{\partial f^i}{\partial x'^i}A_1^\alpha\frac{\partial x'^i}{\partial a_1^\alpha} = 0.$$

Since the jacobian $\left|\dfrac{\partial f(x';a_2)}{\partial x'}\right|$ is not identically zero (§4), we have

$$A_1^\alpha \frac{\partial f^i(x;a_1)}{\partial a_1^\alpha} = 0,$$

and consequently the a's are not essential parameters. This result follows indirectly from the second of (5.2), when we observe that the roles of the parameters a_1^α and a_2^α are interchanged in the set of inverse transformations (cf. 4.15). Consequently (5.2) is established.

Because of the second of (5.2) equations (4.6) can be solved for a_2^α as functions of a_1^α and a_3^α. When these expressions are substituted in (4.7), the latter become identities in the x's, a_1's and a_3's. Consequently we have from (4.7)

$$\frac{\partial f^i(x'; a_2)}{\partial x'^i} \frac{\partial x'^i}{\partial a_1^\alpha} + \frac{\partial f^i(x'; a_2)}{\partial a_2^\beta} \frac{\partial a_2^\beta}{\partial a_1^\alpha} = 0.$$

Because of (4.2) there exist functions $\psi_j^i(x'; a_2)$ defined by

$$\psi_i^k \frac{\partial f^i(x'; a_2)}{\partial x'^i} = \delta_j^k \quad (i, j, k = 1, \cdots, n),$$

where δ_j^k are Kronecker deltas (2.14), and consequently the above equations are equivalent to

$$\frac{\partial x'^i}{\partial a_1^\alpha} = -\psi_j^i(x'; a_2) \frac{\partial f^j(x'; a_2)}{\partial a_2^\beta} \frac{\partial a_2^\beta}{\partial a_1^\alpha}.$$

The quantities $\dfrac{\partial a_2^\beta}{\partial a_1^\alpha}$ are functions of the a_1's and a_3's, but by means of (4.6) are expressible in terms of the a_1's and a_2's. The other quantities in the right-hand member of each of the above equations are functions of the x''s and a_2's. But the left-hand member of each of these equations does not involve the a_2's and consequently the combinations on the right are independent of the a_2's. Consequently if $(a_2^\alpha)_0$ are particular values of the a_2's and we put

$$(5.4) \qquad \xi_b^i(x') \equiv -\left[\psi_j^i \frac{\partial f^j(x'; a_2)}{\partial a_2^b} \right]_{a_2 = (a_2)_0}$$

$$A_\alpha^b(a_1) \equiv \left(\frac{\partial a_2^b}{\partial a_1^\alpha} \right)^*_{a_2 = (a_2)_0} \qquad (b, \alpha = 1, \cdots, r),$$

we have

$$(5.5) \qquad \frac{\partial x'^i}{\partial a^\alpha} = \xi_b^i(x') A_\alpha^b(a) \qquad \left(\begin{matrix} i = 1, \cdots, n; \\ \alpha, b = 1, \cdots, r \end{matrix} \right).$$

Since these equations admit as solutions $f^i(x; a)$ whatever be the x's, the latter may be looked upon as n arbitrary constants, and consequently the system of equations (5.5) is completely integrable. This fact, which is of fundamental importance, will enable us in §6 to find conditions which the ξ's and A's must satisfy.

---

* In what follows indices $a, b, c, \ldots$ take the values $1, \ldots, r$; and $i, j, k, \ldots$, the values $1, \ldots, n$.

We observe that the determinant of $A_\alpha^b(a)$ is not identically zero, that is,

(5.6) $$|A_\alpha^b| \neq 0.$$

Otherwise there would exist a set of functions A^α of the a's such that $A^\alpha A_\alpha^b = 0$, and then from (5.5) we should have

$$A^\alpha \frac{\partial f^i(x; a)}{\partial a^\alpha} = 0,$$

and the a's would not be essential (cf. §3).

Because of (5.6) a set of functions $A_b^\alpha(a)$ are defined by the equations

(5.7) $$A_\alpha^b A_b^\beta = \delta_\alpha^\beta,$$

from which follow also the relations

(5.8) $$A_\alpha^b A_a^\alpha = \delta_a^b.^*$$

By means of the quantities A_b^α equations (5.5) may be written in the form

(5.9) $$\xi_b^i(x') = A_b^\alpha \frac{\partial x'^i}{\partial a^\alpha},$$

from which it follows that there cannot exist a set of functions of the a's, say $\varphi^b(a)$, such that

(5.10) $$\varphi^b(a)\xi_b^i(x') = 0;$$

otherwise the parameters a^α would not be essential (cf. §3). From this it follows that we cannot have

(5.11) $$c^b \xi_b^i(x) = 0,$$

where the c's are constants. Hence the ξ's are linearly independent (constant coefficients); we use this notation frequently to distinguish this condition from that which arises when the rank of the matrix of ξ_a^i is less than r, in which case there are relations of the form (5.11) with the c's functions of the x's (not all constants).

It should be remarked that all the results of this section apply equally well to semi-groups.

---

* The quantities A_α^a and A_a^α are essentially different in character as shown in §7; the reader is advised to bear this in mind constantly.

6. First fundamental theorem. We have observed that the equations (5.5) are completely integrable, consequently the conditions of integrability of these equations must be satisfied identically. The conditions are

(6.1) $$\left(\xi_a^j \frac{\partial \xi_b^i}{\partial x'^i} - \xi_b^j \frac{\partial \xi_a^i}{\partial x'^i}\right) A_\alpha^a A_\beta^b + \xi_b^i \left(\frac{\partial A_\beta^b}{\partial a^\alpha} - \frac{\partial A_\alpha^b}{\partial a^\beta}\right) = 0.$$

By means of the functions A_a^α defined by (5.7) these equations may be written

(6.2) $$\xi_a^j \frac{\partial \xi_b^i}{\partial x'^i} - \xi_b^j \frac{\partial \xi_a^i}{\partial x'^i} = c_{ab}^e \xi_e^i,$$

where

(6.3) $$c_{ab}^e = A_a^\alpha A_b^\beta \left(\frac{\partial A_\alpha^e}{\partial a^\beta} - \frac{\partial A_\beta^e}{\partial a^\alpha}\right).$$

Since (6.2) are identities in the x''s and the a's, and the latter do not enter in the ξ's, we have

$$\frac{\partial c_{ab}^e}{\partial a^\alpha} \xi_e^i = 0.$$

From (6.3) it follows that the c's do not involve the x''s. Consequently we have $\frac{\partial c_{ab}^e}{\partial a^\alpha} = 0$; otherwise we have relations of the form (5.10). Hence we have:

[6.1] *The quantities c_{ab}^e in equations (6.2) and (6.3) are constants.*

They are called the *constants of structure* of the group; some writers call them the *constants of composition*. From (6.2) we observe that

$$c_{ab}^e = -c_{ba}^e,$$

that is c_{ab}^e is *skew-symmetric* in the indices a and b.

In consequence of (5.8) the equations (6.3) can be written in the form

(6.4) $$\frac{\partial A_\alpha^e}{\partial a^\beta} - \frac{\partial A_\beta^e}{\partial a^\alpha} = c_{ab}^e A_\alpha^a A_\beta^b,$$

and also in the form, because of (5.7),

$$(6.5) \qquad A_a^\alpha \frac{\partial A_b^\beta}{\partial a^\alpha} - A_b^\alpha \frac{\partial A_a^\beta}{\partial a^\alpha} = c_{ab}^{\;\;e} A_e^\beta.$$

Equations (6.4) are due to Maurer.*

If we write (4.7) in the form

$$(6.6) \qquad x''^i = f^i(x'; a_2) = f^i(x; a_3),$$

we have analogously to (5.5), according as we use the first or second expression for x''^i,

$$\frac{\partial x''^i}{\partial a_2^\alpha} = \xi_a^i(x'') A_\alpha^a(a_2),$$

$$\frac{\partial x''^i}{\partial a_3^\alpha} = \xi_b^i(x'') A_\alpha^b(a_3).$$

If we substitute the second expression for the x'''s in the first set of these equations and note that the a_3's are functions of the a_1's and a_2's, we have

$$\frac{\partial x''^i}{\partial a_3^\beta} \frac{\partial a_3^\beta}{\partial a_2^\alpha} = \xi_a^i(x'') A_\alpha^a(a_2).$$

From these two sets of equations it follows that

$$(6.7) \qquad \frac{\partial a_3^\beta}{\partial a_2^\alpha} = A_b^\beta(a_3) A_\alpha^b(a_2),$$

otherwise we should have relations of the form (5.10).

Suppose conversely that we have a completely integrable set of equations (5.5) with (5.6) satisfied; then (6.2) and (6.5) hold.

Since the equations (6.5) are of the same form as (6.2), we see on comparing (6.7) with (5.5) that the former is necessarily completely integrable, in consequence of (6.5). These equations admit solutions a_3^β such that $a_3^\beta = a_1^\beta$ when a_2^α have the initial values a_0^α, say

$$(6.8) \qquad a_3^\alpha = \varphi^\alpha(a_1; a_2; a_0), \qquad a_1^\alpha = \varphi^\alpha(a_1; a_0; a_0), \qquad \left| \frac{\partial \varphi}{\partial a_2} \right| \neq 0.$$

Suppose then that a set of solutions of (5.5) is $f^i(x; a)$ such that (4.2) is satisfied, and we put

$$(6.9) \qquad x_3^i = f^i(x; a_3),$$

---

* 1888, 2, p. 117.

where a_3^α have the values (6.8). We have accordingly

$$\frac{\partial x_3^i}{\partial a_2^\alpha} = \frac{\partial x_3^i}{\partial a_3^\beta}\frac{\partial a_3^\beta}{\partial a_2^\alpha} = \xi_a^i(x_3)A_\beta^a(a_3)A_b^\beta(a_3)A_\alpha^b(a_2) = \xi_b^i(x_3)A_\alpha^b(a_2),$$

in consequence of (5.5), (6.7) and (5.8). Since these equations are of the form (5.5), they admit solutions $f^i(\bar{x}; a_2)$, where the $\bar{x}$'s, being independent of the a_2's, are functions of the x's, a_1's and a_0's at most such that

(6.10) $$f^i(x; \varphi(a_1; a_2; a_0)) = f^i(\bar{x}; a_2)$$

are identities in the x's, a_1's and a_2's.

Suppose that the solution $f^i(x; a)$ is such that for a set of values of the a's, say a_0^α, we have

$$f^i(x; a_0) = x^i,$$

and such that $|A_\alpha^a(a_0)| \neq 0$, that is, the values a_0^α determine the identity. If we use these values of a_0^α in the above determination of the functions φ^α (6.8), it follows that when in (6.10) we put $a_2^\alpha = a_0^\alpha$ these identities reduce to $f^i(x; a_1) = \bar{x}^i$. When these expressions for $\bar{x}^i$ are substituted in (6.10), we obtain (5.1). Furthermore, if a_1^α are the values of the parameters of a given transformation, the transformation for the values a_2^α, where the latter are defined by

(6.11) $$\varphi^\alpha(a_1; a_2; a_0) = a_0^\alpha,$$

is the inverse of the given transformation.

The foregoing results may be stated as the following *first fundamental theorem* (cf. §11) of continuous groups:

[6.2] *A necessary condition that a set of n functions $f^i(x; a)$ of n variables x^i and r essential parameters a^α define a group G_r of continuous transformations is that these functions satisfy a system of differential equations of the form* (5.5), *where the determinant of the functions $A_\alpha^a(a)$ is of rank r and the functions ξ_a^i are independent (constant coefficients).*   (S-G)

Conversely, if $f^i(x; a)$ are solutions of a completely integrable system of equations of the form (5.5), *where the A's and ξ's satisfy the conditions stated above, such that for values a_0^α of the a's the determinant of the A's is not zero and*

(6.12) $$f^i(x; a_0) = x^i,$$

these functions define a continuous group of transformations, that is, the inverse of each transformation is a member of the group.*

We shall show that the conditions of the second part of this theorem can be met when we have a completely integrable system (5.5). In fact, if $f^i(x; a)$ are n independent solutions of such a system, that is, such that (4.2) is satisfied and a_0^α are values of the a's for which $|A_\alpha^a(a_0)| \neq 0$ and we write the equations of the transformation T_{a_0} and $T_{a_0}^{-1}$ in the forms

$$x_0^i = f^i(x; a_0), \qquad x^i = \bar{f}^i(x_0; a_0),$$

then the equations of a transformation $T_a T_{a_0}^{-1}$ are

$$x'^i = f^i[\bar{f}(x_0; a_0); a].$$

These expressions evidently are a solution of (5.5) in which x^i are replaced by $\bar{f}^i(x_0; a_0)$. Moreover $T_{a_0} T_{a_0}^{-1} = I$ and consequently these transformation satisfy the conditions of the second part of the above theorem; in fact

$$f^i[\bar{f}(x_0; a_0); a_0] = x_0^i.$$

Hence we have the theorem:

[6.3] *If the functions ξ_a^i and A_α^a satisfy the conditions* (6.2) *and* (6.4), *equations* (5.5) *admit one independent set of solutions defining a group G_r of transformations, for which any given values of the a's such that $|A_\alpha^a| \neq 0$ determine the identity transformation.*†

7. Conditions upon the constants of structure. Transformations of coordinates and parameters. The functions ξ_a^i which enter in equations (5.5) serve to determine r linear operators (§2)

$$(7.1) \qquad X_a f = \xi_a^i(x) \frac{\partial f}{\partial x^i} = \xi_a^i p_i.$$

From (6.2) it follows that

$$(7.2) \qquad (X_a, X_b)f = c_{ab}^e X_e f \qquad (a, b, e = 1, \cdots, r).$$

Applying the Jacobi identity (2.4) to (7.2), we obtain

$$(c_{ab}^e c_{ec}^f + c_{bc}^e c_{ea}^f + c_{ca}^e c_{cb}^f)\xi_f^i = 0.$$

---

* The proof of this theorem is an adaptation of the proof given by *Schur*, 1891, 1, pp. 264–268.

† Cf. *Bianchi*, 1918, 1, p. 85.

Since the ξ's are linearly independent (constant coefficients), we have:

[7.1]   *The constants of structure of a group satisfy the conditions*

(7.3) $$c_{ab}^{e} = -c_{ba}^{e},$$

(7.4) $$c_{ab}^{e}c_{ec}^{f} + c_{bc}^{e}c_{ea}^{f} + c_{ca}^{e}c_{eb}^{f} = 0. \qquad \text{(S-G)}$$

We call these the *Jacobi relations*.

If we subject the x's to a non-singular transformation, such as (4.16) and (4.19), it follows from (5.5) that

$$\frac{\partial y'^{i}}{\partial a^{\alpha}} = \frac{\partial y'^{i}}{\partial x'^{i}}\xi_{a}^{i}(x')A_{\alpha}^{a}(a) = \eta_{a}^{i}(y')A_{\alpha}^{a}(a),$$

where

(7.5) $$\eta_{a}^{i}(y) = \xi_{a}^{i}(x)\frac{\partial y^{i}}{\partial x^{i}}.$$

Two sets of quantities ξ_{a} and η_{a}^{i} related as in (7.5) are said to be the components in the respective coordinate systems of a contravariant vector in V_{n}.* Also we have

(7.6) $$Y_{a}f \equiv \eta_{a}^{i}(y)\frac{\partial f}{\partial y^{i}} = X_{a}f,$$

that is, $X_{a}f$ is a scalar† under transformation of the coordinates.

The functions ξ_{a}^{i} and A_{α}^{a} have been defined by (5.4). Suppose, however, that we define a new set of functions $\xi_{a}'^{i}$ by

(7.7) $$\xi_{a}'^{i}(x) = c_{a}^{b}\xi_{b}^{i}(x),$$

where the c's are constants, such that the determinant $|c_{a}^{b}| \neq 0$. Then the ξ''s are linearly independent (constant coefficients). A set of constants $\bar{c}_{a}^{b}$ is uniquely defined by

(7.8) $$c_{a}^{b}\bar{c}_{b}^{e} = \delta_{a}^{e}, \qquad c_{a}^{b}\bar{c}_{e}^{a} = \delta_{e}^{b}.$$

If then we put

(7.9) $$A'^{a}_{\alpha}(a) = \bar{c}_{b}^{a}A_{\alpha}^{b}(a), \qquad A'^{\alpha}_{a}(a) = c_{a}^{b}A_{b}^{\alpha}(a),$$

we have that (5.5) may be written

$$\frac{\partial x'^{i}}{\partial a^{\alpha}} = \xi_{b}'^{i}(x')A'^{b}_{\alpha}(a),$$

_____

* Cf. 1926, 3, p. 4.
† Cf. 1926, 3, p. 6.

so that we have the generality in the choice of the ξ's and A's set forth in (7.7) and (7.9).

If we effect a change in the parameters a^α by putting

$$(7.10) \qquad a^\alpha = \varphi^\alpha(a'^1, \cdots, a'^r),$$

where $\left|\dfrac{\partial \varphi}{\partial a'}\right| \neq 0$, it follows from (5.4) that

$$\xi_b'^i(x') = \xi_a^i(x')\left(\frac{\partial a_2^a}{\partial a_2'^b}\right)_{a_2' = (a_2')_0}, \qquad A_\beta'^b(a') = A_\alpha^a(a)\frac{\partial a^\alpha}{\partial a'^\beta}\left(\frac{\partial a_2'^b}{\partial a_2^a}\right)_{a_2 = (a_2)_0}.$$

Thus a change in the a's may be said to induce a linear transformation on the functions $\xi_a^i(x')$. However, in view of the generality referred to above we may choose the ξ's for the parameters a'^α so that for the ξ's in the a''s we take $\xi_a^i[f(x; \varphi^1(a'), \ldots, \varphi^r(a'))]$, that is, treat them as scalars under change of the parameters. If then we denote the A's for the new parameters by $A_a'^b(a')$, it follows from (5.5) and similar equations in the parameters a'^α that

$$\xi_b'^i(x')\left[A_\beta'^b(a') - A_\alpha^b(a)\frac{\partial a^\alpha}{\partial a'^\beta}\right] = 0.$$

In consequence of the observation (5.10) it follows that

$$(7.11) \qquad A_\beta'^b(a') = A_\alpha^b(a)\frac{\partial a^\alpha}{\partial a'^\beta} \qquad (\alpha, \beta, b = 1, \cdots, r).$$

Thus the A's obey the law of transformation of covariant vectors[*] in the space S of coordinates a^α, the index b indicating the vector and α the components; they constitute an ennuple of independent covariant vectors in this space S, which is called the *group-space* of G_r. From (5.8) it follows that the quantities A_b^α are the components of the ennuple of contravariant vectors in S, conjugate to the ennuple A_α^b, where the index b indicates the vector and α the component.[†]

We return to the consideration of (7.7) and put $X_a'f = \xi_a'^i(x)p_i$. In place of (7.2) we have

$$(7.12) \qquad (X_a', X_b')f = c_{ab}'^e X_e'f,$$

where

$$(7.13) \qquad c_{ab}'^e = c_a^{a_1}c_b^{b_1}\bar{c}_{e_1}^e c_{a_1b_1}^{e_1}.$$

---

[*] Cf. 1926, 3, 7.

[†] Cf. 1926, 3, p. 15.

Lie calls $X_a f$ for $a = 1, \cdots, r$ the *symbols* of the group; we shall refer to them as the *basis* of the group; the significance of these terms appears more clearly in §§10, 11. When the basis is changed in accordance with (7.7), the new constants of structure are given by (7.13). Two r-parameter groups whose constants are in the relation (7.13) are said to have the *same structure*.

8. Semi-group of inverse transformations. In §4 we saw that the inverses of the transformations of a semi-group G_r form a semi-group $\overline{G}_r$, and that if they coincide G_r is a group. For the present we treat the general case.

For the semi-group $\overline{G}_r$ we have analogously to (5.5) the completely integrable system of equations

$$(8.1) \qquad \frac{\partial x^i}{\partial a^\alpha} = \bar{\xi}^i_b(x)\bar{A}^b_a(a),$$

of which the functions $\bar{f}^i(x'; a)$ indicated in (4.3) are solutions. In this case the equations analogous to (5.1) are, in consequence of (4.15),

$$(8.2) \qquad \bar{f}^i[\bar{f}(x'; a_2); a_1] = \bar{f}^i(x'; \varphi(a_1; a_2)).$$

Consequently in place of (6.7) we have, since the roles of a_1 and a_2 are interchanged,

$$(8.3) \qquad \frac{\partial a^\beta_3}{\partial a^\alpha_1} = \bar{A}^\beta_b(a_3)\bar{A}^b_\alpha(a_1).$$

For any values a^α_0 for which the determinant $|\bar{A}^b_\alpha(a_0)|$ is different from zero a set of constants c^a_b can be found such that

$$c^a_b\bar{A}_\alpha(a_0) = A^a_\alpha(a_0)$$

and in consequence of (7.7) and (7.9) we have that the functions $\bar{\xi}^i_b$ and $\bar{A}^b_\alpha$ can be chosen so that

$$(8.4) \qquad \bar{A}^b_\alpha(a_0) = A^b_\alpha(a_0), \qquad \bar{A}^\alpha_b(a_0) = A^\alpha_b(a_0).$$

When we express the conditions of consistency of (6.7) and (8.3), we obtain

$$A^b_\alpha(a_2)\bar{A}^a_\gamma(a_1)\left(\frac{\partial A^\beta_b(a_3)}{\partial a^\delta_3}\bar{A}^\delta_a(a_3) - \frac{\partial \bar{A}^\beta_a(a_3)}{\partial a^\delta_3}A^\delta_b(a_3)\right) = 0.$$

Since these must hold for all values of the a_1's and a_2's, the quantity in parenthesis must be zero, so that on changing indices we have

$$(8.5) \qquad \frac{\partial A_b^\alpha}{\partial a^\beta} \bar{A}_a^\beta - \frac{\partial \bar{A}_a^\alpha}{\partial a^\beta} A_b^\beta = 0.$$

In order to consider these equations further, we define functions $L_{\beta\gamma}^\alpha$ of the a's by the equations

$$(8.6) \qquad \frac{\partial A_b^\alpha}{\partial a^\gamma} + A_b^\beta L_{\beta\gamma}^\alpha = 0,$$

which because of (5.8) are equivalent to

$$(8.7) \qquad L_{\beta\gamma}^\alpha = -A_\beta^b \frac{\partial A_b^\alpha}{\partial a^\gamma} = A_b^\alpha \frac{\partial A_\beta^b}{\partial a^\gamma}.$$

From (8.7) and (6.5) we have

$$(8.8) \qquad L_{\beta\gamma}^\alpha - L_{\gamma\beta}^\alpha = c_{ab}^e A_\beta^a A_\gamma^b A_e^\alpha.$$

If we put, analogously to (8.6),

$$(8.9) \qquad \frac{\partial \bar{A}_b^\alpha}{\partial a^\gamma} + \bar{A}_b^\beta \bar{L}_{\beta\gamma}^\alpha = 0,$$

we find from (8.5) and (8.7) that

$$(8.10) \qquad \bar{L}_{\beta\gamma}^\alpha = L_{\gamma\beta}^\alpha.$$

Analogously to (6.5) we have

$$(8.11) \qquad \bar{A}_a^\alpha \frac{\partial \bar{A}_b^\beta}{\partial a^\alpha} - \bar{A}_b^\alpha \frac{\partial \bar{A}_a^\beta}{\partial a^\alpha} = \bar{c}_{ab}^e \bar{A}_e^\beta,$$

and similarly to (8.8)

$$(8.12) \qquad \bar{L}_{\beta\gamma}^\alpha - \bar{L}_{\gamma\beta}^\alpha = c_{ab}^e \bar{A}_\beta^a \bar{A}_\gamma^b \bar{A}_e^\alpha.$$

From (8.10), (8.8) and (8.12) we have, on putting $a^\alpha = a_0^\alpha$ and making use of (8.4),

$$(8.13) \qquad \bar{c}_{ab}^e = -c_{ab}^e.$$

Comparing these equations with (7.13), we see that the latter arise from the former, if we put $c_b^a = -\delta_a^b$. Consequently the semi-group of inverse transformations of a semi-group has the same structure as the latter. In fact, the minus sign is a consequence of the choice (8.4).

Since the determinant $|A_b^\alpha|$ is of rank r (§5), a set of functions $\rho_a^b(a)$ are defined by the equations

$$(8.14) \qquad \bar{A}_a^\alpha = \rho_a^b A_b^\alpha,$$

which transform as scalars, when the parameters a^α undergo a general analytic transformation; in particular,

$$(8.15) \qquad \rho_a^b(a_0) = \delta_a^b,$$

as follows from (8.4). When the above expressions for $\bar{A}_a^\alpha$ are substituted in (8.9) and $\bar{L}_{\beta\gamma}^\alpha$ are replaced by $L_{\gamma\beta}^\alpha$, we obtain, in consequence of (8.6),

$$(8.16) \qquad \frac{\partial \rho_a^b}{\partial a^\gamma} = \rho_a^e A_\alpha^b A_e^\beta (L_{\beta\gamma}^\alpha - L_{\gamma\beta}^\alpha);$$

these are reducible by (8.8) and (5.8) to

$$(8.17) \qquad \frac{\partial \rho_a^b}{\partial a^\alpha} = c_{ef}^b \rho_a^e A_\alpha^f.$$

The conditions of integrability of these equations are reducible by means of (8.6) and (8.8) to

$$(c_{hd}^e c_{ec}^b + c_{dc}^e c_{eh}^b + c_{ch}^e c_{ed}^b) \rho_a^h A_\alpha^c A_\beta^d = 0,$$

which are satisfied identically because of the Jacobi relations (7.4). Hence equations (8.17) are completely integrable, and ρ_a^b are the solutions satisfying the initial conditions (8.15). That (8.17) are completely integrable is seen also from the fact that equations (8.9) are completely integrable.

Equations (8.14) may be written in the form

$$(8.18) \qquad \rho_a^b = \bar{A}_a^\alpha A_\alpha^b.$$

Another set of scalars $\bar{\rho}_a^b$ are defined by

$$(8.19) \qquad \bar{\rho}_a^b = \bar{A}_a^b A_a^\alpha,$$

which are such that

$$(8.20) \qquad \bar{\rho}_a^b \rho_c^a = \delta_c^b.$$

From (8.20) and (8.17) we have

$$(8.21) \qquad \frac{\partial \bar{\rho}_b^a}{\partial a^\alpha} = c_{eb}^d \rho_d^a A_\alpha^e.$$

Moreover, from (8.8), (8.10), (8.12) and (8.13) we have

$$(8.22) \qquad c_{ab}^{d} = c_{ef}^{g} \bar{\rho}_{a}^{e} \bar{\rho}_{b}^{f} \rho_{g}^{d}.$$

9. The parameter-groups of a group G_r. In §4 we indicated by

$$(9.1) \qquad T_{a_3} = T_{a_2} T_{a_1}$$

the first condition that the transformations T_a form a group, the parameters being in the relation

$$(9.2) \qquad a_3^{\alpha} = \varphi^{\alpha}(a_1; a_2).$$

We may look upon (9.1) as defining a transformation of the point of coordinates a_1^{α} in the group-space S into the point of coordinates a_3^{α}, the transformation being induced by T_{a_2}. Then (9.2) are the equations of this transformation, a_2^{α} being the parameters. We shall show that these transformations form a semi-group. In fact, if we put

$$a_3'^{\alpha} = \varphi^{\alpha}(a_3; a_2'),$$

we have

$$T_{a_3'} = T_{a_2'} T_{a_3} = T_{a.}(T_{a_2} T_{a_1}) = (T_{a_2'} T_{a_2}) T_{a_1} = T_{a_2''} T_{a_1},$$

where the last expression is a consequence of (9.1), and

$$(9.3) \qquad a_2''^{\alpha} = \varphi^{\alpha}(a_2; a_2').$$

Thus we have shown that equations (9.2) in which the a_2's are parameters define a semi-group of transformations, and furthermore that the functions φ^{α} connecting its parameters are the same as for a semi-group G_r. If G_r is a group, as distinguished from a semi-group, a_0^{α} being the values of the parameters for the identity, then for these values of a_2^{α} equations (9.2) define the identity of the latter, as follows from (4.14). Conversely, if equations (9.2) define a group, then G_r is a group. We remark further that equations (6.7) are the equations for (9.2) analogous to (5.5) for G_r.

In similar manner the symbolical equations

$$T_{a_3}^{-1} = T_{a_1}^{-1} T_{a_2}^{-1}$$

may be looked upon as defining a transformation of the point of coordinates a_2^{α} into that of coordinates a_3^{α} in the group-space S, the transformation being induced by $T_{a_1}^{-1}$; and the equations of the transformation are (9.2) with a_1^{α} as parameters. If now we put

$$a_3'^{\alpha} = \varphi^{\alpha}(a_1'; a_3),$$

we have

$$T_{a_3'}^{-1} = T_{a_1'}^{-1}T_{a_3}^{-1} = (T_{a_1'}^{-1}T_{a_1}^{-1})T_{a_2}^{-1} = T_{a_1'}^{-1}T_{a_2'}^{-1},$$

where

$$a_1''^\alpha = \varphi^\alpha(a_1'; a_1).$$

Consequently equations (9.2), in which a_1^α are the parameters, define a semi-group or a group, according as G_r is a semi-group or group. In this case equations (8.3) are the equations for this group (or semi-group) analogous to (8.1) for $\bar{G}_r$.

When G_r is a group, the groups defined by (9.2) as a_2^α and a_1^α are considered as the parameters are called respectively the *first* and *second parameter-groups* of G_r. Their respective symbols are

$$(9.4) \qquad A_a f = A_a^\alpha \frac{\partial f}{\partial a^\alpha}, \qquad \bar{A}_a f = \bar{A}_a^\alpha \frac{\partial f}{\partial a^\alpha}.$$

From §8 it follows that these two groups have the same structure as G_r (for a semi-group G_r the same structure as G_r and its inverse $\bar{G}_r$ respectively). In consequence of (4.14) and (6.7) we have

$$(9.5) \qquad A_b^\beta(a_1) = A_b^\alpha(a_0)\frac{\partial \varphi^\beta(a_1; a_0)}{\partial a_0^\alpha},$$

and from (8.3)

$$(9.6) \qquad \bar{A}_b^\beta(a_1) = \bar{A}_b^\alpha(a_0)\frac{\partial \varphi^\beta(a_0; a_1)}{\partial a_0^\alpha}.$$

10. One-parameter groups. Suppose that

$$(10.1) \qquad x'^i = f^i(x^1, \cdots, x^n; a) \qquad (i = 1, \cdots, n)$$

are the equations of a one-parameter group G_1 and a_0 is the value of a for the identity, that is,

$$(10.2) \qquad f^i(x; a_0) = x^i.$$

The functions f^i are the solution of a system of ordinary differential equations (cf. (5.5))

$$(10.3) \qquad \frac{dx'^i}{da} = \xi^i(x')A(a)$$

satisfying the initial conditions (10.2). If we define the parameter t by

$$(10.4) \qquad t = \int_{a_0}^a A(a)da,$$

then $t = 0$ yields the identity, and equations (10.3) become

(10.5)
$$\frac{dx'^i}{dt} = \xi^i(x').$$

Corresponding to (4.6) and (4.14) we have

(10.6)
$$a_3 = \varphi(a_1, a_2), \qquad a = \varphi(a_0, a) = \varphi(a, a_0),$$

and likewise

$$t_3 = \theta(t_1, t_2), \qquad t = \theta(t, 0) = \theta(0, t).$$

The function $\varphi(a_1; a_2)$ must be such—in view of the way in which (6.7) was obtained—that we have in this case

(10.7)
$$\frac{\partial a_3}{\partial a_2} = \frac{A(a_2)}{A(a_3)}.$$

If we write the inverse of (10.4) as

(10.8)
$$a = a(t),$$

we must have from (10.6)

$$a(\theta) = \varphi(a(t_1), a(t_2)),$$

as an identity in t_1 and t_2. Differentiating with respect to t_2, we have

$$\frac{da(t_3)}{dt_3} \frac{\partial \theta}{\partial t_2} = \left(\frac{\partial \varphi(a_1, a_2)}{\partial a_2}\right)_{\substack{a_1=a(t_1)\\a_2=a(t_2)}} \frac{da(t_2)}{dt_2}.$$

In consequence of (10.7) and (10.4) this reduces to $\dfrac{\partial \theta}{\partial t_2} = 1$. Hence because of the initial condition we have

(10.9)
$$t_3 = t_1 + t_2.$$

This result follows also from equations (10.5). In fact, any solution of (10.5) satisfies the equations

$$\frac{dx'^1}{\xi^1} = \cdots = \frac{dx'^n}{\xi^n}.$$

The general solution of these equations is given by $n - 1$ equations

(10.10)
$$\varphi^\alpha(x'^1, \cdots, x'^n) = c^\alpha \quad (\alpha = 1, \cdots, n - 1),$$

where the φ's are independent and the c's are constants. Without loss of generality we may assume them solved for $x'^1, \ldots, x'^{n-1}$ as functions of x'^n and the c's. When these are substituted in $\xi^n(x')$ yielding $\bar{\xi}^n(x'^n, c^1, \ldots, c^{n-1})$, and we put

$$t + c^n = \int \frac{dx'^n}{\bar{\xi}^n} = \bar{\varphi}^n(x'^n, c^1, \cdots, c^{n-1}) = \varphi^n(x'^1, \cdots, x'^n),$$

φ^n being the value of $\bar{\varphi}^n$ when the c^α are replaced by φ^α, then in addition to (10.10) we have

(10.11) $$\varphi^n(x'^1, \cdots, x'^n) = c^n + t.$$

Since $x'^i = x^i$ for $t = 0$, we have as another form of (10.1)

(10.12) $$\varphi^i(x') = \varphi^i(x) + \delta_n^i t \qquad (i = 1, \cdots, n),$$

from which it follows that the law of composition of the parameters is (10.9).

If we effect the change of coordinates

(10.13) $$y^i = \varphi^i(x^1, \cdots, x^n) \qquad (i = 1, \cdots, n),$$

equations (10.12) become

(10.14) $$y'^i = y^i + \delta_n^i t \qquad (i = 1, \cdots, n),$$

which is a *translation*. Two groups in the same number of variables, which are transformable into one another by a non-singular transformation of coordinates are said to be *equivalent*. Hence we have:

[10.1] *Any one parameter group of transformations is equivalent to a one-parameter group of translations.*

It is important to remark that this equivalence may obtain only in a limited domain of t. Thus if we consider the group of rotations in the euclidean plane, for which the operator is (cf. Ex. 4, p. 42)

(10.15) $$x^1 p_2 - x^2 p_1,$$

the functions φ^i, referred to above, are in this case

$$\varphi^1 = \sqrt{(x^1)^2 + (x^2)^2} \equiv r \qquad \varphi^2 = \cos^{-1}\frac{x^1}{r}.$$

Consequently (10.14) in this case refers to a rotation and thus is closed after t takes values from 0 to 2π, whereas a translation is open in the sense that t may take on values without limit and yield new points.

We observe that the functions $\varphi^\alpha(x')$ in (10.10) are solutions of the equations

(10.16) $$Xf \equiv \xi^i p_i = 0,$$

and that $X\varphi^n = 1$. Hence for the transformation (10.13) we have in consequence of (7.5)

$$\eta^\alpha(y) = 0, \quad \eta^n(y) = 1 \quad (\alpha = 1, \cdots, n - 1),$$

so that in the y's equations (10.5) are

$$\frac{dy'^\alpha}{dt} = 0, \quad \frac{dy'^n}{dt} = 1,$$

of which the solution is (10.14).

Incidentally we have established the theorem:

[10.2] *If $\xi^i(x)$ are the components of a contravariant vector, there exists a coordinate system in which the components of the vector are δ_n^i.*

Equations (10.10) define the congruence of integral curves of equations (10.5), one through each point. Equations (10.12) give the curve through the point $P(x)$ in terms of the parameter t. They are called the *trajectories* of the group, each of which is described by a point as the latter undergoes a continuous transformation of the group.

We return to the consideration of equations (10.5) and seek their solution as power series in t. Evidently we have

(10.17) $$x'^i = x^i + t\xi^i(x) + \frac{t^2}{2!}\xi^i(x)\frac{\partial \xi^i(x)}{\partial x^j} + \cdots.$$

Making use of the notation in (10.16) and indicating by $X^m f$ the result of operating with X m times on f, these expressions may be written

(10.18) $$x'^i = x^i + tXx^i + \frac{t^2}{2!}X^2x^i + \cdots + \frac{t^m}{m!}X^mx^i + \cdots.$$

Moreover any function $f(x'^1, \ldots, x'^n)$ regular in the domain of x^i is expressible in the form

(10.19) $$f(x') = f(x) + tXf + \cdots + \frac{t^m}{m!}X^mf + \cdots.$$

For, we have

$$f(x') = f(x) + \left(\frac{\partial f}{\partial x'^i}\frac{dx'^i}{dt}\right)_{t=0} t + \cdots$$

and the result follows from (10.5).

If the series (10.18) are convergent for $0 \leqslant t \leqslant t_1$, the equations so defined, say

$$(10.20) \qquad\qquad x'^i = F^i(x; t),$$

are equivalent to (10.1) for the range of values of a as given by (10.8) for these values of t. This may mean that only a portion of the trajectory through $P(x)$, as defined by (10.1), is given by (10.18). By a transformation of coordinates as (10.13) equations (10.20) take the form (10.14) for the limited values of t. From this result it is seen that if $P(x')$ is a point on the segment of the trajectory referred to above, then on applying (10.20) to P' we get another segment. Hence by sufficient repetition of (10.20) we obtain the trajectory through $P(x)$ in so far as it is defined by (10.1).

For the rotation in the plane with the symbol (10.15) we have that (10.18) become

$$x'^1 = x^1 - x^2 t - x^1 \frac{t^2}{2!} + x^2 \frac{t^3}{3!} + \cdots = x^1 \cos t - x^2 \sin t,$$

$$x'^2 = x^2 + x^1 t - x^2 \frac{t^2}{2!} - x^1 \frac{t^3}{3!} + \cdots = x^1 \sin t + x^2 \cos t.$$

Consequently (10.18) holds for the whole trajectory in this case.

If in (10.17) we replace t by an infinitesimal δt and neglect powers of δt higher than the first, we obtain

$$(10.21) \qquad\qquad x'^i = x^i + \xi^i(x)\,\delta t,$$

which is called the *infinitesimal transformation* of the group G_1. By the integration of the infinitesimal transformation in the sense of the preceding paragraph the group G_1 is obtained. Lie was the first to undertake a systematic study of the construction of continuous groups by means of infinitesimal transformations in this sense.

11. Sub-groups G_1 of a group G_r. If in equations (4.1) of a group G_r the a's are expressed in terms of a single parameter t, say

$$(11.1) \qquad\qquad a^\alpha = a^\alpha(t),$$

we obtain the equations,

$$(11.2) \qquad x'^i = f^i(x; a^1(t), \cdots, a^r(t)) \equiv F^i(x; t),$$

of ∞^1 transformations. The question arises under what conditions these transformations form a group G_1. For this to be the case there must be a relation

$$t_3 = \theta(t_1, t_2)$$

such that from (4.6) we have

$$a^\alpha(t_3) = \varphi^\alpha(a^1(t_1), \cdots, a^r(t_1); a^1(t_2), \cdots, a^r(t_2)) \equiv$$
$$\varphi^\alpha(a(t_1); a(t_2)).$$

From these equations we have by differentiation with respect to t_2

$$\frac{da^\alpha(t_3)}{dt_3}\frac{\partial\theta}{\partial t_2} = \left(\frac{\partial\varphi^\alpha}{\partial a_2^\beta}\right)_{\substack{a_1^\alpha = a^\alpha(t_1) \\ a_2^\alpha = a^\alpha(t_2)}}\frac{da^\beta(t_2)}{dt_2},$$

which in consequence of (6.7) become

$$(11.3)\qquad \frac{da^\alpha(t_3)}{dt_3}\frac{\partial\theta}{\partial t_2} = A_b^\alpha(a(t_3))A_\beta^b(a(t_2))\frac{da^\beta(t_2)}{dt_2},$$

where

$$A_\beta^b(a(t)) \equiv A_\beta^b(a^1(t), \cdots, a^r(t)).$$

If we put

$$(11.4)\qquad \psi^b(t) \equiv A_\alpha^b(a(t))\frac{da^\alpha(t)}{dt},$$

equations (11.3) are reducible in consequence of (5.7) and (5.8) to

$$\psi^b(t_3)\frac{\partial\theta}{\partial t_2} = \psi^b(t_2).$$

From these equations it follows that

$$\frac{\psi^a(t_3)}{\psi^b(t_3)} = \frac{\psi^a(t_2)}{\psi^b(t_2)} \qquad (a, b = 1, \cdots, r).$$

Since these relations must hold for all values of t_2, we have that each term is a constant, and hence we have

$$\psi^a(t) = e^a\psi(t),$$

where the e's are constants, and equations (11.4) become

$$A_\alpha^b(a(t))\frac{da^\alpha(t)}{dt} = e^b\psi(t).$$

If we define a new parameter t' by $t' = \int_{t_0}^t \psi(t)dt$, where t_0 is such that $a^\alpha(t_0) = a_0^\alpha$, the latter giving the identity, then the above equations in terms of the new parameter which we now call t assume the form

$$(11.5)\qquad \frac{da^\beta}{dt} = e^b A_b^\beta(a).$$

From the choice of t it follows that, when $t = 0$, the transformation is the identity. Hence the desired functions $a^\alpha(t)$ are the solutions of (11.5) such that $a^\alpha(0) = a_0^\alpha$. By means of (11.5) equations (11.3) reduce to

$$(11.6) \qquad \frac{\partial \theta}{\partial t_2} = 1.$$

From (4.14) it follows that $\theta(0, t_2) = t_2$ and $\theta(t_1, 0) = t_1$. Hence the solution of (11.6) is

$$(11.7) \qquad t_3 = t_1 + t_2.$$

Conversely, if e^a in (11.5) are any set of constants and the functions $a^\alpha(t)$ are the solutions such that $a^\alpha(0) = a_0^\alpha$, then (11.3) are satisfied, if we make use of (11.7). Consequently

$$a^\alpha(t_1 + t_2) - \varphi^\alpha(a(t_1); a(t_2))$$

is at most a function of t_1. But this expression vanishes when $t_2 = 0$, whatever be t_1, and consequently it vanishes identically. Hence if the functions $a^\alpha(t)$ are substituted in (11.2), the resulting equations define a one-parameter group for each set of values of the e's and in each case for $t = 0$ we have the identity.

Suppose then that we have the solution of (11.5) involving the e's as parameters and t, and satisfying the initial condition

$$a^\alpha(0) = a_0^\alpha.$$

Then using (5.5) we have

$$(11.8) \qquad \frac{dx'^i}{dt} = \xi_a^i(x') A_\alpha^a(a) \frac{da^\alpha}{dt} = e^a \xi_a^i(x') \equiv \xi^i(x').$$

The integrals, as power series in t, of these equations such that $x'^i = x^i$ when $t = 0$, are (cf. §10)

$$(11.9) \qquad x'^i = x^i + tXx^i + \cdots + \frac{t^m}{m!} X^m x^i + \cdots,$$

where

$$(11.10) \qquad Xf = \xi^i(x)p_i = e^a \xi_a^i(x)p_i.$$

Because of the second expressions for Xf in (11.10) equations (11.9) may be written

(11.11) $x'^i = x^i + te^a X_a x^i + \cdots$

$$+ \frac{t^m}{m!} e^{a_1} \cdots e^{a_m} X_{a_1} \cdots X_{a_m} x^i + \cdots.$$

From the form of these expressions it follows that the a's as functions of the e's and t, to which we have referred, are in fact functions of the form

(11.12) $a^\alpha = \psi^\alpha(u^1, \cdots, u^r), \qquad u^a = e^a t;$

the expressions for the ψ's as infinite series are obtained in §12.

The values of t for which the series in (11.11) converge depends evidently upon the values of the e's and the x's.   In each case and for the appropriate domain of t equations (11.11) may be written in the form

(11.13) $x'^i = F^i(x; u).$

Evidently when the expressions (11.12) are substituted in (11.2), we get (11.13).   Hence for the values of t for which the series in (11.11) converge, for given values of the e's and x's, equations (11.11) define transformations of the given group G_1, and these transformations, in the terminology of Lie, are generated by the corresponding infinitesimal transformation

(11.14) $x'^i = x^i + e^a \xi_a^i(x) \delta t;$

in the sense of §10.

For given values of the x's in (11.11) there are values of $u^a = e^a t$, in general limited, for which the series can be inverted, and we have u^a as functions of x^i and x'^i.   Consequently for any point $P'(x')$ sufficiently near $P(x)$ there exists a transformation which sends P into P'.

If in the equations $x'^i = f^i(x; a)$, we give the a's a set of values for which these equations have a meaning, we get the point $P'(x')$ into which $P(x)$ is transformed by the corresponding transformation. If these values of the a's are such that the u's are defined by (11.12) (that is, if it is possible to solve these equations for the u's as functions of the a's with these values), then a sub-group G_1 is defined and the trajectory of this group through $P(x)$ passes through $P'(x')$.   In this case, as seen in §10, $P'(x')$ can be obtained from $P(x)$ by the transformations (11.11), for the corresponding values of the e's or by a product of such transformations.

There are cases, however, for which the values of the a's giving P' are such that it is not possible to find a group G_1 whose trajectory through P passes through P'. The question has been raised in these cases as to whether the passage from P to P' can be accomplished by a set of trajectories, so that P is transformed into a point P_1 by means of one set of e's, P_1 into P_2 by another set and so on. It has been shown that, if $P(x)$ and $P'(x')$ are points of a connected manifold, then P can be transformed into P' in this manner, that is, by the product of transformations of a number of groups of order one.* It is stated sometimes that a group G_r consists of the totality of transformations of the groups G_1 generated by the infinitesimal transformations with the symbols $e^a X_a f$, where the e's are arbitrary constants, $X_a f$ being called the *generators* of G_r. We have seen that the above results should be stated as follows:

[11.1] *The transformations of a continuous group G_r, whose symbols are $X_a f$, consist of the transformations of the groups G_1 generated by the infinitesimal transformations with the symbols $e^a X_a f$, where the e's are arbitrary constants, or of the products of such transformations.*

A continuous group satisfying these conditions is called a *Lie* group. In view of the preceding discussion we have:

[11.2] *A necessary and sufficient condition that the group G_1 generated by the infinitesimal transformation with the symbol Xf be a sub-group of a G_r with the symbols $X_1 f$, . . . , $X_r f$ in the same coordinate system is that Xf be a linear homogeneous combination (constant coefficients) of the latter.*

And as a corollary of this we have:

[11.3] *A necessary and sufficient condition that two sets of operators $X_a f$ and $X'_a f$, expressed in the same coordinate system, be the symbols of the same group is that*

$$(11.15) \qquad\qquad X'_a f = c_a^b X_b f,$$

where the c's are constants such that the determinant $|c_a^b|$ is not zero.

If we denote by e'^a the constants for the symbols $X'_a f$ yielding the same infinitesimal transformation as $e^a X_a f$, then from (11.15) and (7.8) we have

$$(11.16) \qquad\qquad e'^a = \bar{c}_b^a e^b.$$

---

* *Schreier, 1925, 2, p. 18, 19.*

In terms of the parameters a^α appearing in (11.2) an infinitesimal transformation is defined by

$$(11.17) \quad x'^i = f^i(x; a_0 + \delta a_0) = x^i + \left(\frac{\partial f^i}{\partial a^\alpha}\right)_{a=a_0} \delta a_0^\alpha =$$
$$x^i + \xi_a^i(x) A_\alpha^a(a_0) \delta a_0^\alpha,$$

in consequence of (5.5). From this result and (11.5) we have:

[11.4] *When the finite equations of a group are given, the vectors ξ^i of a general infinitesimal transformation of the group are obtained from the equations*

$$x^i + \xi^i \delta t = f^i(x; a_0^1 + e^1 \delta t, \cdots, a_0^r + e^r \delta t),$$

where the e's are constants, on neglecting second and higher powers of δt.

Consider now the transformation determined by the values $a^\alpha + \delta a^\alpha$, when (11.2) gives the transformation for a^α; the equations are

$$x''^i = f^i(x; a + \delta a) = x'^i + \xi_a^i(x') A_\alpha^a(a) \delta a^\alpha.$$

Comparing this result with (11.17), we see that if we determine quantities $\delta_1 a_0^\alpha$ by the equations

$$A_\alpha^a(a_0) \delta_1 a_0^\alpha = A_\beta^a(a) \delta a^\beta,$$

the above equations become

$$x''^i = x'^i + \xi_a^i(x') A_\alpha^a(a_0) \delta_1 a_0^\alpha,$$

and consequently

$$(11.18) \qquad T_{a+\delta a} = T_{a_0 + \delta_1 a_0} T_a.$$

We obtain the same result from (4.6), (4.14) and (9.5); thus

$$a^\alpha + \delta a^\alpha = \varphi^\alpha(a; a_0 + \delta_1 a_0) = \varphi^\alpha(a; a_0) + \frac{\partial \varphi^\alpha(a; a_0)}{\partial a_0^\beta} \delta_1 a_0^\beta$$
$$= a^\alpha + A_a^\alpha(a) A_\beta^a(a_0) \delta_1 a_0^\beta.$$

In similar manner we have from (9.6)

$$a^\alpha + \delta a^\alpha = \varphi^\alpha(a_0 + \delta_2 a_0; a) = \varphi^\alpha(a_0; a) + \frac{\partial \varphi^\alpha(a_0; a)}{\partial a_0^\beta} \delta_2 a_0^\beta$$
$$= a^\alpha + \bar{A}_a^\alpha(a) \bar{A}_\beta^a(a_0) \delta_2 a_0^\beta.$$

Consequently for $\delta_2 a_0^\alpha$ defined by these equations we have

(11.19) $$T_{a+\delta a} = T_a T_{a_0 + \delta_2 a_0}.$$

Since $T_{a+\delta a}$ is a transformation in the neighborhood of T_a, we have:

[11.5] *A transformation in the neighborhood of T_a of a group G_r may be obtained by multiplying T_{a_0} either on the left or on the right by a suitable infinitesimal transformation of G_r.*

Exercises

1. The symbol of the G_1 of *dilatations*, or *homothetic* transformations, whose equations are $x'^i = a x^i$ is

$$Xf = x^i p_i.$$

2. Show that each of the following sets of equations defines a one-parameter group, and find the symbol and the trajectories in each case:

(i) $x' = ax,$ $y' = \dfrac{1}{a} y$;

(ii) $x' = a^m x,$ $y' = a^n y$ (m, n integers);

(iii) $x' = \dfrac{x}{1 - ax},$ $y' = \dfrac{y}{1 - ay}.$

3. Find the finite equations of the one parameter groups whose symbols are the following:

(i) $x^2 p + xyq$ $\left(p = \dfrac{\partial f}{\partial x},\; q = \dfrac{\partial f}{\partial y} \right)$;

(ii) $(x - y)p + (x + y)q$;

(iii) $x \dfrac{\partial f}{\partial y} - y \dfrac{\partial f}{\partial x} + m \dfrac{\partial f}{\partial z}.$

4. For the group G_3 of motions in the plane defined by the equations

$$x'^1 = a^1 + x^1 \cos a^3 - x^2 \sin a^3,$$
$$x'^2 = a^2 + x^1 \sin a^3 + x^2 \cos a^3,$$

the equations (5.5) are satisfied by the functions

$$\left\| \begin{matrix} \xi_1^1 & \xi_2^1 & \xi_3^1 \\ \xi_1^2 & \xi_2^2 & \xi_3^2 \end{matrix} \right\| = \left\| \begin{matrix} 1 & 0 & x'^2 \\ 0 & 1 & -x'^1 \end{matrix} \right\|,$$

$$\left\| \begin{matrix} A_1^1 & A_1^2 & A_1^3 \\ A_2^1 & A_2^2 & A_2^3 \\ A_3^1 & A_3^2 & A_3^3 \end{matrix} \right\| = \left\| \begin{matrix} 1 & 0 & 0 \\ 0 & 1 & 0 \\ a^2 & -a^1 & -1 \end{matrix} \right\|.$$

The corresponding symbols of the group are

$$X_1 f = p_1, \qquad X_2 f = p_2, \qquad X_3 f = x^2 p_1 - x^1 p_2,$$

and the constants of structure are

$$c_{12}^1 = c_{12}^2 = c_{12}^3 = 0; \quad c_{13}^1 = c_{13}^3 = 0, \quad c_{13}^2 = -1; \quad c_{23}^1 = 1, \quad c_{23}^2 = c_{23}^3 = 0.$$

5. In terms of the angles of Euler the equations of the G_3 of rotations in euclidean 3-space are

$$x'^1 = (\cos a^1 \sin a^2 \sin a^3 + \cos a^2 \cos a^3)x^1$$
$$+ (\cos a^1 \cos a^2 \sin a^3 - \sin a^2 \cos a^3)x^2 + \sin a^1 \sin a^3 x^3,$$
$$x'^2 = (\cos a^1 \sin a^2 \cos a^3 - \cos a^2 \sin a^3)x^1$$
$$+ (\cos a^1 \cos a^2 \cos a^3 + \sin a^2 \sin a^3)x^2 + \sin a^1 \cos a^3 x^3,$$
$$x'^3 = - \sin a^1 \sin a^2 x^1 - \sin a^1 \cos a^2 x^2 + \cos a^1 x^3.$$

The equations (5.5) for the G_3 are satisfied by

$$\begin{Vmatrix} \xi_1^1 & \xi_1^2 & \xi_1^3 \\ \xi_2^1 & \xi_2^2 & \xi_2^3 \\ \xi_3^1 & \xi_3^2 & \xi_3^3 \end{Vmatrix} = \begin{Vmatrix} 0 & x'^3 & -x'^2 \\ -x'^3 & 0 & x'^1 \\ x'^2 & -x'^1 & 0 \end{Vmatrix},$$

$$\begin{Vmatrix} A_1^1 & A_1^2 & A_1^3 \\ A_2^1 & A_2^2 & A_2^3 \\ A_3^1 & A_3^2 & A_3^3 \end{Vmatrix} = \begin{Vmatrix} \cos a^3 & -\sin a^3 & 0 \\ -\sin a^1 \sin a^3 & -\sin a^1 \cos a^3 & -\cos a^1 \\ 0 & 0 & 1 \end{Vmatrix}.$$

The corresponding symbols of the group are

$$X_1 f = x^3 p_2 - x^2 p_3, \quad X_2 f = x^1 p_3 - x^3 p_1, \quad X_3 f = x^2 p_1 - x^1 p_2,$$

and the constants of structure are

$$c_{12}^3 = c_{23}^1 = c_{31}^2 = 1, \quad c_{ab}^a = -c_{ba}^a = 0 \quad (a, b = 1, 2, 3; \ a \text{ not summed}).$$

6. Find the symbols of the group G_3 of rotations of Ex. 5 in terms of polar coordinates.

7. The vectors ξ^i of the general infinitesimal transformation of the *linear*, or *affine*, group

$$x'^i = a_j^i x^j + a^i$$

are (cf. theorem [11.4])

$$\xi^i = \alpha^i + \alpha_j^i x^j,$$

where the α's are arbitrary constants; also find the constants of structure.

8. Derive the ξ's for the group of motions of euclidean 3-space as a special case of Ex. 7.

9. The vectors ξ^i of the general infinitesimal transformation of the group G_8 of projective transformations of the plane

$$x'^i = \frac{a_j^i x^j + a^i}{b_j x^j + 1}$$

are

$$\xi^i = \alpha^i + \alpha_j^i x^j + \alpha_j x^j x^i,$$

where the α's are arbitrary constants; also find the constants of structure.

10. Show that the equations of the infinitesimal transformations $T_{a+\delta a} T_a^{-1}$ and $T_a^{-1} T_{a+\delta a}$ are respectively

$$x'^i = x^i + \left(\frac{\partial f^i(y;\,a)}{\partial a^\alpha}\right)_{y=\bar{f}(x;a)} \delta a^\alpha\ ,$$

$$x'^i = x^i - \left(\frac{\partial \bar{f}^i(y;\,a)}{\partial a^\alpha}\right)_{y=f(x;a)} \delta a^\alpha\ .$$

Bianchi, 1918, 1, pp. 71, 72.

11. Show that the symbol of the transform (§4) of the infinitesimal transformation with the symbol Xf by the infinitesimal transformation with the symbol Yf is

$$Xf + \delta t(X,\,Y)f.$$

12. Canonical parameters. If we apply to equations (6.7) considerations similar to those which led from (5.5) to (11.8) with the aid of (11.5), we see that in fact equations (11.5) are the differential equations whose integral curves are the trajectories in the group-space S of the first parameter group. We can therefore write down the integrals of (11.5) in a form analogous to (11.11), namely

$$(12.1)\quad a'^\alpha = a^\alpha + te^a A_a a^\alpha + \cdots$$
$$+ \frac{t^m}{m!} e^{a_1} \cdots e^{a_m} A_{a_1} \cdots A_{a_m} a^\alpha + \cdots,$$

where we have put

$$A_a(a) = A_a^\alpha(a)\frac{\partial f}{\partial a^\alpha}.$$

In particular, the solution of (11.5) such that $a^\alpha = a_0^\alpha$ when $t = 0$, which was used in §11, is given by

$$(12.2)\quad a^\alpha = a_0^\alpha + te^a A_a a_0^\alpha + \cdots$$
$$+ \frac{t^m}{m!} e^{a_1} \cdots e^{a_m} A_{a_1} \cdots A_{a_m} a_0^\alpha + \cdots$$

We shall write the latter in another form by making use of the functions $\Gamma_{\beta\gamma}^\alpha$ defined by

$$(12.3)\qquad\qquad \Gamma_{\beta\gamma}^\alpha = \frac{1}{2}(L_{\beta\gamma}^\alpha + L_{\gamma\beta}^\alpha),$$

the functions $L_{\beta\gamma}^\alpha$ being defined by (8.7); thus $\Gamma_{\beta\gamma}^\alpha$ is the symmetric part of $L_{\beta\gamma}^\alpha$. In fact, if equations (11.5) are differentiated with respect to t, the resulting equations are reducible by (11.5) and (8.6) to

$$(12.4)\qquad\qquad \frac{d^2 a^\alpha}{dt^2} + \Gamma_{\beta\gamma}^\alpha \frac{da^\beta}{dt}\frac{da^\gamma}{dt} = 0.$$

Thus the trajectories of the first parameter group are paths of S determined by the linear connection $\Gamma^\alpha_{\beta\gamma}$.*

Differentiating (12.4) and making use of these equations in the reduction, we obtain

$$\frac{d^3 a^\alpha}{dt^3} + \Gamma^\alpha_{\beta\gamma\delta}\frac{da^\beta}{dt}\frac{da^\gamma}{dt}\frac{da^\delta}{dt} = 0,$$

where

$$\Gamma^\alpha_{\beta\gamma\delta} = \frac{1}{3}P\left(\frac{\partial\Gamma^\alpha_{\beta\gamma}}{\partial a^\delta} - \Gamma^\alpha_{\epsilon\gamma}\Gamma^\epsilon_{\beta\delta} - \Gamma^\alpha_{\beta\epsilon}\Gamma^\epsilon_{\gamma\delta}\right),$$

where P indicates the sum of terms obtained by permuting the subscripts cyclically. Continuing this process, we have

$$\frac{d^m a^\alpha}{dt^m} + \Gamma^\alpha_{\beta_1\cdots\beta_m}\frac{da^{\beta_1}}{dt}\cdots\frac{da^{\beta_m}}{dt} = 0,$$

where

$$\Gamma^\alpha_{\beta_1\cdots\beta_m} = \frac{1}{m}P\left[\frac{\partial\Gamma^\alpha_{\beta_1\cdots\beta_{m-1}}}{\partial a^{\beta_m}} - \sum_1^{m-1}\epsilon\,\Gamma^\alpha_{\beta_1\cdots\beta_{\epsilon-1}\gamma\beta_{\epsilon+1}\cdots\beta_{m-1}}\Gamma^\gamma_{\beta_\epsilon\beta_m}\right].$$

Because of these results and (11.5) the expansions

$$a^\alpha = a_0^\alpha + \left(\frac{da^\alpha}{dt}\right)_0 t + \frac{1}{2}\left(\frac{d^2 a^\alpha}{dt^2}\right)_0 t^2 + \cdots$$

may be written in the form

$$(12.5)\quad a^\alpha = a_0^\alpha + u^a A_a^\alpha(a_0) - \frac{1}{2}(\Gamma^\alpha_{\beta\gamma}A_a^\beta A_b^\gamma)_0 u^a u^b - \cdots$$
$$- \frac{1}{m!}(\Gamma^\alpha_{\beta_1\cdots\beta_m}A_{a_1}^{\beta_1}\cdots A_{a_m}^{\beta_m})_0 u^{a_1}\cdots u^{a_m} + \cdots,$$

where

$$(12.6)\qquad\qquad u^a = e^a t.$$

If we put

$$(12.7)\qquad y^\alpha = u^a A_a^\alpha(a_0),\qquad u^a = y^\alpha A_\alpha^a(a_0),$$

in consequence of (12.6) and (11.5) we have

$$\left(\frac{dy^\alpha}{dt}\right)_0 = e^a A_a^\alpha(a_0) = \left(\frac{da^\alpha}{dt}\right)_0,$$

* 1927, 1, p. 14.

and hence

$$(12.8) \qquad y^\alpha = \left(\frac{da^\alpha}{dt}\right)_0 t.$$

Moreover, the expressions (12.5) become

$$(12.9) \quad a^\alpha = a_0^\alpha + y^\alpha - \frac{1}{2}(\Gamma_{\beta\gamma}^\alpha)_0 y^\beta y^\gamma - \cdots$$

$$- \frac{1}{m!}(\Gamma_{\beta_1 \cdots \beta_m}^\alpha)_0 y^{\beta_1} \cdots y^{\beta_m} + \cdots .$$

If in place of the a's we had started with another set of coordinates a'^α, corresponding to (12.8) we should have

$$y'^\alpha = \left(\frac{da'^\alpha}{dt}\right)_0 t = \left(\frac{\partial a'^\alpha}{\partial a^\beta}\right)_0 y^\beta.$$

Because of this result and (7.11) we have

$$y'^\alpha A_\alpha'^a(a_0') = y^\alpha A_\alpha^a(a_0),$$

and consequently, as follows from (12.7), the quantities u^a are unchanged when the coordinates a^α undergo a general analytic transformation.

If we take linear combinations with constant coefficients of the vectors A_α^a and of A_a^α as in (7.9), it follows from (8.7) that the coefficients $L_{\beta\gamma}^\alpha$ are not changed. Accordingly the coordinates y^α, defined by (12.9), are not affected by such changes of the vectors A_α^a, but the quantities u^a are changed. In fact, from (7.9) and (12.7) we have, on denoting the new coordinates by u'^a,

$$(12.10) \qquad u'^a = \bar{c}_{\,b}^{\,a} u^b.$$

We call the u^a *canonical parameters*. Equations (12.10) are in agreement with (11.16).

If we denote by U_α^a the components in the u's of the vector of components A_α^a in the a's, we have

$$(12.11) \qquad U_\alpha^a = A_\beta^a \frac{\partial a^\beta}{\partial u^\alpha}.$$

Then from (12.5), on observing that $u^\alpha = 0$ when $a^\alpha = a_0^\alpha$, we have

$$(12.12) \qquad U_\alpha^a(0) = \delta_\alpha^a, \qquad U_a^\alpha(0) = \delta_a^\alpha.$$

Analogously to the equations (11.5) we have

$$\frac{du^\alpha}{dt} = e^a U_a^\alpha(u).$$

From these equations and (12.6) it follows that

$$e^\alpha = e^a U_a^\alpha,$$

and consequently we have

(12.13) $\qquad u^a U_a^\alpha = u^\alpha, \qquad u^\alpha U_\alpha^a = u^a.$

We observe that, if a_0^α are the coordinates of any point in the group-space S, then in terms of the coordinates y^α defined by (12.9) the equations of the paths through the point are given by (12.8). Evidently the coordinates y^α are determined for the domain about the point for which (12.9) are convergent and such that no two paths through the point meet again in the domain. They are in fact the normal coordinates with this point as origin of the space with the linear connection determined by the quantities $\Gamma_{\beta\gamma}^\alpha$.*

Although the results of the preceding development are of use in the theory of continuous groups, the ideas of the latter have not been used in their development and on reviewing these results we have:

[12.1] *Given an ennuple of vectors $A_a^\alpha(a)$ in a space of coordinates a^α and define as in* (8.7) *a linear connection; any point for which the determinant* $|A_\alpha^a|$ *is not zero may be taken as the origin of a set of canonical parameters u^a, defined by* (12.5), *where $\Gamma_{\beta\gamma}^\alpha$ are the symmetric parts of the coefficients of the linear connection; in this system of coordinates the components U_a^α of the vectors satisfy the equations* (12.13); *the coordinates u^a are unchanged by a general analytic transformation of the a's, but undergo a linear homogeneous transformation with constant coefficients, when the vectors are subjected to such a transformation.*

These coordinates u^α have been used in other theories recently.†

For the values of t for which the right-hand members of (12.5) converge these equations may be written in the form (11.12), and

---

* 1927, 1, p. 59.

† Cf. *Thomas*, 1930, 4; *Michal*, 1929, 3.

these in turn lead to (11.13). If equations (11.11), written in the form

$$(12.14) \quad x'^i = x^i + u^a X_a x^i + \cdots$$
$$+ \frac{1}{m!} u^{a_1} \cdots u^{a_m} X_{a_1} \cdots X_{a_m} x^i + \cdots$$

converge for these values of t, then (11.13) are equivalent to (11.11).

13. Abelian groups. If all the constants of composition of a group are zero for a given basis of the group, the same is true for every basis, as follows from (7.13). In this case we have from (7.2) that $(X_a, X_b)f = 0$. In §29 it is shown that, when and only when this condition is satisfied, any two transformations, T_a and T_b, are commutative, that is, that $T_b T_a = T_a T_b$. If all the transformations of a group possess this property, the group is said to be *Abelian*. Before proceeding to the discussion of certain canonical forms for non-Abelian groups, we shall develop canonical expressions for the equations of Abelian groups.

For an Abelian group equations (6.4) become

$$\frac{\partial A_\alpha^a}{\partial a^\beta} - \frac{\partial A_\beta^a}{\partial a^\alpha} = 0,$$

so that there exist functions $A^a(a^1, \ldots, a^r)$ such that

$$(13.1) \qquad\qquad A_\alpha^a = \frac{\partial A^a}{\partial a^\alpha}$$

and the jacobian $\left| \dfrac{\partial A}{\partial a} \right| \neq 0$.

If we introduce a new set of parameters u^α defined by

$$(13.2) \qquad\qquad u^\alpha = A^\alpha(a^1, \cdots, a^r),$$

and denote by $U_\alpha^a(u)$ and $U_a^\alpha(u)$ the functions of the components in the u's of the vectors of components A_α^a and A_a^α in the a's, we have in consequence of (13.1) and (7.11)

$$(13.3) \qquad U_\alpha^a(u) = \delta_\alpha^a, \qquad U_a^\alpha(u) = \delta_a^\alpha.$$

We observe that these results are in keeping with (12.13), but this is the only case in which the U's have the particular values (13.3), as follows from the equations in the u's analogous to (6.4). The fundamental equations (5.5) are in this case

(13.4) $$\frac{\partial x'^i}{\partial u^a} = \xi_a^i(x').$$

The corresponding equations (6.2) are in this case, on dropping primes for the present,

(13.5) $$\xi_a^j \frac{\partial \xi_b^i}{\partial x^j} - \xi_b^j \frac{\partial \xi_a^i}{\partial x^j} = 0.$$

We choose the coordinate system in accordance with theorem [10.2] so as to have for the vector ξ_1^i of an Abelian group

(13.6) $$\xi_1^i = \delta_1^i.$$

If in (13.5) we put $a = 1$ and $b = 2, \ldots, r$, it follows that the vectors ξ_b^i are independent of x^1. The rank q of the matrix

(13.7) $$M = \|\xi_a^i\| \quad (i = 1, \cdots, n; a = 1, \cdots, r)$$

for general values of the coordinates x^i is called its *generic* rank; evidently $q \leqslant r$ and also $q \leqslant n$. We consider first the case when $q = r$. If $\xi_2^1 \not\equiv 0$, the equation $\xi_2^i p_i = 0$ admits $n - 1$ independent solutions

$$\varphi^1 = x^1 + \psi^1(x^2, \cdots, x^n), \qquad \varphi^\sigma(x^2, \cdots, x^n)$$
$$(\sigma = 3, \cdots, n);$$

if $\xi_2^1 = 0$, we may take $\psi^1 = 0$. Denote by $\varphi^2(x^2, \ldots, x^n)$ a solution of $\xi_2^i p_i = 1$. Then the equations $x'^i = \varphi^i$ define a non-singular transformation of coordinates. In this coordinate system, which we now call x^i, we have from (7.5) that (13.6) hold and $\xi_2^i = \delta_2^i$, and then from (13.5) that the other ξ's are independent of x^1 and x^2. Then we take the equation $\xi_3^i p_i = 0$ and get solutions

$$\varphi^1 = x^1 + \psi^1(x^3, \cdots, x^n), \qquad \varphi^2 = x^2 + \varphi^2(x^3, \cdots, x^n),$$
$$\varphi^\sigma(x^3, \cdots, x^n) \quad (\sigma = 4, \cdots, n)$$

and $\varphi^3(x^3, \ldots, x^n)$ a solution of $\xi_3^i p_i = 1$, and effect the transformation of coordinates $x'^i = \varphi^i$. Continuing this process, we have ultimately a coordinate system for which the components of the ξ's are

(13.8) $$\xi_a^i = \delta_a^i \quad (a = 1, \cdots, r; i = 1, \cdots, n; r \leqslant n).$$

From (13.4) we have as the finite equations of the group

(13.9) $$x'^\alpha = x^\alpha + u^\alpha, \qquad x'^\sigma = x^\sigma$$
$$(\alpha = 1, \cdots, r; \sigma = r + 1, \cdots, n; r \leqslant n).$$

If the generic rank q of (13.7) is less than r, by renumbering the subscripts of the ξ's, we have that the matrix $\|\xi_h^i\|$ for $h = 1, \cdots, q$ is of rank q. Then we have in any coordinate system

(13.10) $\xi_p^i = \varphi_p^h(x)\xi_h^i$ $(h = 1, \cdots, q; p = q + 1, \cdots, r)$.

Proceeding with this case in a manner similar to the preceding, we have in consequence (13.10)

(13.11) $\xi_h^i = \delta_h^i, \qquad \xi_p^h = \psi_p^h(x^{q+1}, \cdots, x^n), \qquad \xi_p^s = 0$

$$\begin{pmatrix} i = 1, \cdots, n; h = 1, \cdots, q; \\ p = q + 1, \cdots, r; s = q + 1, \cdots, n \end{pmatrix}.$$

In this case equations (13.4) are

$$\frac{\partial x'^h}{\partial u^\alpha} = \delta_\alpha^h, \qquad \frac{\partial x'^h}{\partial u^\sigma} = \psi_\sigma^h(x'^{q+1}, \cdots, x'^n), \qquad \frac{\partial x'^s}{\partial u^a} = 0$$

$$\begin{pmatrix} h, \alpha = 1, \cdots, q; a = 1, \cdots, r; \\ s = q + 1, \cdots, n; \sigma = q + 1, \cdots, r \end{pmatrix}.$$

The integral of these equations is

(13.12) $x'^h = x^h + u^h + u^\sigma\psi_\sigma^h(x^{q+1}, \cdots, x^n), \qquad x'^s = x^s$

$$\begin{pmatrix} h = 1, \cdots, q; \sigma = q + 1, \cdots, r; \\ s = q + 1, \cdots, n \end{pmatrix}.$$

In consequence of (13.3) the equations (6.7) in terms of the parameters u^α are

$$\frac{\partial u_3^\alpha}{\partial u_2^\beta} = \delta_\beta^\alpha,$$

and consequently the analogue of (6.8) is

$$u_3^\alpha = u_1^\alpha + u_2^\alpha.$$

This result is in agreement with (13.9) and (13.12).

It should be remarked that the preceding results must be thought of as conditioned by the extent of the domain to which they apply. Thus in specifying the existence of integrals by the aid of which these forms were obtained, we have not gone into the nature of these integrals, which question clearly bears directly on the above remark.

14. The vectors U_α^a. Making use of equations (6.4) in canonical parameters, namely

$$(14.1) \qquad \frac{\partial U_\alpha^a(u)}{\partial u^\beta} - \frac{\partial U_\beta^a(u)}{\partial u^\alpha} = c_{bd}^a U_\alpha^b(u) U_\beta^d(u),$$

we have in consequence of (12.6) and (12.13)

$$\frac{dU_\alpha^a(et)}{dt} = \frac{\partial U_\alpha^a(u)}{\partial u^\beta} e^\beta = \frac{1}{t}\left[\frac{\partial U_\beta^a(u)}{\partial u^\alpha} u^\beta + c_{bd}^a U_\alpha^b(u) U_\beta^d(u) u^\beta\right]$$

$$= \frac{1}{t}(\delta_\alpha^a - U_\alpha^a(et)) + c_{bd}^a U_\alpha^b(et) e^d.$$

If then we put

$$(14.2) \qquad \theta_\alpha^a(e;t) = t U_\alpha^a(et),$$

the preceding equations become

$$(14.3) \qquad \frac{d\theta_\alpha^a}{dt} = \delta_\alpha^a + c_{bd}^a \theta_\alpha^b e^d,$$

and in consequence of equations (14.1) we have

$$(14.4) \qquad \frac{\partial \theta_\alpha^a}{\partial e^\beta} - \frac{\partial \theta_\beta^a}{\partial e^\alpha} = c_{bd}^a \theta_\alpha^b \theta_\beta^d.$$

The solutions of (14.3) for which $(\theta_\alpha^a)_{t=0} = 0$ are of the form

$$(14.5) \quad \theta_\alpha^a = t\left[\delta_\alpha^a + \frac{1}{2}c_{\alpha\beta_1}^a e^{\beta_1} t + \frac{1}{3!}c_{\alpha\beta_1}^{\gamma_1} c_{\gamma_1\beta_2}^a e^{\beta_1} e^{\beta_2} t^2 + \cdots\right.$$

$$\left. + \frac{1}{(r+1)!}c_{\alpha\beta_1}^{\gamma_1} c_{\gamma_1\beta_2}^{\gamma_2} \cdots c_{\gamma_{r-1}\beta_r}^a e^{\beta_1} \cdots e^{\beta_r} t^r + \cdots\right],$$

and consequently

$$(14.6) \quad U_\alpha^a(u) = \delta_\alpha^a + \frac{1}{2}c_{\alpha\beta_1}^a u^{\beta_1} + \cdots$$

$$+ \frac{1}{(r+1)!}c_{\alpha\beta_1}^{\gamma_1} c_{\gamma_1\beta_2}^{\gamma_2} \cdots c_{\gamma_{r-1}\beta_r}^a u^{\beta_1} \cdots u^{\beta_r} + \cdots.$$

Because of (7.3) these expressions agree with (12.13).

If $\left|c_{\alpha\beta}^\gamma\right| < c$ for all values of α, β, γ from $1, \ldots, r$, and we put $u = \sum_\alpha |u^\alpha|$, we have that the series on the right in (14.6) is less than the series

$$1 + \frac{1}{2}cu + \frac{1}{3!}rc^2u^2 + \cdots + \frac{1}{(n+1)!}r^{n-1}(cu)^n + \cdots,$$

whose sum is $(e^{cru} - 1)/cur^2 + 1 - \frac{1}{r}*$, and consequently the series (14.6) are convergent.

If G_r is a group for which the identity is determined by a_0^α and the vectors $\bar{A}_a^\alpha$ are chosen so that (8.4) are satisfied, it follows from (12.5) that the canonical parameters with respect to the ennuple $\bar{A}_a^\alpha$ are the same as with respect to the ennuple A_a^α. This is a consequence also of the fact that the symmetric part of the coefficients $L_{\beta\gamma}^\alpha$ is the same as of $\bar{L}_{\beta\gamma}^\alpha$ in consequence of (8.10). Accordingly, if U_α^a and $\bar{U}_\alpha^a$ are the components in the u's of the respective vectors A_a^α and $\bar{A}_a^a$, we have

(14.7) $\quad u^a\bar{U}_\alpha^\alpha = u^a, \qquad u^a\bar{U}_a^\alpha = u^\alpha, \qquad \bar{U}_\alpha^a(0) = \delta_\alpha^a, \qquad \bar{U}_a^\alpha(0) = \delta_a^\alpha.$

Since the functions $\bar{U}_\alpha^a(u)$ are defined by (14.6) with the c's replaced by the $\bar{c}$'s, in consequence of (8.13) we have

(14.8) $$\bar{U}_\alpha^a(u) = U_\alpha^a(-u),$$

and because of (5.7)

(14.9) $$\bar{U}_a^\alpha(u) = U_a^\alpha(-u).$$

If analogously to (8.14) we put

(14.10) $$\bar{U}_a^\alpha(u) = \sigma_a^b U_b^\alpha(u),$$

then the σ's are equal to the ρ's in (8.14) under the transformation of the a's into the u's. In place of (8.17) we have

(14.11) $$\frac{\partial\sigma_a^b}{\partial u^\alpha} = c_{ef}^b\sigma_a^e U_\alpha^f,$$

and in consequence of (12.13)

(14.12) $$\frac{d\sigma_a^b}{dt} = c_{ef}^b\sigma_a^e e^f.$$

From (12.12) and (14.7) we have $\sigma_a^b(0) = \delta_a^b$. The solutions of (14.12) satisfying this condition are of the form

---

* *Schur*, 1891, 1, p. 270.

$$(14.13) \quad \sigma_a^b = \delta_a^b + c_{af_1}^b u^{f_1} + \frac{1}{2} c_{e_1f_1}^b c_{af_2}^{e_1} u^{f_1} u^{f_2}$$

$$+ \frac{1}{3!} c_{e_1f_1}^b c_{e_2f_2}^{e_1} c_{af_3}^{e_2} u^{f_1} u^{f_2} u^{f_3} + \cdots .$$

From (14.10) and (14.9) we have

$$(14.14) \qquad U_a^\alpha(-u) U_\alpha^b(u) = \sigma_a^b(u)^*,$$

and also

$$(14.15) \qquad U_\alpha^b(u) = \sigma_a^b(u) U_\alpha^a(-u).$$

When canonical parameters are used, we indicate by

$$(14.16) \qquad u_3^\alpha = \psi^\alpha(u_1; u_2)$$

the equations analogous to (4.6), and we have that the ψ's are such that (cf. (4.14))

$$(14.17) \qquad u_1^\alpha = \psi^\alpha(u_1; 0), \qquad u_2^\alpha = \psi^\alpha(0; u_2).$$

From the equations (cf. (6.7) and (8.3))

$$(14.18) \qquad \frac{\partial u_3^\alpha}{\partial u_2^\beta} = U_a^\alpha(u_3) U_\beta^a(u_2), \qquad \frac{\partial u_3^\alpha}{\partial u_1^\beta} = \overline{U}_a^\alpha(u_3) \overline{U}_\beta^a(u_1)$$

and (14.17) we have

$$(14.19) \quad U_a^\alpha(u_1) = \left(\frac{\partial \psi^\alpha(u_1; u_2)}{\partial u_2^a} \right)_{u_2^\alpha = 0}, \qquad \overline{U}_a^\alpha(u_2) = \left(\frac{\partial \psi^\alpha(u_1; u_2)}{\partial u_1^a} \right)_{u_1^\alpha = 0},$$

which are the analogues of (9.5) and (9.6).

15. The second and third fundamental theorems. Suppose that in a V_n of coordinates x^i we have r sets of contravariant vectors ξ_a^i linearly independent (constant coefficients) such that the equations (6.2) are satisfied, where the c's are constants; as previously shown these constants must be such that (7.3) and (7.4) hold. We shall show that there exist functions A_α^a of parameters $a^1, \ldots, a^r$ such that the corresponding set of equations (5.5) are completely integrable, and consequently their solutions define a group.

To this end we consider the differential equations (14.3), that is,

$$(15.1) \qquad \frac{d\theta_\alpha^a}{dt} = \delta_\alpha^a + c_{df}^a \theta_\alpha^d e^f,$$

---

* Cf. *Schur*, 1891, 1, p. 275.

where the e's are a set of r parameters. The solutions of these equations which vanish when $t = 0$ are given as power series by (14.5), from which it is seen that these solutions are of the form

$$\theta_\alpha^a = t f_\alpha^a(e^1 t, \cdots, e^r t).$$

We have from (15.1)

$$\frac{d}{dt}\left(\frac{\partial \theta_\alpha^a}{\partial e^\beta}\right) = c_{b\beta}^{\;a}\theta_\alpha^b + c_{df}^{\;a}e^f \frac{\partial \theta_\alpha^d}{\partial e^\beta},$$

$$\frac{d}{dt}(c_{bd}^{\;a}\theta_\alpha^b\theta_\beta^d) = c_{\alpha d}^{\;a}\theta_\beta^d + c_{b\beta}^{\;a}\theta_\alpha^b + \theta_\alpha^b\theta_\beta^d e^f(c_{bf}^{\;h}c_{hd}^{\;a} + c_{fd}^{\;h}c_{hb}^{\;a})$$

$$= c_{\alpha d}^{\;a}\theta_\beta^d + c_{b\beta}^{\;a}\theta_\alpha^b + \theta_\alpha^b\theta_\beta^d e^f c_{bd}^{\;h}c_{hf}^{\;a},$$

the latter being a consequence of (7.3) and (7.4). Hence we have

$$\frac{d}{dt}\left(\frac{\partial \theta_\alpha^a}{\partial e^\beta} - \frac{\partial \theta_\beta^a}{\partial e^\alpha} - c_{bd}^{\;a}\theta_\alpha^b\theta_\beta^d\right) = c_{hf}^{\;a}e^f\left(\frac{\partial \theta_\alpha^h}{\partial e^\beta} - \frac{\partial \theta_\beta^h}{\partial e^\alpha} - c_{bd}^{\;h}\theta_\alpha^b\theta_\beta^d\right).$$

From (14.5) we have that θ_α^a and $\dfrac{\partial \theta_\alpha^a}{\partial e^\beta}$ vanish when $t = 0$. Consequently when $t = 0$, the right-hand members of these equations are zero, and from the theory of differential equations it follows that for all values of t we have

$$\frac{\partial \theta_\alpha^a}{\partial e^\beta} - \frac{\partial \theta_\beta^a}{\partial e^\alpha} = c_{bd}^{\;a}\theta_\alpha^b\theta_\beta^d.$$

If we put $\theta_\alpha^a = tU_\alpha^a(u)$, where $u^\alpha = e^\alpha t$, we obtain equations (14.1), and $U_\alpha^a(u)$ are given by (14.6), which series have been shown to be convergent.[*] Since $U_\alpha^a(0) = \delta_\alpha^a$, the determinant $|U_\alpha^a|$ is of rank r. Hence we have the following second fundamental theorem of Lie:

[15.1] *If r sets of functions $\xi_a^i(x)$ satisfy the conditions (6.2) and they are linearly independent (constant coefficients), there exist functions $U_\alpha^a(u)$, whose determinant is of rank r, such that the equations*

$$\frac{\partial x'^i}{\partial u^\alpha} = \xi_a^i(x')U_\alpha^a(u)$$

are completely integrable; the solutions $f^i(x; u)$ of these equations such that $f^i(x; 0) = x^i$[†] define a group G_r of transformations consisting of

---

[*] Cf. also *Engel*, 1891, 2, pp. 308–311.

[†] That such solutions exist has been shown in §6.

the transformations of the groups G_1 generated by the ∞^{r-1} infinitesimal transformations $e^a X_a f$ or of products of such transformations.

In the foregoing proof no use has been made of the functions ξ_a^i except as giving the constants c_{ab}^e and thence the conditions (7.3) and (7.4). What has been shown in fact is that, if a set of constants satisfy these conditions, the functions U_α^a defined by (14.6) satisfy equations (14.1), that

$$u^\alpha U_\alpha^a = u^a, \qquad U_\alpha^a(0) = \delta_\alpha^a$$

and consequently the determinant $|U_\alpha^a(u)|$ is of rank r. Furthermore the functions $U_a^\alpha(u)$, satisfy the equations

$$U_a^\alpha \frac{\partial U_b^\beta}{\partial u^\alpha} - U_b^\alpha \frac{\partial U_a^\beta}{\partial u^\alpha} = c_{ab}^e U_e^\beta.$$

Consequently the equations

$$\frac{\partial y'^\alpha}{\partial u^\beta} = U_a^\alpha(y') U_\beta^a(u)$$

are completely integrable, and the solutions

$$y'^\alpha = f^\alpha(y; u)$$

such that $y'^\alpha = y^\alpha$ when $u = 0$, define a group G_r of transformations. This is the same as the first parameter group when the c's arise from a given group G_r. Hence we have the following *third fundamental theorem* of Lie:

[15.2] *Any set of constants c_{ab}^e, which satisfy the conditions (7.3) and (7.4) are the constants of structure of a group.*

It should be remarked that the preceding argument applies only to the domain about the origin for which the determinant of U_α^a is not zero *

Exercises

1. Show that the inverse of $T_{a_0 + \delta a_0}$ is $T_{a_0 - \delta a_0}$ where a_0^α determine the identity and that the inverses of $T_{a + \delta a} T_a^{-1}$ and $T_a^{-1} T_{a + \delta a}$ are $T_{a - \delta a} T_a^{-1}$ and $T_a^{-1} T_{a - \delta a}$ respectively.

2. Show that

$$T_{a_2 + \delta a_2} T_{a_1 + \delta a_1} = T_{a_2} T_{a_1},$$

---

* Cf. *Cartan*, 1930, 1, p. 17.

if δa_1^α and δa_2^α are chosen so that

$$\frac{\partial \varphi^\beta(a_1; a_2)}{\partial a_1^\alpha} \delta a_1^\alpha + \frac{\partial \varphi^\beta(a_1; a_2)}{\partial a_2^\alpha} \delta a_2^\alpha = 0,$$

and that in this case $T_{a_2}^{-1}T_{a_2+\delta a_2}$ and $T_{a_1+\delta a_1}T_{a_1}^{-1}$ are inverses of one another, and consequently (Ex. 1)

$$T_{a_1+\delta a_1}T_{a_1}^{-1} = T_{a_2}^{-1}T_{a_2-\delta a_2}.$$

3. Given two sets of transformations

$$x'^i = f_1^i(x^1, \cdots, x^n; a_1^1, \cdots, a_1^r), \qquad x'^i = f_2^i(x^1, \cdots, x^n; a_2^1, \cdots, a_2^r),$$

in which the a's are essential, and denote by T_{a_1} and T_{a_2} transformations of the two sets; in order that $T_{a_2}T_{a_1}$ be of the form

$$x''^i = f_3^i(x^1, \cdots, x^n; a_3^1, \cdots, a_3^r),$$

where the a_3's are essential, in which case

$$a_3^\alpha = \psi^\alpha(a_1; a_2),$$

it is necessary that the functions f_1^i satisfy a set of equations of the form

$$\frac{\partial x'^i}{\partial a_1^\alpha} = \xi_a^i(x')A_\alpha^a(a_1).$$

Then show that a necessary and sufficient condition that the above conditions be met is that

$$T_{a_1} = S_{b_1}T_{(a_1)_0}, \qquad T_{a_2} = T_{(a_2)_0}S_{b_2},$$

where S_{b_1} and S_{b_2} are transformations of a group G_r, and $T_{(a_1)_0}$ and $T_{(a_2)_0}$ are particular transformations of the two sets.

Bianchi, 1918, 1, p. 118.

4. If the r parameters in the equations $y^i = f^i(x; a^1, \cdots, a^r)$ are essential, the maximum number in the sequence (3.7), say μ_s, is equal to r and consequently the $s + 1$ system of equations

$$y^i = f^i, \qquad y^i{}_{,i_1} = f^i{}_{,i_1}, \cdots, y^i{}_{,i_1 \cdots i_s} = f^i{}_{,i_1 \cdots i_s}$$

can be solved for the a's; when the a's are eliminated from these equations, q equations

$$(1) \qquad F_\alpha(x, y, y_{,i_1}, \cdots, y_{,i_1 \cdots i_s}) = 0 \qquad (\alpha = 1, \cdots, q)$$

result, where $q = \epsilon_0 + \epsilon_1 + \cdots + \epsilon_s - r$, ϵ_t being the number of equations involving the t th derivatives and $\epsilon_0 = n$. Then $y^i{}_{,i_1 \cdots i_{s+1}}$ are expressible in the form

$$(2) \qquad y^i{}_{,i_1 \cdots i_{s+1}} = \Phi^i{}_{j_1 \cdots i_{s+1}}(x, y, y_{,i_1}, \cdots, y_{,i_1 \cdots i_s}).$$

Thus (1) and (2) constitute the differential equations arising from $y^i = f^i(x; a)$ by the elimination of the a's, and the latter is the general integral of the system (1).

5. If the equations $y^i = f^i(x; a)$ in Ex. 4 define a group, and if $\varphi^i(x)$ and $\psi^i(x)$ are solutions of the system (1), so also are $\psi^i(\varphi(x))$ solutions. The system (1) is called the *differential equations of definition* of the group.

6. For the general affine group $x'^i = a^i_j x^j + a^i$ the equations of definition (Ex. 5) are

$$\frac{\partial^2 x'^i}{\partial x^j \partial x^k} = 0 \qquad\qquad (i, j, k = 1, \cdots, n).$$

7. A particular case of Ex. 4 is afforded by the set of equations

$$\xi^i = e^a \xi^i_a,$$

where ξ^i_a are the vectors of a group G_r. In this case equations (1) are linear and homogeneous in ξ^i and their derivatives of order 1 to s, the coefficients being functions of the x's. These equations admit as solutions each set ξ^i_a and also $\xi^i_a \frac{\partial \xi^j_b}{\partial x^i} - \xi^i_b \frac{\partial \xi^j_a}{\partial x^i}$. The corresponding equations (1) are called the *equations of definition of the infinitesimal transformations of the group*.

Lie-Engel, 1888, 1, vol. 1, p. 185.

8. Show that if a system of differential equations linear and homogeneous in ξ^i and its derivatives to a finite order possesses the property that when ξ^i_1 and ξ^i_2 are any two sets of solutions so also is $\xi^i_1 \frac{\partial \xi^j_2}{\partial x^i} - \xi^i_2 \frac{\partial \xi^j_1}{\partial x^i}$, and the general solution ξ^i involves a finite number, r, of arbitrary constants, then $\xi^i \frac{\partial f}{\partial x^i}$ is the general symbol of a group G_r.

Lie-Engel, 1888, 1, vol. 1, p. 188.

9. The equations of definition of the infinitesimal transformation of a group follow from the equations of definition (Ex. 4 (1)) of the group, if y^i are replaced by $x^i + \xi^i \delta t$ and the coefficient of δt is equated to zero.

Bianchi, 1918, 1, p. 128.

10. Find the equations of definition of the infinitesimal transformation of the affine group (Ex. 6) and show that

$$\xi^i = e^i + e^i_j x^j \qquad\qquad (i, j = 1, \cdots, n),$$

where the e's are constants.

11. The group of motions in euclidean n-space in cartesian coordinates is defined by

$$x'^i = a^i + a^i_j x^j,$$

where the a's are constants subject to the conditions

$$a^i_j a^i_k = \delta_{jk}.$$

Show that the equations of definition (Ex. 4) are

$$\frac{\partial x'^i}{\partial x^j} \frac{\partial x'^i}{\partial x^k} = \delta_{jk}.$$

12. For the group of motions in euclidean n-space (Ex. 11) the equations of definition of the infinitesimal transformations are

$$\frac{\partial \xi^i}{\partial x^j} + \frac{\partial \xi^j}{\partial x^i} = 0 \qquad\qquad (i, j = 1, \cdots, n),$$

of which the integrals are

$$\xi^i = b^i + b^i_j x^j, \qquad b^i_j = -b^j_i.$$

Consequently the symbols are

$$p_i, \qquad x^i p_j - x^j p_i \qquad\qquad (i, j = 1, \cdots, n).$$

13. When a group of transformations is defined not by a single set of equations as (4.1), but by $m (>1)$ sets of equations, say

$$x'^i = f^i_{(s)}(x; a) \qquad\qquad (s = 1, \cdots, m),$$

in each of which there enter r essential parameters a^α_s, the equations are said to define a *mixed group*. Show that in this case each set of functions $f^i_{(s)}$ satisfy a set of equations

$$\frac{\partial f^i_{(s)}}{\partial a^\alpha_s} = \xi^i_a(f_{(s)}) A^a_{(s)\alpha}(a_s),$$

where the ξ^i_a are the same functions for each value of s, but not the $A_{(s)}$'s.

Lie-Engel, 1888, 1, vol. 1, p. 314.

14. If the functions $\xi^i_a(x)$ in Ex. 13 are such that $X_a f = \xi^i_a p_i$ are the symbols of a group by theorem [6.3], and if a^α_{s0} are values for which the determinant $|A_{(s)\alpha}{}^a(a_{s0})|$ is not zero, then $T^{-1}_{(s)a_0} T_{(s)a}$ are transformations of the mixed group. Since the identity is in one set say $T_{(1)a}$, this set is the group of symbols $X_a f$, and

$$T_{(s)a} = T_{(s)a_0} T_{(1)a}.$$

15. Show that

$$x' = x \cos a - y \sin a, \qquad y' = x \sin a + y \cos a,$$
$$x' = x \cos a + y \sin a, \qquad y' = x \sin a - y \cos a$$

define a mixed group; that the symbol is $y\dfrac{\partial f}{\partial x} - x\dfrac{\partial f}{\partial y}$; and that if T_{20} denotes the second transformation when $a = 0$, then

$$T_{2a} = T_{20} T_{1a}.$$

16. In order that m sets of equations define a mixed group it is necessary that the same number of essential parameters enter into each set of equations.

Lie-Engel, 1888, 1, vol. 1, p. 319.

CHAPTER II

PROPERTIES OF GROUPS. DIFFERENTIAL EQUATIONS

16. Sub-groups of a G_r. We have seen that for each set of
values of the constants e^a in $e^a X_a f$, the latter is the symbol of a
sub-group G_1 of the given G_r. The question arises whether $m(<r)$
sets of independent constants e_t^a for $a = 1, \cdots, r$ and $t = 1, \cdots,$
$m(<r)$ can be found so that

$$(16.1) \qquad\qquad Y_t f = e_t^a X_a f$$

are the symbols of group of order m, and thus a sub-group of G_r.
Since the constants e_t^a are independent, their matrix $\|e_t^a\|$ must be
of rank m. Also we must have

$$(16.2) \qquad\qquad (Y_t, Y_u)f = \gamma_{tu}^v Y_v f \qquad (t, u, v = 1, \cdots, m),$$

where the γ's are constants. Because of (16.1), (16.2) and (7.2)
we have

$$(16.3) \qquad\qquad e_v^c \gamma_{tu}^v = e_t^a e_u^b c_{ab}^c.$$

The problem reduces accordingly to the consistency of these
equations and is an algebraic one. For each set of values of t
and u, these equations must admit a solution. Since the matrix
$\|e_t^a\|$ is by hypothesis of rank m, it means that the augmented
matrix, for given values of t and u, namely

$$\|e_v^c, e_t^a e_u^b c_{ab}^c\| \quad (c = 1, \cdots, r; \; v = 1, \cdots, m)$$

must be of rank m. Thus all these matrices for values of t and
u from 1 to m must be of rank m; that is, the equations obtained
by equating to zero all the determinants of order $m + 1$ of these
$\frac{1}{2}m(m - 1)$ augmented matrices must be consistent in the e's, the
latter being subject to the condition that the matrix $\|e_t^a\|$ be of
rank m.

If these conditions are satisfied and the basis $X_a f$ is chosen so
that $X_1 f, \ldots, X_m f$ are symbols of the sub-group, then

$$(16.4) \qquad\qquad (X_t, X_u)f = c_{tu}^v X_v f \qquad (t, u, v, = 1, \cdots, m).$$

Consequently we have

$$(16.5) \qquad\qquad c_{tu}^{\ z} = 0 \qquad \begin{pmatrix} t,\, u = 1,\, \cdots,\, m; \\ z = m+1,\, \cdots,\, r \end{pmatrix}.$$

As an example of sub-groups we have that for the group of motions in euclidean 3-space, the translations form an Abelian sub-group G_3, and the rotations about a point form a sub-group G_3.

We shall establish the theorem:

[16.1]   *The basis of a group G_2, or of a sub-group G_2, can be chosen so that*

$$(16.6) \qquad\qquad (X_1,\, X_2)f = X_2 f \text{ or } (X_1,\, X_2)f = 0.$$

For a given basis we have

$$(X_1,\, X_2)f = a X_1 f + b X_2 f.$$

If $a \neq 0$, $b = 0$, we replace $X_2 f$ by $-a X_1 f$ and $X_1 f$ by $X_2 f$; if $a = 0$, $b \neq 0$ we replace $X_1 f$ by $b X_1 f$; if $a \neq 0$, $b \neq 0$, we replace $X_1 f$ by $b X_1 f$ and $X_2 f$ by $\dfrac{1}{b} X_2 f - a X_1 f$.

Let $u^a X_a f$ be the symbol of a G_1 of a G_r, and find under what conditions $v^a X_a f$ is the symbol of a G_1, so that the two form a sub-group G_2 of G_r. For this to be the case it is necessary and sufficient that

$$(16.7) \qquad\qquad (u^a X_a,\, v^b X_b)f = \sigma u^a X_a f + \rho v^b X_b f.$$

In consequence of (7.2) we have

$$(16.8) \qquad\qquad \eta_b^e(u) v^b - \sigma u^e - \rho v^e = 0,$$

where

$$(16.9) \qquad\qquad \eta_b^e(u) = c_{ab}^{\ e} u^a.$$

Since $\eta_b^e(u) u^b = 0$ in consequence of (7.3), we have that the rank of the matrix $\|\eta_b^e\|$ is less than r. Hence in order that (16.7) with $\sigma = \rho = 0$ be satisfied by v^a different from u^a, it is necessary that the rank of this matrix be less than $r - 1$, and we have:

[16.2]   *A necessary and sufficient condition that $u^a X_a f$ be the symbol of a G_1 contained in an Abelian sub-group G_2 of a G_r is that the rank of the matrix of $c_{ab}^{\ e} u^a$ be less than $r - 1$.*

In order that $\rho = 0$, $\sigma \neq 0$ in (16.7), in which case we may take $\sigma = 1$, as in theorem [16.1], it follows from (16.8) that the augmented matrix $\|\eta_b^e, \ u^e\|$ must be of the same rank as the matrix $\|\eta_b^e\|$, and that if the rank is $r - p$, then certain p of the v's may be chosen arbitrarily and the others are determined.

In order that $\rho \neq 0$ in (16.7), we may take $\sigma = 0$ as in the case of theorem [16.1] and equations (16.8) become

$$(16.10) \qquad\qquad (\eta_b^e - \rho\delta_b^e)v^b = 0.$$

From theorem [16.2] it follows that, if zero is a multiple root of the determinant equation

$$(16.11) \qquad\qquad |\eta_b^e - \delta_b^e\rho| = 0,$$

there are values of v^a other than u^a satisfying (16.10) and these lead to Abelian sub-groups G_2. If ρ is a non-zero root of (16.11), v^a determined by (16.10) are different from u^a, and we have a G_2 such that

$$(u^a X_a, \ v^b X_b)f = \rho v^a X_a f.$$

As a result of the preceding investigation we have:

[16.3] *Any sub-group G_1 of a G_r is contained in at least one sub-group G_2 of G_r.*

17. Absolute and relative invariants of a group. A function $F(x)$ which is unaltered by all the transformations of a group G_r, that is, such that $F(x') \equiv F(x)$ is called an *absolute invariant* of the group. For a sub-group G_1 of transformations the trajectories are given by (11.8), that is,

$$\frac{dx'^i}{dt} = e^a \xi_a^i(x'),$$

where the e's are definite constants. A necessary and sufficient condition that F be an absolute invariant of this G_1 is that

$$F(x') = F(x)$$

be an identity when the x''s are replaced by the solutions of the above equations determined by the x's as the values when $t = 0$. Hence we must have

$$\frac{dF(x')}{dt} = e^a \xi_a^i(x') \frac{\partial F(x')}{\partial x'^i} = 0.$$

If F is to be an absolute invariant for G_r, this condition must hold for all values of the e's and we have:

[17.1] *A necessary and sufficient condition that a function $F(x^1, \ldots, x^n)$ be an absolute invariant of a G_r is that*

$$(17.1) \qquad\qquad X_a F = 0 \qquad\qquad (a = 1, \cdots, r).$$

If the rank of the matrix

$$(17.2) \qquad\qquad M = \|\xi_a^i\|$$

is q, then q of the equations (17.1) are independent, say

$$(17.3) \qquad\qquad X_\sigma f = 0 \qquad\qquad (\sigma = 1, \cdots, q).$$

In consequence of (7.2) and the fact that the other symbols are expressible in terms of those in (17.3) the system (17.1) is complete. Hence if $q < n$, there are $n - q$ independent absolute invariants, and none when $q = n$.

Suppose that a function $F(x)$ is not an invariant of the group with symbols $X_a f$, but that it is invariant for the transformations with the p linearly independent (constant coefficients) symbols

$$Y_l f = \lambda_l^a X_a f \qquad\qquad (l = 1, \cdots, p),$$

where the λ's are constants, and only for these symbols. If we take as basis these p symbols and $r - p$ of the $X_a f$ so that the whole set be linearly independent (constant coefficients), we have

$$(Y_l, Y_m)f = \gamma_{lm}^k Y_k f + \gamma_{lm}^t X_t f \quad \begin{pmatrix} k, l, m = 1, \cdots, p; \\ t = p + 1, \cdots, r \end{pmatrix}.$$

When $f = F$, we obtain from these identities the equations

$$\gamma_{lm}^t X_t F = 0,$$

from which it follows that $\gamma_{lm}^t = 0$, otherwise we should have further symbols for which F is an invariant. Hence we have:

$$(Y_l, Y_m)f = \gamma_{lm}^k Y_k f,$$

and consequently by the second fundamental theorem the transformations with the symbols $Y_a f$ form a group, which is a *sub-group* of G_r. This result may be obtained also by observing that the product of any two transformations applied to $F(x)$ leaves it invariant and consequently is a member of the same set. We call this sub-group the *sub-group of the function $F(x)$.* From the results at the beginning of this section it follows that the number of independent functions invariant for a sub-group is equal to n minus the rank of the matrix of the symbols of the sub-group.

Suppose that we have p independent functions $F_1, \ldots, F_p$ such that

$$(17.4) \qquad X_a F_\alpha = 0 \qquad \begin{pmatrix} a = 1, & \cdots, r; \\ \alpha = 1, & \cdots, p \end{pmatrix}$$

not identically, but for the values of the x's for which

$$(17.5) \qquad F_\alpha = 0.$$

If we change the coordinates so that equations (17.5) are in the new coordinate system

$$(17.6) \qquad x^\alpha = 0 \qquad (\alpha = 1, \cdots, p),$$

then from (17.4) we have ξ_a^α equal to zero when $x^1, \ldots, x^p$ are zero; we indicate this by

$$(17.7) \qquad (\xi_a^\alpha)_0 = 0.$$

If now we take the sub-set

$$\frac{dx^\alpha}{dt} = e^a \xi_a^\alpha(x)$$

of equations (11.8) and treat the quantities $x^{p+1}, \ldots, x^n$ entering therein as parameters, we have that the solutions of these equations which have the initial values (17.6) are in fact (17.6), in consequence of (17.7) and the theory of systems of ordinary differential equations. Hence we have:

[17.2] *If $X_\alpha f$ are the symbols of a G_r and there exist a set of independent functions $F_1, \ldots, F_p$ such that $X_a F_\alpha = 0$ for points of the variety defined by $F_1 = \cdots = F_p = 0$, this variety is transformed into itself by each transformation of G_r.*

In this case the functions F_α are said to define a *relative invariant* of the group.

Theorems [17.1] and [17.2] are established immediately, if we consider only infinitesimal transformations of the group, so that we have proved that invariance under the infinitesimal transformations of a group implies invariance under the finite transformations generated by them. We show in §51 that the same principle holds in the case of another type of invariant.

18. Ordinary and singular points.   Order of a transformation at a point.   Consider the matrix

$$(18.1) \qquad\qquad M = \|\xi_a^i\|$$

of the components of the vectors of a group G_r. Let q be the rank of M for general values of the x's, that is, q is the *generic* rank (§13). A point for whose coordinates the rank is q we call an *ordinary point* of the transformation, and a point for which the rank is less than q a *singular* point.

In order to distinguish between types of singular points, we proceed as follows.   If the equations obtained by equating to zero all the elements of (18.1) are consistent, we denote by L_0 the locus of points whose coordinates satisfy these equations and say that these points are singular of *order zero*.

If there is such a locus it consists of the points satisfying a set of independent and consistent equations, say

$$F_1 = 0, \; \cdots \;, F_s = 0.$$

If $s = n$, L_0 consists of one or more isolated points.   If $s < n$, then L_0 is a sub-variety of $n - s$ dimensions.   For example, in the case of the group of rotations in euclidean 3-space (Ex. 5, p. 43) the origin is the locus L_0.

If the equations obtained by equating to zero all the minors of order two of M are consistent, we denote by L_1 the locus of points, whose coordinates satisfy these equations and do not belong to L_0, if any, and call these points of *order one*.   Proceeding in this manner we have the possible loci $L_0, \ldots, L_{q-1}$ of singular points. The above remarks concerning the nature of L_0 apply to each of the loci $L_1, \ldots, L_{q-1}$.   Thus for any point in L_ρ the rank of the matrix is ρ.   In this sense the locus of the ordinary points is denoted by L_q.   Thus L_q consists of all the points of V_n other than those in $L_0, \ldots, L_{q-1}$, if any.

If the components ξ^i of a transformation are regular in the neighborhood of a point P_0 of coordinates x_0^i, and they are expressed in the form

$$(18.2) \qquad \xi^i(x) = \xi^i(x_0) + \left(\frac{\partial \xi^i}{\partial x^i}\right)_0 (x^i - x_0^i) + \cdots,$$

we say that the transformation is of *order zero* at P_0, when not all of the $\xi^i(x_0)$ are zero; that it is of *order one* when all of the $\xi^i(x_0)$ are zero but not all of the $\left(\dfrac{\partial \xi^i}{\partial x^j}\right)_0$, and so on.

If the rank of M at P_0 is q_0, then in the equations

$$(18.3) \qquad\qquad b^a \xi_a^i(x_0) = 0,$$

q_0 of the b's are expressible linearly and homogeneously in terms of the remaining $r - q_0$. Hence there are $r - q_0$ linearly independent transformations

$$(18.4) \qquad\qquad b^a X_a f$$

of order greater than zero at P_0 and q_0 of order zero.

Each of the infinitesimal transformations of order greater than zero leaves the point P_0 invariant, as follows from (10.21). If two such transformations are applied successively, the point P_0 is unaltered and consequently the resulting transformation is of the same kind. Consequently all the infinitesimal transformations of order greater than zero generate a group of order $r - q_0$, which is a subgroup of G_r; it is called the *sub-group of stability* of P_0.

In order to find expressions for the generators of this sub-group, we assume that the indices in (18.1) have been arranged so that the matrix $\|\xi_p^i(x_0)\|$ for $p = 1, \cdots, q_0$ is of rank q_0. Then we have

$$(18.5) \qquad \xi_s^i(x_0) = a_s^p \xi_p^i(x_0) \qquad \begin{pmatrix} p = 1, \cdots, q_0; \\ s = q_0 + 1, \cdots, r \end{pmatrix},$$

where the a's are constants. Hence the symbols

$$(18.6) \qquad\qquad X_s f - a_s^p X_p f,$$

being linearly independent and of order greater than zero at P_0, generate the sub-group of stability of P_0 of order $r - q_0$.

From the foregoing considerations it follows that, when the rank of M is r, an ordinary point does not admit a subgroup of stability other than the identity. If the rank q of M is less than r, then there is a sub-group of stability for each point; it is of order $r - q$ for an ordinary point and $r - p$ for a singular point of order p.

Exercises

1. The symbols of the general projective group G_8 in x and y, if we put $p = \dfrac{\partial f}{\partial x}$, $q = \dfrac{\partial f}{\partial y}$, are (cf. Ex. 9, p. 43)

$$p, \quad q, \quad xp, \quad yp, \quad xq, \quad yq, \quad x^2 p + xyq, \quad xyp + y^2 q;$$

show that the following are the symbols of sub-groups:

$$
\begin{array}{llllll}
\text{(i)} & p, & q, & xp, & yp, & xq, & yq; \\
\text{(ii)} & xp, & yp, & xq, & yq, & x^2p + xyq, & xyp + y^2q; \\
\text{(iii)} & p, & q, & xq, & xp - yq, & yp; \\
\text{(iv)} & xq, & xp - yq, & yp, & x^2p + xyq, & xyp + y^2q; \\
\text{(v)} & p, & q, & xp, & xq, & yq.
\end{array}
$$

Lie-Scheffers, 1893, 1, p. 272.

2. Show that the general projective group in the plane does not have a sub-group of order 7.

Lie-Scheffers, 1893, 1, p. 267.

3. The G_3 with the symbols

$$
p + xq, \qquad xp + 2yq, \qquad (x^2 - y)p + xyq
$$

leaves the parabola $x^2 - 2y = 0$ invariant.

4. Show that p, q, yq, xp are the symbols of a G_4, that $p + yq, q$ and $p + yq, yq$ are the symbols of sub-groups G_2, and these are the only sub-groups G_2 of which $p + yq$ is one of the symbols.

Lie-Scheffers, 1893, 1, p. 554.

5. If the ranks of M (18.1) and

$$
M_1 = \left\| \xi_a^i; \frac{\partial \xi_a^i}{\partial x^j} \right\|, \qquad M_2 = \left\| \xi_a^i; \frac{\partial \xi_a^i}{\partial x^j}, \frac{\partial^2 \xi_a^i}{\partial x^j \partial x^k} \right\|, \cdots
$$

for an ordinary point $P(x_0)$ are $q_0, q_0 + q_1, q_0 + q_1 + q_2, \cdots$, there are q_0 independent transformations of order zero at P, q_1 of order one, q_2 of order two and so on.   There is an M_s such that

(i)
$$
q_0 + q_1 + \cdots + q_s = r.
$$

6. If $X_1 f$ and $X_2 f$ are of orders h_1 and h_2 at $P(x_0)$, $(X_1, X_2)f$ is of order $h_1 + h_2 - 1$.

7. For a G_r in one variable the sequence (i) of Ex. 5 is $0, 1, \cdots, r - 1$, from which and Ex. 6 it follows that r is at most equal to three.

8. If $X_1 f, X_2 f, X_3 f$ are symbols of transformations of a G_3 on one variable of order zero, one and two at an ordinary point, by a suitable change of base one has

(i)
$$
(X_1, X_2)f = X_1 f, \qquad (X_1, X_3)f = 2X_2 f, \qquad (X_2, X_3)f = X_3 f.
$$

If the coordinate x is chosen so that $X_1 f = p$, the symbols of G_3 are

(ii)
$$
p, \qquad xp, \qquad x^2 p.
$$

and the finite equation of the group is

$$
x' = \frac{ax + b}{cx + d},
$$

that is, the projective group on the line; show also that the first two symbols of (ii) define the subgroup of affine transformations on the line.

Bianchi, 1918, 1, p. 374.

9. Find the varieties L_0, L_1, . . . , (§18) for the group in x, y, z whose matrix is

$$\begin{Vmatrix} y & -x & 0 \\ x & y & z \\ x^2 & xy & xz \\ xy & y^2 & yz \\ xz & yz & z^2 \end{Vmatrix}$$

and interpret the result geometrically.

10. For the group with the symbols

$$X_1f = p_1 + 2x^1p_2 + 3x^2p_3, \quad X_2f = x^1p_1 + 2x^2p_2 + 3x^3p_3, \quad X_3f = p_2 + 3x^1p_3$$

the function $3x^1x^2 - x^3 - 2(x^1)^3$ is a relative invariant, and an absolute invariant of the sub-group with the symbols X_1f and X_3f.

19. Invariant varieties. Two points are said to be *equivalent* under a given group, if they are transformable into one another by one or more transformations of the group. A sub-space V_m of V_n is called an *invariant variety* for the group, if all the points equivalent to each point of V_m lie in V_m. Thus if equations (17.1) admit s independent solutions, say F_1, . . . , F_s, then the equations

$$F_t = a_t \qquad (t = 1, \cdots, s),$$

where the a's are constants, define an invariant V_m, where

$$m = n - s,$$

and any one of these equations defines an invariant variety of order $n - 1$.

Consider a group G_r $(r > 1)$ and let the rank of the matrix M (18.1) at a given point $P_0(x_0)$ be $q_0 < n$. Let the indices of the symbols of G_r be so arranged that the rank of the matrix of the symbols X_ef $(e = 1, \cdots, q_0)$ is q_0 at P_0. Then the transformations with the symbols a^eX_ef, where the a's are arbitrary constants, are of order zero at P_0. The points equivalent to P_0 under G_r lie on the trajectories through P_0 of the G_1's determined by these symbols. The locus of the points is a variety of dimension q_0 and is the variety of least dimension containing points equivalent to P_0 under the group; it is called the *minimum invariant variety* for P_0.

Making use of the fact that the sub-group of stability of P_0 is of order $r - q_0$ (§18), we shall show that the rank of the matrix M at each point equivalent to P_0 is q_0. For let P be such a point and T

the transformation which sends P_0 into P, T^{-1} its inverse and $\overline{T}$ any transformation of stability of P_0; then

$$T\overline{T}T^{-1}(P) = P.$$

Since each $\overline{T}$ determines in this manner a transformation leaving P fixed, the sub-group of stability of P is at least of the same order as for P_0, and on reversing the process we have that it is of the same order. From this result it follows that, if P_0 is an ordinary point (§18), the points equivalent to it are ordinary points, and if P_0 is a singular point its equivalents are singular points of the same order.

If P_0 is an ordinary point ($q_0 = q < n$), there are $n - q$ absolute invariants F_σ and the minimum invariant variety for P_0 is given by the equations

$$(19.1) \qquad F_\sigma(x) = F_\sigma(x_0) \qquad (\sigma = 1, \cdots, n - q),$$

and these varieties constitute L_q as defined in §18. Furthermore, from the above results it follows that this is the minimum invariant variety for each of its points. Although equations (19.1) define an invariant variety for each P_0, it is the minimum variety for P_0 only in case P_0 is an ordinary point.

Consider, for example, the group of rotations in euclidean 3-space (Ex. 5, p. 43). The origin is the only singular point and is of order zero. The generic rank of the matrix is two and the function $x^2 + y^2 + z^2$ is the only absolute invariant of the group, so that the minimum invariant variety for each ordinary point is a sphere.

If P_0 is a singular point, that is, $q_0 < q$, the coordinates of any point equivalent to P_0 satisfy the equations obtained by equating to zero all the determinants of M of order $q_0 + 1$. The variety defined by these equations is an invariant variety of G_r. If we exclude from this variety all points, if any, for which the rank of M is less than q_0, and thus obtain the locus L_{q_0} of §18, the points of the minimum invariant variety of each point of L_{q_0} lie in L_{q_0}. However, the variety for a given point does not necessarily coincide with L_{q_0}.

The coordinates x^i of the points equivalent to P_0 are given by

$$(19.2) \qquad x^i = f^i(x_0; a),$$

when the a's take all possible values. In consequence of (5.5) the matrix

(19.3)
$$\left\| \frac{\partial f(x_0;\ a)}{\partial a} \right\|$$

is the product of the matrix $\|\xi_\alpha^i(x)\|$ and the non-singular matrix $\|A_\alpha^a\|$, and therefore is of the same rank as M at the point of coordinates x^i.* Equations (19.2) define the minimum invariant variety of the point P_0 and the order of this variety is equal to the rank of the matrix (19.3). If q_0 is the rank of this matrix, then q_0 of the equations (19.2) can be solved for q_0 of the a's, and when these are substituted in the remaining equations and the other $r - q$ a's are given a set of possible values, we obtain the equations

(19.4)
$$\psi_\rho(x;\ x_0) = 0 \qquad (\rho = 1,\ \cdots,\ n - q_0),$$

defining the minimum invariant variety of P_0. If $q_0 = q$, equations (19.4) are necessarily equivalent to (19.1).

We have just seen that the equations of invariant varieties may be obtained in several ways. Let us assume that the independent equations

(19.5)
$$F_1 = \cdots = F_p = 0$$

define an invariant variety. By definition the trajectories through each point in the variety lie in it. Then from (10.19) it follows that it is necessary that the equations

(19.6)
$$X_a F_\alpha = 0 \qquad \begin{pmatrix} a = 1,\ \cdots,\ r; \\ \alpha = 1,\ \cdots,\ p \end{pmatrix}$$

be satisfied by the coordinates of each point in the variety. If these equations are satisfied identically, the functions F_α are absolute invariants, and we have the case discussed in connection with equations (19.1). If (19.6) are satisfied because of (19.5), the functions F_α are relative invariants as shown in §17, and these are the only conditions to be satisfied to insure that equations (19.5) define an invariant variety. Evidently the second case arises only when all the points of this variety are singular. Recapitulating these results we have:

[19.1] *Let q be the rank of the matrix M (18.1) for a G_r. If the minors of M of order $p(<q)$ when equated to zero are consistent, the resulting equations define an invariant variety of G_r, which includes the minimum invariant variety for any point of it; whether $q = n$ or*

---

* Cf. *Bôcher*, 1907, 1, p. 79.

$q < n$, *the points of these invariant manifolds, if they exist, are singular points.   If $q < n$, there are minimum invariant varieties of ordinary points and these are defined by* (19.1).

20. The group induced in an invariant variety. Let V_m be an invariant variety, whose equations are given in the parametric form

$$(20.1) \qquad x^i = \psi^i(y^1, \cdots, y^m).$$

Since each of the vectors ξ^i_a at any point of V_m is tangential to V_m, they are expressible in the form

$$(20.2) \qquad \xi^i_a = \eta^\sigma_a \frac{\partial x^i}{\partial y^\sigma} \qquad\qquad (\sigma = 1, \cdots, m),$$

where the η's are functions of the y's.*   From (20.2) we have

$$(20.3) \qquad X_a f = \xi^i_a \frac{\partial f}{\partial x^i} = \eta^\sigma_a \frac{\partial f}{\partial y^\sigma} = Y_a f.$$

It may be that at all points of V_m one or more equations of the form

$$(20.4) \qquad c^a \xi^i_a = 0,$$

where the c's are constants, are satisfied.   In this case the G_1 with the symbol $c^a X_a f$ leaves V_m point-wise invariant.   If there are p such independent relations (20.4), there is a sub-group G_p of G_r which leaves V_m point-wise invariant.   Since the jacobian matrix $\left\| \dfrac{\partial x}{\partial y} \right\|$ is of rank m, from (20.4) and (20.2) we have

$$(20.5) \qquad c^a \eta^\sigma_a = 0.$$

Hence the symbols $Y_a f$ are not independent (constant coefficients), and in fact there are only $r - p$ of them thus independent.   These define a group Γ_{r-p} which is said to be induced in V_m by G_r.

Conversely, if p relations (20.5) hold, then (20.4) follow from (20.2) and there is a sub-group G_p of G_r which leaves V_m point-wise invariant.   Hence we have:

[20.1] *If V_m is an invariant variety for a G_r and there is a sub-group G_p which leaves V_m point-wise invariant, the group induced in V_m is a Γ_{r-p}; and conversely.*

---

* Cf. 1926, 3, p. 46.

If $Y_\rho f$ for $\rho = 1, \cdots, r - p$, are independent (constant coefficients) and we have

$$(20.6) \qquad Y_\tau f = d_\tau^\rho Y_\rho f \quad \begin{pmatrix} \rho = 1, \cdots, r - p; \\ \tau = r - p + 1, \cdots, r \end{pmatrix},$$

then from (7.2) and (20.3) we have

$$(20.7) \quad (Y_{\rho_1}, Y_{\rho_2})f = (c_{\rho_1\rho_2}^{\rho_3} + c_{\rho_1\rho_2}^{\tau} d_\tau^{\rho_3})Y_{\rho_3}f \; (\rho_1, \rho_2, \rho_3 = 1, \cdots, r - p)$$

as the equations of composition of the induced group.

If V_m is the minimum invariant variety for each of its points, then the sub-group of stability of each point is of order $r - m$, but only in case there is a sub-group of stability which holds for every point is the induced group of order less than r. Consider, for example, the group of rotations in 3-space about the origin. The spheres with centers at the origin are the minimum invariant varieties, but there is no sub-group of stability holding for all the points of any one of these spheres.

When in particular the equations (20.1) are taken in the form

$$(20.8) \qquad x^\tau = \psi^\tau(x^1, \cdots, x^m) \quad (\tau = m + 1, \cdots, n),$$

we have from (20.2)

$$\xi_a^i = \eta_a^\sigma \delta_\sigma^i \qquad (\sigma = 1, \cdots, m).$$

Hence we have:

[20.2] *When the equations of an invariant variety V_m are given in the form* (20.8), *the symbols $Y_a f$ of the induced group are of the form*

$$(20.9) \qquad Y_a f = \xi_a^\sigma(x^1, \cdots, x^m, \psi^{m+1}, \cdots, \psi^n)\frac{\partial f}{\partial x^\sigma}$$

$$\begin{pmatrix} a = 1, \cdots, r; \\ \sigma = 1, \cdots, m \end{pmatrix}.*$$

21. Transitive and intransitive groups. A group is said to be *transitive* when every two ordinary points of the space are equivalent under the group (§19); otherwise the group is *intransitive*. For example, the group of translations in euclidean 3-space is transitive and the group of rotations is intransitive. From the consideration of the finite equations of a group G_r it is seen that

---

* Cf. *Bianchi*, 1918, 1, pp. 164, 165.

for a group to be transitive we must have $r \geqslant n$ and the rank of the matrix (19.3) must be n; when $r = n$, the group is called *simply transitive*, otherwise *multiply transitive*.*

For an intransitive group there are invariant varieties of order less than n, otherwise by a combination of transformations any ordinary point could be transformed into any other ordinary point. If q is the generic rank of the matrix M (18.1) and $q = n$, there are no absolute invariants of the group (§17) and consequently the invariant variety of an ordinary point is the space itself, so that the group is transitive. But if $q < n$ there are absolute invariants of the group, and the group is intransitive, since there are invariant varieties for ordinary points. Hence we have:

[21.1]   *A necessary and sufficient condition that a group G_r be transitive is that $r \geqslant n$ and that the generic rank of the matrix M be n.*

For example, the first and second parameter groups of a G_r are simply transitive on r variables (the a's) (cf. §9). Also from §19 it follows that the group induced on an invariant variety is transitive, when it is the minimum invariant variety for its points, but otherwise it is intransitive.

A transitive group may have invariant varieties, the points of which necessarily are singular. For example, the group G_3 of Ex. 10, p. 67 is transitive and the equation

$$3x^1x^2 - x^3 - 2(x^1)^3 = 0$$

defines an invariant variety.

We shall develop in the remainder of this section certain equations which will be used in the subsequent development. We recall from §6 the fundamental equations

$$(21.1) \qquad \xi_a^\alpha \frac{\partial \xi_b^\beta}{\partial x^\alpha} - \xi_b^\alpha \frac{\partial \xi_a^\beta}{\partial x^\alpha} = c_{ab}^c \xi_c^\beta \qquad \left(\begin{array}{l} a,\, b,\, c = 1,\, \cdots ,\, r; \\ \alpha,\, \beta = 1,\, \cdots ,\, n \end{array} \right).$$

If the generic rank q of the matrix M (18.1) is less than r, we assign the subscripts of ξ_a^α so that the matrix $\|\xi_h^\alpha\|$ for $h = 1,\, \cdots ,\, q$ is of rank q and put

$$(21.2) \qquad \xi_p^\alpha = \varphi_p^h \xi_h^\alpha \qquad (h = 1,\, \cdots ,\, q;\, p = q + 1,\, \cdots ,\, r),$$

---

*Some writers use the term multiply transitive in the sense in which we use the term k-fold transitive in Ex. 14, p. 108.

where the φ's are functions of the x's. Then equations (21.1) may be written

$$(21.3) \qquad \xi_a^\alpha \frac{\partial \xi_b^\beta}{\partial x^\alpha} - \xi_b^\alpha \frac{\partial \xi_a^\beta}{\partial x^\alpha} = (c_{ab}^h + c_{ab}^p \varphi_p^h)\xi_h^\beta.$$

If we take $b = p$ in (21.3) and replace ξ_p^β by its expression (21.2), the result is equivalent, in consequence of equations of the form (21.3) when b is replaced by h, to

$$(21.4) \qquad X_a \varphi_p^h = \Phi_{ap}^h,$$

where

$$(21.5) \qquad \Phi_{ap}^h = c_{ap}^h + c_{ap}^s \varphi_s^h - \varphi_p^g(c_{ag}^h + c_{ag}^s \varphi_s^h)$$
$$\begin{pmatrix} a = 1, \cdots, r; \\ h, g = 1, \cdots, q; \\ p, s = q + 1, \cdots, r \end{pmatrix}.$$

When a in (21.4) takes the values $1, \ldots, q$, we have

$$(21.6) \qquad X_k \varphi_p^h = \Phi_{kp}^h \qquad \begin{pmatrix} h, k = 1, \cdots, q; \\ p = q + 1, \cdots, r \end{pmatrix}.$$

In consequence of (21.2) and (21.6) when a in (21.4) takes the values $q + 1, \cdots, r$, the latter are reducible to

$$(21.7) \qquad \varphi_s^k \Phi_{kp}^h = \Phi_{sp}^h \qquad \begin{pmatrix} h, k = 1, \cdots, q; \\ p, s = q + 1, \cdots, r \end{pmatrix},$$

which are fundamental identities connecting the Φ's. Hence equations (21.4) are equivalent to (21.6) and (21.7).

When $q < n$, that is, when G_r is intransitive, the equations $X_a f = 0$ form a complete system of q independent equations, and consequently admit $n - q$ independent solutions, say ψ^σ. Without loss of generality we may assume that the determinant $\left| \dfrac{\partial \psi^t}{\partial x^u} \right| \neq 0$ for $t, u = q + 1, \cdots, n$. If we effect the transformation

$$x'^\lambda = x^\lambda, \qquad x'^\sigma = \psi^\sigma \qquad (\lambda = 1, \cdots, q; \sigma = q + 1, \cdots, n),$$

and note that the ξ's transform as contravariant vectors, that is,

$$\xi_a'^\alpha = \xi_a^\beta \frac{\partial x'^\alpha}{\partial x^\beta},$$

then in the new coordinate system, which we call x^i (dropping primes), we have

(21.8) $\xi_a^\sigma = 0$ $(a = 1, \cdots, r; \sigma = q + 1, \cdots, n)$.

In this coordinate system the equations

$$x^{q+1} = a^{q+1}, \cdots, x^n = a^n$$

for particular values of the constants a^i define an invariant variety V_q.   Corresponding to equations (20.1) we have in this case

$$x^\lambda = x^\lambda, \qquad x^\sigma = a^\sigma \qquad (\lambda = 1, \cdots, q; \sigma = q + 1, \cdots, n),$$

and from (20.2) it follows that $\eta_a^\lambda = \xi_a^\lambda$.   Hence when in ξ_a^λ we put $x^\sigma = a^\sigma$ for $\sigma = q + 1, \cdots, n$, we obtain the vectors of the induced group in the invariant variety.   This is in accordance with (20.9).

The matrix $\|\xi_h^\alpha\|$, for $h = 1, \cdots, q$, being assumed to be of rank q, since (21.8) holds, we have that a set of functions ξ_μ^h for $\mu = 1, \cdots, q$ are uniquely defined by

(21.9) $\xi_\mu^h \xi_h^\lambda = \delta_\mu^\lambda, \qquad \xi_\lambda^h \xi_l^\lambda = \delta_l^h$ $(\lambda, \mu, h, l = 1, \cdots, q).$*

In consequence of (21.8) equations (21.3) are equivalent to

(21.10) $\xi_h^\lambda \dfrac{\partial \xi_l^\mu}{\partial x^\lambda} - \xi_l^\lambda \dfrac{\partial \xi_h^\mu}{\partial x^\lambda} = (c_{hl}^m + c_{hl}^p \varphi_p^m) \xi_m^\mu$

$$\begin{pmatrix} \lambda, \mu, h, l, m = 1, \cdots, q; \\ p = q + 1, \cdots, r \end{pmatrix}$$

and (21.6), which because of (21.9) may be written

(21.11) $\dfrac{\partial \varphi_p^h}{\partial x^\lambda} = \Phi_{lp}^h \xi_\lambda^l$ $\begin{pmatrix} \lambda, h, l = 1, \cdots, q; \\ p = q + 1, \cdots, r \end{pmatrix}.$

When $q = n$, that is, when the group is transitive, the results of the preceding paragraph hold on replacing q by n.

We define functions $\Lambda_{\beta\mu}^\alpha$ by

(21.12) $\Lambda_{\beta\mu}^\lambda = \xi_h^\lambda \dfrac{\partial \xi_\mu^h}{\partial x^\beta} = -\xi_\mu^h \dfrac{\partial \xi_h^\lambda}{\partial x^\beta}, \qquad \Lambda_{\beta\mu}^\sigma = 0$

$$\begin{pmatrix} \lambda, \mu, h = 1, \cdots, q; \\ \beta = 1, \cdots, n; \\ \sigma = q + 1, \cdots, n \end{pmatrix}.$$

---

* Note in this connection the foot-note bearing on equations (5.8).

From (21.12) and (21.9) we have

(21.13)
$$\frac{\partial \xi_h^\lambda}{\partial x^\beta} + \xi_h^\mu \Lambda_{\beta\mu}^\lambda = 0$$

and

(21.14)
$$\frac{\partial \xi_\lambda^h}{\partial x^\beta} - \xi_\mu^h \Lambda_{\beta\lambda}^\mu = 0.$$

From (21.12) and (21.10) we obtain

(21.15)
$$\Lambda_{\mu\nu}^\lambda - \Lambda_{\nu\mu}^\lambda = (c_{hc}^i + c_{hc}^p \varphi_p^i)\,\xi_\nu^h \xi_\mu^c \xi_i^\lambda$$
$$\begin{pmatrix} \lambda, \, \mu, \, \nu, \, h, \, c, \, j = 1, \, \cdots, \, q; \\ p = q + 1, \, \cdots, \, r \end{pmatrix}$$

Since the determinant $|\xi_h^\lambda|$ is different from zero, the conditions of integrability of (21.13) are

(21.16)
$$\frac{\partial \Lambda_{\alpha\mu}^\lambda}{\partial x^\beta} - \frac{\partial \Lambda_{\beta\mu}^\lambda}{\partial x^\alpha} + \Lambda_{\alpha\mu}^\pi \Lambda_{\beta\pi}^\lambda - \Lambda_{\beta\mu}^\pi \Lambda_{\alpha\pi}^\lambda = 0$$
$$\begin{pmatrix} \alpha, \, \beta = 1, \, \cdots, \, n; \\ \lambda, \, \mu, \, \pi = 1, \, \cdots, \, q \end{pmatrix}$$

These are necessarily identities as may be verified by substitution from (21.12). If we replace β in (21.16) by ν for $\nu = 1, \, \cdots, \, q$ and subtract from these equations the corresponding ones obtained by interchanging μ and ν, we obtain

$$\frac{\partial \Lambda_{\alpha\mu}^\lambda}{\partial x^\nu} - \frac{\partial \Lambda_{\alpha\nu}^\lambda}{\partial x^\mu} = \frac{\partial}{\partial x^\alpha}(\Lambda_{\nu\mu}^\lambda - \Lambda_{\mu\nu}^\lambda) + \Lambda_{\alpha\pi}^\lambda(\Lambda_{\nu\mu}^\pi - \Lambda_{\mu\nu}^\pi) - \Lambda_{\alpha\mu}^\pi \Lambda_{\nu\pi}^\lambda + \Lambda_{\alpha\nu}^\pi \Lambda_{\mu\pi}^\lambda.$$

If we put

(21.17)
$$\Lambda_{\alpha\beta\gamma}^\epsilon = \frac{\partial \Lambda_{\alpha\gamma}^\epsilon}{\partial x^\beta} - \frac{\partial \Lambda_{\alpha\beta}^\epsilon}{\partial x^\gamma} + \Lambda_{\alpha\gamma}^\delta \Lambda_{\delta\beta}^\epsilon - \Lambda_{\alpha\beta}^\delta \Lambda_{\delta\gamma}^\epsilon$$
$$(\alpha, \, \beta, \, \gamma, \, \delta, \, \epsilon = 1, \, \cdots, \, n),$$

the preceding equations may be written, in consequence of the second set of (21.12) if $q < n$,

$$\Lambda_{\alpha\mu\nu}^\lambda = \frac{\partial}{\partial x^\alpha}(\Lambda_{\mu\nu}^\lambda - \Lambda_{\nu\mu}^\lambda) + \Lambda_{\alpha\pi}^\lambda(\Lambda_{\mu\nu}^\pi - \Lambda_{\nu\mu}^\pi) + \Lambda_{\alpha\nu}^\pi(\Lambda_{\pi\mu}^\lambda - \Lambda_{\mu\pi}^\lambda)$$
$$+ \Lambda_{\alpha\mu}^\pi(\Lambda_{\nu\pi}^\lambda - \Lambda_{\pi\nu}^\lambda).$$

In consequence of (21.13), (21.14) and (21.15) these equations are reducible to

$$(21.18) \quad \Lambda^\lambda_{\alpha\mu\nu} = c^p_{hi}\frac{\partial\varphi^j_p}{\partial x^\alpha}\xi^h_\nu\xi^i_\mu\xi^\lambda_j \quad \begin{pmatrix} \alpha = 1, \cdots, n; \, p = q+1, \cdots, r; \\ \lambda, \mu, \nu, h, i, j = 1, \cdots, q \end{pmatrix}.$$

When $q = r$, there are no functions φ^j_p and consequently

$$(21.19) \qquad\qquad \Lambda^\lambda_{\alpha\mu\nu} = 0 \quad \begin{pmatrix} \alpha = 1, \cdots, n; \\ \lambda, \mu, \nu = 1, \cdots, r; \, r = q \end{pmatrix}.$$

22. Equivalent groups. Two groups G_r and H_r, whose finite equations are

$$(22.1) \quad \begin{aligned} \bar{x}^\alpha &= f^\alpha(x^1, \cdots, x^n; a^1, \cdots, a^r), \\ \bar{x}'^\alpha &= h^\alpha(x'^1, \cdots, x'^n; a'^1, \cdots, a'^r) \quad (\alpha = 1, \cdots, n), \end{aligned}$$

are said to be *equivalent* when there exists a set of r independent functions $\varphi^\alpha(a)$ such that when we put in the above equations $a'^\alpha = \varphi^\alpha(a)$ for $\alpha = 1, \cdots, r$, there exists a non-singular transformation of coordinates which transforms either set of equations (22.1) into the other; some writers call two such groups *similar*. From the results of §7 we have that by a suitable choice of bases of the two groups the constants of structure are the same, that is, the two groups have the same structure.   Hence we have:

[22.1] *A necessary condition that two r-parameter groups in the same number of variables be equivalent is that they have the same structure.*

Suppose that $\xi_a(x)$ and $\xi'_a(x')$ are the vectors of the bases of two groups G_r and H_r respectively, such that the constants of structure are the same.   In order that the two groups be equivalent it is necessary and sufficient that there exist a non-singular transformation

$$(22.2) \qquad\qquad x'^\alpha = \psi^\alpha(x),$$

such that

$$(22.3) \qquad\qquad \xi'^\alpha_a = \xi^\beta_a\frac{\partial x'^\alpha}{\partial x^\beta}.$$

A necessary condition is that the two matrices $\|\xi\|$ and $\|\xi'\|$ be of the same rank.

We consider first the case where the generic rank q of the matrix $\|\xi\|$ is equal to r. If $r < n$, the coordinates x^α and x'^α can be chosen (cf. §21) so that

$$(22.4) \qquad \xi_a^\sigma = \xi_a'^\sigma = 0 \quad (a = 1, \cdots, r; \sigma = q + 1, \cdots, n).$$

From (22.3) for $\alpha = r + 1, \cdots, n$ it follows that

$$(22.5) \qquad \frac{\partial x'^\sigma}{\partial x^h} = 0 \quad (\sigma = q + 1, \cdots, n; h = 1, \cdots, q).$$

Hence x'^σ are functions $\varphi^\sigma(x^{r+1}, \ldots, x^n)$. Since these are independent, we may take them as new coordinates, $x^{r+1}, \ldots, x^n$ without affecting (22.4). Consequently in all generality we have

$$(22.6) \qquad x'^\sigma = x^\sigma \qquad (\sigma = q + 1, \cdots, n).$$

When α in (22.3) takes the values 1 to $q(=r)$, these equations may be written

$$(22.7) \qquad \frac{\partial x'^\lambda}{\partial x^\mu} = \xi_a'^\lambda(x')\xi_\mu^a(x) \quad (\lambda, \mu, a = 1, \cdots, r),$$

where

$$(22.8) \qquad \xi_\mu^a \xi_a^\lambda = \delta_\mu^\lambda, \qquad \xi_\lambda^a \xi_b^\lambda = \delta_b^a.$$

If in the functions $\xi_a'^\lambda(x')$ in equations (22.7) we replace x'^σ for $\sigma = q + 1, \cdots, n$ by x^σ in accordance with (22.6), we have a system of differential equations in the independent variables $x^1, \ldots, x^q$ with $x^{q+1}, \ldots, x^n$ as parameters. With the aid of (21.13) and (21.14) and similar equations in the ξ''s we obtain from (22.7)

$$(22.9) \qquad \frac{\partial^2 x'^\lambda}{\partial x^\mu \partial x^\nu} = -\Lambda_{\pi\tau}'^\lambda \xi_h'^\tau \xi_\mu^h \xi_l'^\pi \xi_\nu^l + \xi_h'^\lambda \xi_\mu^h \Lambda_{\nu\mu}^\tau$$

$$\left(\begin{matrix} \lambda, \mu, \nu, \pi, \tau = 1, \cdots, q; \\ h = 1, \cdots, q \end{matrix} \right).$$

Since $q = r$, equations (21.15) are in this case

$$\Lambda_{\mu\nu}^\lambda - \Lambda_{\nu\mu}^\lambda = c_{hi}^j \xi_\nu^h \xi_\mu^i \xi_j^\lambda \qquad (h, i, j = 1, \cdots, r).$$

Because of these relations it follows from (22.9) that the conditions of integrability of (22.7) are satisfied identically. Hence a solution is determined by taking for initial values of $x'^1, \ldots, x'^r$ arbitrary functions of $x^{r+1}, \ldots, x^n$. If these are chosen so that the jacobian of the solution with respect to x^1, . . . , x^r is different from zero, these functions and (22.6) define a non-singular transformation of G_r into H_r. Hence we have:

[22.2]   *Two intransitive groups G_r and H_r in the same number of variables, with the same constants of structure and whose matrices M are of rank r, are equivalent; the equations of transformation involve r arbitrary functions.*

If $r = n$, that is, if the group is simply transitive, we have in place of (22.4), (22.6) and (22.7) only equation (22.7) in which λ, μ, $a = 1$, . . . , n. As before these equations are completely integrable, and their solution involves n arbitrary constants, the initial values of the x''s. Hence we have:

[22.3]   *Two simply transitive groups in the same number of variables and with the same constants of structure are equivalent, and the equations of transformation of one into the other involve n arbitrary constants.*

We consider next the case when $q < r$. For the group G_r we choose the ξ's so that $\|\xi_h^\alpha\|$ for $h = 1, \cdots, q$ is of rank q, then the matrix $\|\xi_h'^\alpha\|$ for $h = 1, \cdots, q$ must be of rank q, and analogously to (21.2) we put

$$(22.10) \qquad \xi_p'^\alpha = \varphi_p'^h \xi_h'^\alpha \qquad (p = q + 1, \cdots, r).$$

From these equations and (22.3) we have

$$(22.11) \qquad \varphi_p'^h(x') = \varphi_p^h(x) \qquad \begin{pmatrix} h = 1, \cdots, q; \\ p = q + 1, \cdots, r \end{pmatrix}.$$

There are $q(r - q)$ of these equations. Evidently in order that the two groups be equivalent, it is necessary that these equations be consistent, and that it be impossible to eliminate the x''s and get a relation between the x's, and vice-versa. We shall show that this condition is also sufficient.

If $q < n$, we assume that the coordinates x' and x are chosen so that (22.4) and (22.6) hold. As in the case when $q = r$, we get the system of equations (22.7) in which now λ, μ, $a = 1, \cdots, q$. When in the ξ''s in these equations and in (22.11) we replace x'^σ for $\sigma = q + 1, \cdots, n$ by x^σ, we have a mixed system (§1) of differential equations in x'^1, . . . , x'^q in the independent variables x^1, . . . , x^q which system involves x^{q+1}, . . . , x^n as parameters. As in the preceding case we have (22.9), and in consequence of

(21.15) the conditions of integrability of (22.7) are satisfied because of (22.11).

From (21.11) we have

$$(22.12) \qquad \frac{\partial \varphi_p^h}{\partial x^\lambda} = \Phi_{lp}^h \xi_\lambda^l, \qquad \frac{\partial \varphi_p'^h}{\partial x'^\lambda} = \Phi_{lp}'^h \xi_\lambda'^l.$$

If equations (22.11) are differentiated with respect to x^μ for $\mu = 1, \cdots, q$, the resulting equations are reducible in consequence of (22.6), (22.7) and (22.12) to

$$(\Phi_{lp}'^h - \Phi_{lp}^h)\xi_\mu^l = 0,$$

which are satisfied because of (22.11), as is evident from (21.5).

Thus in applying the general theory of mixed systems (§1) to the present case, we see that equations (22.11) are the system F_0, but all the systems $F_1, \ldots,$ are consequences of F_0. Hence there exists a solution, if equations (22.11) are consistent, and do not lead to relations between the x's alone or the x''s alone. If they are consistent, a number s of the functions $\varphi_p'^h$ are independent, say $\varphi'^1, \ldots, \varphi'^s$, and s must be equal to or less than q, because (22.6) also must hold. Consequently the jacobian of these φ''s and x'^σ must be of rank $n - q + s$, and therefore the rank of $\left\| \dfrac{\partial \varphi'^\alpha}{\partial x'^\lambda} \right\|$ for $\alpha = 1, \cdots, s$ and $\lambda = 1, \cdots, q$ is s. Hence the equations $\varphi'^\alpha = \varphi^\alpha$ can be solved for s of $x'^1, \ldots, x'^q$ as functions of $x'^{q+1}, \ldots, x'^n, x^1, \ldots, x^n$; there is no loss in generality in assuming that these are $x'^1, \ldots, x'^s$. When in these we replace x'^σ for $\sigma = q + 1, \cdots, n$ by x^σ, we have

$$(22.13) \qquad x'^\alpha = \psi^\alpha(x'^{s+1}, \cdots, x'^q; x^1, \cdots, x^n)$$
$$(\alpha = 1, \cdots, s).$$

These are the analogue of (1.8) in which $x^{q+1}, \ldots, x^n$ enter as parameters. Then we have to solve the complete system (analogue of (1.11))

$$(22.14) \qquad \frac{\partial x'^\rho}{\partial x^\lambda} = f_\lambda^\rho(x'^{s+1}, \cdots, x'^q; x^1, \cdots, x^n)$$
$$\left(\begin{matrix} \rho = s + 1, \cdots, q; \\ \lambda = 1, \cdots, q \end{matrix} \right),$$

obtained from (22.7) on replacing x'^1, . . . , x'^s by the expressions from (22.13) and x'^σ by x^σ for $\sigma = q + 1, \cdots , n$. The solution of (22.14) involves $q - s$ arbitrary functions of the parameters x^{q+1}, . . . , x^n.

If $q = n$, we do not have (22.4) and (22.6), and hence (22.7) and (22.11) involve all the x's as independent variables, but the above process applies just the same.

Hence we have the following theorem established by Lie in a different manner:[*]

[22.4] *Two r-parameter groups in the same number of variables, with the same constants of structure and for which the generic ranks of the matrices $\|\xi\|$ and $\|\xi'\|$ are less than r, are equivalent, when and only when these respective ranks are equal (say q), each pair of corresponding minors of order q have the same rank, and the corresponding set of equations (22.11) are consistent and do not lead to a relation between the variables of either set.*

23. Imprimitive and primitive groups. In §19 it was seen that each ordinary point of an intransitive group lies in an invariant variety which is transformed into itself by all transformations of the group, and that there is a set of such invariant varieties, one through each ordinary point [cf. (19.1)]. Although there is no such state of affairs for transitive groups, there are transitive groups for which there exist a set of varieties, one through each ordinary point, such that if a point of one variety is transformed into a point of another, each point of the former goes into a point of the latter. A simple example of this is afforded by any family of parallel planes in case of the group of translations of euclidean 3-space; the same is true of a congruence of parallel lines under these transformations, and thus we see that for a given group there may be more than one set of varieties possessing this property. Lie[†] called a group possessing this property *imprimitive,* and the corresponding varieties a *system of imprimitivity;* a group not possessing the property he called *primitive.*   For example, the group of motions in the euclidean plane is primitive, because any point and direction at it is transformable into any point and any direction at it, and if there were a system of imprimitivity each member through a point would have a given direction.

---

[*] *Lie-Engel,* 1888, 1, p. 354; *Eisenhart,* 1932, 4.

[†] 1888, 1, vol. 1, p. 220.

A system of imprimitivity of varieties of dimension p, if it exists, is defined by a set of equations

$$(23.1) \qquad F_\mu = c_\mu \qquad (\mu = 1, \cdots, n - p),$$

where the F's are independent and the c's are arbitrary constants, such that

$$(23.2) \qquad F_\mu(f(x; a)) = \Phi_\mu(F_1(x), \cdots, F_{n-p}(x); a),$$

where the Φ's are such functions that these equations are identities in the x's and a's. A natural way to handle this problem is to consider the complete system of partial differential equations satisfied by the functions F_μ. Since the F's are independent by hypothesis, the equations

$$b^i \frac{\partial F_\mu}{\partial x^i} = 0 \qquad (\mu = 1, \cdots, n - p; i = 1, \cdots, n)$$

in the b's admit p independent sets of solutions, b_α^i. Consequently the equations

$$(23.3) \qquad B_\alpha f = b_\alpha^i p_i = 0 \qquad (\alpha = 1, \cdots, p)$$

form a complete system, whose solutions are $F_1, \ldots, F_{n-p}$ and any function of the latter. Consequently the functions $F_\mu(f(x; a))$ must be solutions of (23.3) for all values of the a's. Hence we have:

[23.1] *A necessary and sufficient condition that a group admit a system of imprimitivity of varieties of dimension p is that there exist a complete system (23.3) such that if $F_\mu(x)$ for $\mu = 1, \cdots, n - p$ is a set of independent solutions, then $F_\mu(f(x; a))$ also are solutions.*

If the group is intransitive, the equations

$$\xi_\alpha^i p_i = 0$$

form a complete system, and any system of imprimitivity is obtained from the system of invariant varieties (§19).

Suppose that the equations of the group are taken in the form

$$x'^i = x^i + tXx^i + \frac{t^2}{2}X^2x^i + \cdots,$$

where

$$Xf = e^a X_a f,$$

then as seen in §10

$$(23.4) \qquad F_\mu(x') = F_\mu(x) + tXF_\mu + \frac{t^2}{2}X^2F_\mu + \cdots .$$

Since these expressions must satisfy (23.3) when $F_\mu(x)$ do for all values of t and all values of e^a, it follows that it is necessary that X_aF_μ be solutions of (23.3), that is,

$$(23.5) \qquad X_aF_\mu = \Phi_{a\mu}(F_1, \cdots, F_{n-p}).$$

When these conditions are satisfied, we have

$$X_bX_aF_\mu = \frac{\partial\Phi_{a\mu}}{\partial F_\nu}X_bF_\nu = \Phi_{a\mu b}(F_1, \cdots, F_{n-p}).$$

Consequently $X^2F_\mu, \ldots, X^mF_\mu$ are solutions of (23.3), and therefore conditions (23.5) are necessary and sufficient that the expression (23.4) satisfy equations (23.3).   Hence we have:

[23.2] *A necessary and sufficient condition that a system of equations* (23.1) *defines a system of imprimitivity of a group G_r with the symbols X_af for $a = 1, \cdots, r$ is that X_aF_μ be functions of the F's.*

If the functions F_μ are solutions of a complete system (23.3), in consequence of (23.5) we have

$$(X_a, B_\alpha)F_\mu = X_a(B_\alpha F_\mu) - B_\alpha(X_a F_\mu) = 0,$$

and consequently the F's are solutions of the equations $(X_a, B_\alpha)f = 0$. Hence we have:[*]

[23.3] *A necessary and sufficient condition that a group G_r with symbols $X_1f, \ldots, X_rf$ admit a system of imprimitivity of varieties of dimension p is that there exist a complete system* (23.3) *such that*

$$(23.6) \quad (X_a, B_\alpha)f = \lambda_{a\alpha}^\beta(x)B_\beta f, \quad (a = 1, \cdots, r; \alpha, \beta = 1, \cdots, p),$$

where the λ's are at most functions of the x's.

When these conditions are satisfied we say that the complete system *admits* the group.

In order to express these conditions in another form, we assume that we effect a transformation of coordinates so that in the new coordinate system the equations of the system of imprimitivity are

---

[*] *Cf. Bianchi*, 1918, 1, p. 184.

$x^\mu = $ const. $(\mu = p + 1, \cdots, n)$ then $b_\alpha^\mu = 0$, and the complete system (23.3) becomes $b_\alpha^\beta p_\beta = 0$ $(\alpha, \beta = 1, \cdots, p)$. Since the rank of $\|b_\alpha^\beta\|$ is p, there is no loss in generality in taking the system (23.3) in the form

$$p_\alpha = 0 \qquad\qquad (\alpha = 1, \cdots, p),$$

that is, $b_\alpha^i = \delta_\alpha^i$. Then from equations (23.6) we have

$$\frac{\partial \xi_a^\alpha}{\partial x^\beta} = \lambda_{a\beta}^\alpha \qquad \frac{\partial \xi_a^\mu}{\partial x^\alpha} = 0 \qquad \begin{pmatrix} \alpha, \beta = 1, \cdots, p; \\ \mu = p + 1, \cdots, n \end{pmatrix}.$$

Hence the functions ξ_a^μ must be independent of $x^1, \ldots, x^p$. This follows also from (23.5). Consequently we have:

[23.4] *A necessary and sufficient condition that a group G_r have a system of imprimitivity of varieties of dimension p is that the functions $\xi_a^i(x)$ be such that there exist a system of coordinates x'^i for which the expressions*

$$\xi_a^i \frac{\partial x'^\mu}{\partial x^i} \qquad (\mu = p + 1, \cdots, n)$$

are independent of $x'^1, \ldots, x'^p$.

For the group G_3 of translations of a euclidean 3-space the conditions of this theorem are satisfied, the cartesian coordinates being the system x'^i. In this case each of the equations

$$x^i = \text{const.} \qquad\qquad (i = 1, 2, 3)$$

is a system of imprimitivity, and any two define such a system. Moreover, if the b's in (23.3) are constants, we have

$$(B_\rho, X_a)f = 0,$$

and (23.3) gives a system of parallel planes, if $\rho = 1, 2$, and a congruence of parallel lines, if $\rho = 1$.

24. Systatic and asystatic groups. When the generic rank q of M (18.1) is less than r, we have equations (21.2). If $P_0(x_0)$ is an ordinary point, the generators of the sub-group of stability are

$$(24.1) \qquad X_p f - \varphi_p^h(x_0)X_h f \qquad (h = 1, \cdots, q; p = q + 1, \cdots, r).$$

If $n - \rho$ of the functions φ_p^h are independent, the equations

$$(24.2) \qquad\qquad \varphi_p^h(x) = \varphi_p^h(x_0),$$

define a variety V_ρ for all of whose points the sub-group of stability is the same as for P_0. Conversely if $P_1(x_1)$ has the same sub-group of stability as P_0, then as follows from (24.1)

$$[\varphi_p^h(x_1) - \varphi_p^h(x_0)]X_h f = 0,$$

and consequently P_1 is in V_ρ defined by (24.2).

Following Lie* we say that a group is *systatic*, when the sub-group of stability of an ordinary point P is also the sub-group of stability of every point in a continuous variety containing P, called the *systatic variety* for P. When a group does not possess this property, it is *asystatic*. When $q = r$, the sub-group of stability of an ordinary point is the identity (§18); consequently the group is systatic and V_n is the systatic variety. When $q < r$, the asystatic groups are those for which n of the functions φ_p^h are independent. When $n - \rho$ of the functions φ_p^h are independent, the group is systatic, and equations (24.2) define the systatic variety for an ordinary point; evidently there are $\infty^{n-\rho}$ such systatic varieties.

Suppose that T is any transformation of stability of P_0 and S a transformation of G_r not in the sub-group, so that $S(P_0) = P_1$, where P_1 is a point not in V_ρ defined by (24.2). Then STS^{-1} is a transformation of stability of P_1, and of all the points $S(P)$, where P is any point in V_ρ, that is, of all points in the variety V_ρ' into which V_ρ is transformed by S. Conversely if T' is a transformation of stability of the points in V_ρ' then $S^{-1}T'S$ is a transformation of stability of the points in V_ρ. Thus V_ρ' is a systatic variety for G_r. Hence we have:

[24.1]   *When a group G_r is systatic and the generic rank of the matrix $\|\xi\|$ is less than r, the systatic varieties form a system of imprimitivity.*

Exercises

1. Show that $x^2 + y^2 = c^2 z^2$, where c is an arbitrary constant, are invariant varieties for the G_6 of Ex. 9, p. 67 and find the induced group on each by putting

$$x = cu \cos v, \qquad y = cu \sin v, \qquad z = cu.$$

2. If $X_a f$ and $Y_a f$ are the symbols of two simply transitive groups in n variables x^i and y^i respectively and they have the same constants of structure, the equations $X_a f + Y_a f = 0$ form a complete system in the x's and y's, and admit n independent solutions $f_i(x, y)$. The equations $f_i(x, y) = c_i$, where the

---

* 1888, 1, vol. 1, p. 501.

c's are arbitrary constants, define the most general transformation of $X_a f$ into $Y_a f$, and vice-versa.

Bianchi, 1918, 1, p. 261.

3. Find the functions $\Lambda_{\alpha\beta}^{\gamma}$ for the group G_3 of rotations in euclidean 3-space with the symbols $x^i p_j - x^j p_i$. $(i, j = 1, 2, 3)$.

4. Show that there are no multiply-transitive Abelian groups.

5. Find the group induced by the group of Ex. 10, p. 67 on the invariant variety (the ruled surface of Cayley)

$$x^1 = u, \qquad x^2 = v, \qquad x^3 = 3uv - 2u^3,$$

and the system of imprimitivity of the induced group.

6. Show that the group of rotations in euclidean 3-space is systatic and that the straight lines through the origin are the systatic varieties, and systems of imprimitivity.

7. Show that the concentric spheres with the origin for center are systems of imprimitivity of the G_4

$$x^i p_j - x^j p_i, \qquad x^i p_i \qquad\qquad (i, j = 1, 2, 3)$$

8. Show that the groups

$$\begin{aligned}&\text{(i)} \quad p, \qquad q, \qquad xq, \qquad yq\,;\\ &\text{(ii)} \quad q, \qquad xq, \qquad yq\end{aligned}$$

are asystatic.

25. Differential equations admitting linear operators.
In deriving theorem [23.3] no use has been made of the fact that the operators $X_a f$ are the symbols of a group G_r, so that we have the theorem:

[25.1] *If* $X_1 f, \ldots, X_r f$ *are linear operators, a necessary and sufficient condition that* $X_a \theta$ *be solutions of a complete system*

$$(25.1) \qquad\qquad A_\alpha f = a_\alpha^i p_i = 0 \qquad\qquad (\alpha = 1, \cdots, p),$$

when θ *is a solution of this system other than a constant, is that*

$$(25.2) \qquad\qquad (X_a, A_\alpha) f = \lambda_{a\alpha}^{\beta} A_\beta f \quad \begin{pmatrix} a = 1, \cdots, r; \\ \alpha, \beta = 1, \cdots, p \end{pmatrix},$$

where the λ's *are at most functions of the* x's.

When these conditions are satisfied, we say that the complete system *admits the set of operators,* or that it *admits the one-parameter groups* of continuous transformations generated by these operators in the sense of §10.

From the Jacobi identity (2.4) applied to the operators $X_a f$, $X_b f$ and $A_\alpha f$, we obtain in consequence of (25.2)

$$(25.3) \quad ((X_a, \ X_b), \ A_\alpha)f = (X_a \lambda_{b\alpha}^\beta - X_b \lambda_{b\alpha}^\beta + \lambda_{b\alpha}^\gamma \lambda_{a\gamma}^\beta - \lambda_{a\alpha}^\gamma \lambda_{b\gamma}^\beta) A_\beta f.$$

Since this is of the form (25.2) we have:

[25.2] *If a complete system of linear homogeneous partial differential equations admits the operators $X_a f$ and $X_b f$, it admits also the commutator $(X_a, X_b)f$.*

If the operators $X_a f$ are the symbols of a group G_r, no new operators are obtained as a result of this theorem.   We shall consider the more general case of theorem [25.1].

Since the system (25.1) is complete by hypothesis, we have

$$(25.4) \qquad\qquad (A_\alpha, \ A_\beta)f = \sigma_{\alpha\beta}^\gamma A_\gamma f.$$

If we put

$$(25.5) \qquad\qquad Xf = \mu^a X_a f + \nu^\alpha A_\alpha f,$$

then

$$(25.6) \quad (X, \ A_\beta)f = (\mu^a \lambda_{a\beta}^\gamma + \nu^\alpha \sigma_{\alpha\beta}^\gamma - A_\beta \nu^\gamma) A_\gamma f - A_\beta \mu^a X_a f.$$

As a first consequence of these equations we have:

[25.3] *If the complete system (25.1) admits the operators $X_a f$, it admits also $\mu^a X_a f + \nu^\alpha A_\alpha f$, where the ν's are any functions of the x's and the μ's are any constants or solutions of (25.1).*

Suppose that we have a set of r operators admitted by the complete system (25.1); that $r + p \leq n$ and that the matrix

$$(25.7) \qquad\qquad \|a_1^i, \ \ldots, \ a_p^i, \ \xi_1^i, \ \ldots, \ \xi_r^i\|$$

is of rank $r + p$; then we say that the operators are *independent*. If we have another operator (25.5) admitted by the complete system, the right-hand member of (25.6) must be a linear combination of the Af's.   But since the matrix (25.7) is of rank $r + p$, it follows that the μ's are solutions of the complete system or constants. Hence we have:

[25.4] *If the complete system (25.1) admits r independent operators $X_a f$ in the sense that the matrix (25.7) is of rank $r + p$, a necessary and sufficient condition that the complete system admit an operator*

$\mu^a X_a f + \nu^\alpha A_\alpha f$ is that the μ's be solutions of the complete system, or constants.

If we have r independent operators, and we apply the commutator to each pair, we obtain new operators in accordance with theorem [25.2], which are either independent of the given operators or are expressible in the form (25.5). If we obtain independent ones, we add them to the set and continue with the commutator, as in §2, until we have finally $q(\geq r)$ independent operators in the above sense, and $q + p \leq n$. Then for any two of this set we have

$$(25.8) \qquad (X_a, X_b)f = \mu_{ab}^c X_c f + \nu_{ab}^\alpha A_\alpha f \quad (a, b = 1, \cdots, r)$$

and in consequence of theorem [25.4] each μ which is not a constant is a non-trivial solution of the system (25.1). Thus in cases when the $X_a f$ are not the symbols of a group G_q, we may obtain one or more non-trivial solutions, and by theorem [25.1], if θ is a solution so also are $X_a\theta$, and if the latter are not functions of θ, we operate again with X_a obtaining new solutions and so on.

As a special case of theorem [25.1] we have:

[25.5] *A necessary and sufficient condition that an equation*

$$(25.9) \qquad Af = a^i p_i = 0$$

admit a set of operators $X_1 f, \ldots, X_r f$ *is that*

$$(25.10) \qquad (X_a, A)f = \lambda_a Af \quad (a = 1, \cdots, r);$$

then if θ *is a solution of* (25.9), *so also are* $X_a\theta$.

When this condition is satisfied, we proceed in accordance with a theorem analogous to theorem [25.2] to obtain additional operators, if possible, until we have the maximum independent set, in the sense that the matrix

$$\|a^i, \xi_1^i, \ldots, \xi_r^i\|$$

is of rank $r + 1$. Then in place of (25.8) we have

$$(25.11) \qquad (X_a, X_b)f = \mu_{ab}^c X_c f + \nu_{ab} Af,$$

and the μ's are constants or solutions of (25.9).

Thus when r independent operators are known, there is the possibility of obtaining solutions by the direct processes giving (25.11). Furthermore, if $r + 1$ is less than n, equation (25.9) and the equations $X_a f = 0$ form a complete system, in consequence of

(25.10) and (25.11), and there are $n - (r + 1)$ independent solutions of the system and consequently of (25.9), which are found by integrating a system of the form (1.1).

If p independent solutions of (25.9) are known, say $\theta^1, \ldots, \theta^p$, we may designate the x's so that the jacobian $\left| \dfrac{\partial \theta^\alpha}{\partial x^\beta} \right|$ for $\alpha, \beta = 1, \cdots, p$ is not zero.   Hence if we effect the transformation of coordinates

$$x'^\alpha = \theta^\alpha, \qquad x'^\sigma = x^\sigma \quad (\alpha = 1, \cdots, p; \sigma = p + 1, \cdots, n),$$

in the new coordinate system we have $a'^\alpha = 0$, and consequently equation (25.9) becomes

$$a'^\sigma \frac{\partial f}{\partial x'^\sigma} = 0,$$

an equation in $n - p$ variables, but with $x'^1, \ldots, x'^p$ possibly entering as parameters.

In the extreme case when $n - 1$ independent solutions are known, the equation is reducible to $p_1 = 0$.   Evidently this equation admits the $n - 1$ independent operators p_α for $\alpha = 2, \cdots, n$, and these form an Abelian group.   Hence we have:

[25.6] *A linear homogeneous partial equation of the first order in n variables admits an Abelian group of order n − 1.*

In accordance with theorem [25.3] for the equation (25.9) a set of independent operators is given by

$$(25.12) \qquad \overline{X}_a f = X_a f + \nu_a A f,$$

where the ν's are any functions of the x's.   In consequence of (25.10) and (25.11) we have

$$\overline{\xi}_a^i \frac{\partial \overline{\xi}_b^j}{\partial x^i} - \overline{\xi}_b^i \frac{\partial \overline{\xi}_a^j}{\partial x^i} = \mu_{ab}^c \overline{\xi}_c^j + \overline{\nu}_{ab} a^j,$$

where

$$\overline{\nu}_{ab} = X_a \nu_b - X_b \nu_a + \nu_a A \nu_b - \nu_b A \nu_a + \nu_b \lambda_a - \nu_a \lambda_b - \nu_c \mu_{ab}^c.$$

If a^h is one of the a's which is not zero, and we choose the quantities ν_a so that $\xi_a^h + \nu_a a^h = 0$, then $\overline{\xi}_a^h = 0$ in each operator $\overline{X}_a f$.   Hence if in the above equations we put $j = h$, we find that $\bar{\nu}_{ab} = 0$, and consequently

$$(25.13) \qquad (\overline{X}_a, \overline{X}_b)f = \mu_{ab}^{c}X_cf.$$

If the μ's are constants, $\overline{X}_af$ are the symbols of a group G_r in whose equations x^h enters as a parameter. Accordingly we have:

[25.7] *If an equation $Af = 0$ admits r independent operators X_af, either it admits a group G_r of infinitesimal transformations, or solutions of the equation are obtained by direct processes.*

Consider the case when $n = 2$ and there is one operator Xf. If we eliminate λ from the two equations

$$(25.14) \qquad \xi^i\frac{\partial a^j}{\partial x^i} - a^i\frac{\partial \xi^j}{\partial x^i} = \lambda a^j \qquad (j = 1, 2),$$

which are the equations (25.10) in this case, we obtain

$$(25.15) \qquad a^2\left(\xi^i\frac{\partial a^1}{\partial x^i} - a^i\frac{\partial \xi^1}{\partial x^i}\right) - a^1\left(\xi^i\frac{\partial a^2}{\partial x^i} - a^i\frac{\partial \xi^2}{\partial x^i}\right) = 0.$$

This may be written in the form

$$(25.16) \qquad \frac{\partial}{\partial x^1}(a^1M) + \frac{\partial}{\partial x^2}(a^2M) = 0,$$

where

$$(25.17) \qquad M = \frac{1}{\xi^1a^2 - \xi^2a^1}.$$

Consequently M is an integrating factor of the equation

$$(25.18) \qquad a^2dx^1 - a^1dx^2 = 0,$$

and a solution of it and consequently of (25.9) is given by a quadrature.

Conversely, if an integrating factor M of (25.18) is known and any functions ξ^1 and ξ^2 are chosen to satisfy (25.17), equation (25.16) becomes (25.15), and from the latter (25.14) follows. Evidently if ξ^1 and ξ^2 are functions satisfying (25.17), so also are $\xi^1 + \varphi a^1$ and $\xi^2 + \varphi a^2$, where φ is an arbitrary function of x^1 and x^2. Hence we have:

[25.8] *If an equation*

$$(25.19) \qquad a^1p_1 + a^2p_2 = 0$$

admits an operator, its solution is given by a quadrature; if an integrating factor of the corresponding equation (25.18) *is known, an infinity of operators can be found directly each of which is admitted by* (25.19).

If M_1 is a second integrating factor of (25.18), and we denote by ξ_1^1 and ξ_1^2 two functions satisfying the corresponding equation (25.17), then, since we have necessarily $\xi_1^i = \rho\xi^i + \sigma a^i$, it follows that $\rho = M/M_1$, and from theorem [25.4] that ρ is a solution of (25.19).

We apply theorems [16.1] and [25.8] to the determination of canonical forms for the symbols of a two-parameter group in two variables.   We denote by s the rank of the matrix of the two symbols.   There are four cases to be considered.

1° $(X_1,X_2)f = X_1f$, $s = 2$. Since $X_1f = 0$ admits X_2f, a solution of the former can be found by a quadrature, and then a solution of $X_1f = 1$ by a quadrature (cf. §10). Taking these as new coordinates x and y, we have $X_1f = q$ and the above relation becomes $\dfrac{\partial\xi_2}{\partial y}p + \dfrac{\partial\eta_2}{\partial y}q \equiv q$.   Consequently

$$\xi_2 = \varphi_1(x), \qquad \eta_2 = \varphi_2(x) + y,$$

where $\varphi_1 \not\equiv 0$ since $s = 2$.   If we put

$$x' = \psi(x) \equiv e^{\int \frac{dx}{\varphi_1}}, \qquad y' = y - \psi\int\frac{\varphi_2}{\varphi_1\psi}dx,$$

in the new variables we have

$$(25.20) \quad X_1f = q, \qquad X_2f = xp + yq, \qquad (X_1, X_2)f = X_1f.$$

2° $(X_1,X_2)f = X_1f$, $s = 1$. Since $X_2f = \rho X_1f$, we have $X_1\rho = 1$. Hence if a solution of $\eta_1dx - \xi_1dy = 0$ is found, coordinates can be obtained without quadrature in terms of which $X_1f = q$. Now $X_2f = \rho q$ and from the above relation it follows that

$$\rho = \varphi(x) + y.$$

In the variables $x' = x$, $y' = y + \varphi(x)$ the symbols have the form

$$(25.21) \quad X_1f = q, \qquad X_2f = yq, \qquad (X_1, X_2)f = X_1f.$$

3° $(X_1, X_2)f = 0$, $s = 2$. In consequence of theorem [25.8], as in case 1°, coordinates can be found by quadratures so that $X_1 f = q$. Then we have $X_2 f = \varphi_1(x)p + \varphi_2(x)q$, where $\varphi_1 \not\equiv 0$ since $s = 2$. If we put

$$x' = \psi(x) \equiv \int \frac{dx}{\varphi_1}, \qquad y' = y - \int \frac{\varphi_2}{\varphi_1} dx,$$

in the new variables we have

(25.22) $\qquad X_1 f = q, \qquad X_2 f = p, \qquad (X_1, X_2)f = 0.$

4° $(X_1, X_2)f = 0$, $s = 1$. Since $X_2 f = \rho X_1 f$, ρ is a solution of $X_1 f = 0$, and, as in case 1°, coordinates can be found by a quadrature in terms of which $X_1 f = q$. Since $s = 1$, we have then $X_2 f = \varphi_2(x)q$. If we put $x' = \varphi_2(x)$, $y' = y$, in the new variables we have

(25.23) $\qquad X_1 f = q, \qquad X_2 f = xq, \qquad (X_1, X_2)f = 0.$

Hence we have:

[25.9] *The basis of a G_2 and the coordinates can be chosen so that the symbols assume one of the canonical forms* (25.20), (25.21), (25.22), (25.23); *the determination of the coordinates for* (25.21) *requires the solution of an ordinary differential equation of the trajectories and the others require quadratures only.* *

We have just seen that the cases when the solution of an ordinary equation (25.18) reduces to a quadrature are those for which the corresponding equation (25.19) admits an operator and conversely. This is the simplest general example of the relation which exists between the solution of ordinary differential equations and the theory of continuous groups, the problem which was studied extensively by Lie. He pointed out the fact that most of the ordinary differential equations which can be integrated by known methods admit certain continuous groups of transformations, and that the knowledge of the latter aided in their integration. Many of these methods are special but the theory of Lie gives a unifying principle. In the next sections we shall develop this theory further. However, it is not our intention to give an extensive treatment of this subject, and for such treatment refer the reader to other treatises.†

---

* *Lie-Scheffers*, 1891, 3, p. 425; *Dickson*, 1924, 1, p. 363; *Franklin*, 1928, 1, p. 119.

† Cf. *Lie*, 1891, 3; *Cohen*, 1911, 1; *Dickson*, 1924, 1; *Engel* and *Faber*, 1932, 1.

26. Extended groups. Ordinary differential equations of the first order. If for the sake of brevity we denote by x_1^i the differential dx^i, it follows from the finite equations of a group G_r, namely

$$(26.1) \qquad x'^i = f^i(x; a),$$

that

$$(26.2) \qquad x_1'^i = \frac{\partial f^i}{\partial x^j} x_1^j.$$

Equations (26.1) and (26.2) define a transformation in the $2n$ variables x^i and x_1^i involving r parameters; the values a_0^α for which (26.1) becomes the identity yield the identity in (26.2) also. It is readily shown that equations (26.1) and (26.2) define a group G_r in these $2n$ variables which is called the *extended group* of G_r. In fact, from the equations (4.7), namely

$$(26.3) \qquad f^i(f(x; a_1); a_2) = f^i(x; a_3),$$

it follows that

$$\frac{\partial f^i(x'; a_2)}{\partial x'^i} \frac{\partial x'^i}{\partial x^k} = \frac{\partial f^i(x'; a_2)}{\partial x'^i} \frac{\partial f^i(x; a_1)}{\partial x^k} = \frac{\partial f^i(x; a_3)}{\partial x^k}.$$

Consequently we have

$$x_1''^i = \frac{\partial f^i(x'; a_2)}{\partial x'^i} x_1'^i = \frac{\partial f^i(x; a_3)}{\partial x^i} x_1^j,$$

which establishes the group property.

From (26.2) and (5.5) we have

$$\frac{\partial x_1'^i}{\partial a^\alpha} = \frac{\partial^2 f^i}{\partial x^j \partial a^\alpha} x_1^j = \frac{\partial \xi_a^i(x')}{\partial x'^k} \frac{\partial x'^k}{\partial x^j} x_1^j A_\alpha^a(a) = \frac{\partial \xi_a^i(x')}{\partial x'^k} x_1'^k A_\alpha^a(a).$$

Hence if we put

$$(26.4) \qquad \xi_{1a}^i(x; x_1) = \frac{\partial \xi_a^i(x)}{\partial x^k} x_1^k,$$

we have

$$(26.5) \qquad \frac{\partial x_1'^i}{\partial a^\alpha} = \xi_{1a}^i(x'; x_1') A_\alpha^a(a).$$

For the extended group equations (5.5) and (26.5) together in $2n$ variables play the role of (5.5) for the given group. Since the vectors $A_\alpha^a(a)$ are the same for the extended group, we have from (6.3):

[26.1] *The constants of structure of the extended group are the same as for the given group.*

Since the parameter groups are determined by the vectors A_α^a, we have also:

[26.2] *The first and second parameter groups of the extended group are the same as for the given group.*

Analogously to (6.2) we have (6.2) and

$$(26.6) \quad \xi_a^i(x)\frac{\partial \xi_{1b}^i(x;x_1)}{\partial x^j} - \xi_b^j\frac{\partial \xi_{1a}^i}{\partial x^j} + \xi_{1a}^k\frac{\partial \xi_{1b}^i}{\partial x_1^k} - \xi_{1b}^k\frac{\partial \xi_{1a}^i}{\partial x_1^k} = c_{ab}^e\xi_{1e}^i.$$

The symbols of the extended group are

$$(26.7) \quad X_{(1)a}f = \xi_a^i\frac{\partial f}{\partial x^i} + \xi_{1a}^i\frac{\partial f}{\partial x_1^i}.$$

In consequence of (6.2) and (26.6) we have

$$(26.8) \quad (X_{(1)a}, X_{(1)b})f = c_{ab}^e X_{(1)e}f,$$

which result follows also from theorem [26.1].

In accordance with the theory of §17 any function of the x's and x_1's which satisfies the equations

$$(26.9) \quad X_{(1)a}f = 0$$

is called an *absolute invariant* of the extended group. Furthermore, if a set of p independent functions $F_1, \ldots, F_p$ of the x's and x_1's are such that

$$X_{(1)a}F_\alpha = 0 \qquad (\alpha = 1, \cdots, p)$$

not identically but for values of the x's and x_1's for which $F_\alpha(x;x_1) = 0$, we say that these functions define a *relative invariant* of the extended group.

Consider, for example, the p independent Pfaffian equations

$$(26.10) \quad \lambda_i^\alpha x_1^i = 0 \qquad (\alpha = 1, \cdots, p; i = 1, \cdots, n),$$

where the λ's are functions of the x's. From (26.7) and (26.4) we have

$$X_{(1)a}\lambda_i^\alpha x_1^i = \left(\xi_a^j\frac{\partial\lambda_i^\alpha}{\partial x^j} + \frac{\partial\xi_a^j}{\partial x^i}\lambda_j^\alpha\right)x_1^i.$$

Since the right-hand member is linear in the x_1's, in order that $\lambda_i^\alpha x_1^i$ be a relative invariant, it is necessary and sufficient that

$$(26.11) \qquad \xi_a^j\frac{\partial\lambda_i^\alpha}{\partial x^j} + \frac{\partial\xi_a^j}{\partial x^i}\lambda_j^\alpha = \sigma_{a\beta}\lambda_i^\beta,$$

where the σ's are functions of the x's.

In order that a Pfaffian form $\lambda_i^\alpha x_1^i$ be an *absolute invariant*, the σ's in (26.11) must be zero. We consider this case. Suppose that the generic rank q of the matrix $\|\xi_a^i\|$ is $r(\leq n)$. Using the results of §21 we replace equations (26.11) for $\sigma_{a\beta} = 0$ by

$$(26.12) \qquad \frac{\partial\lambda_i^\alpha}{\partial x^l} - \lambda_m^\alpha\Lambda_{il}^m = 0 \qquad \left(\begin{matrix} i = 1, \cdots, n; \\ l, m = 1, \cdots, q \end{matrix}\right).$$

Expressing the condition of integrability of these equations, we obtain

$$(26.13) \qquad \lambda_h^\alpha\Lambda_{ilm}^h = 0 \qquad \left(\begin{matrix} h, l, m = 1, \cdots, q; \\ i = 1, \cdots, n \end{matrix}\right),$$

where Λ_{ilm}^h is defined by (21.17). Moreover, Λ_{ilm}^h are zero, when the rank of $\|\xi_a^i\|$ is r (cf. (21.19)). Hence equations (26.12) are completely integrable. If $r < n$, each set of solutions is determined by n arbitrary functions of $x^{r+1}, \ldots, x^n$, taken as initial values; if $r = n$, each set of solutions is determined by n arbitrary constants taken as initial values. Hence we have:

[26.3] *If the generic rank of the matrix $\|\xi_a^i\|$ of a group G_r is $r(\leq n)$, there exist Pfaffian forms $\lambda_i x_1^i$ which are absolute invariants of the extended group; the λ's involve n arbitrary functions when $r < n$, and n arbitrary constants when $r = n$.*

We consider next the case where the generic rank q of the matrix $\|\xi_a^i\|$ is less than r, and assume that the coordinates are chosen and the ξ's are designated so that the results of §21 obtain. From (26.11) for $a = 1, \cdots, q$ we have a set of equations of the form

(26.12). When a in (26.11) takes the values $q + 1, \cdots, r$, the resulting equations are reducible by means of (21.2) and (26.12) to

$$(26.14) \qquad \lambda^\alpha_i \xi^i_h \frac{\partial \varphi^h_p}{\partial x^i} = 0.$$

When these equations are satisfied, the conditions (26.13) are satisfied in consequence of (21.18). Hence the problem of determining Pfaffian forms when $q < r$, reduces to the solution of (26.12) with the finite conditions (26.14); that is, these are the equations F_0 of §1. Moreover, the set F_1 are those arising from (26.14) by differentiation and reduction by means of (26.12) and (21.13). By the processes of §1 we determine in any case the generality of such absolute invariants of the extended group.

We return to the consideration of equations (26.11) for the case of the single equation (25.18) and one operator Xf. Then we have

$$\xi^i \frac{\partial \lambda_i}{\partial x^j} + \frac{\partial \xi^i}{\partial x^j} \lambda_i = \rho \lambda_i, \qquad \lambda_1 = a^2, \qquad \lambda_2 = -a^1.$$

Eliminating ρ from these two equations, we obtain equation (25.15). In consequence of the preceding results concerning this equation, we have:

[26.4] *If a Pfaffian form $\lambda_i x^i_1$ is a relative invariant of the extended group of a G_1 in two variables x^1 and x^2, the equation $\lambda_i x^i_1 = 0$ can be integrated by a quadrature.*

Hence we say that the equation (25.18) admits the group G_1, if the left-hand member of the former is a relative invariant of the extended group of G_1.

In like manner we say that any differential equation

$$F(x, y, x_1, y_1) = 0,$$

homogeneous in x_1 and y_1 admits the group G_1, if F is an invariant, absolute or relative, of the extended group. In order to give this problem another form we write the equation thus

$$(26.15) \qquad f(x, y, y') = 0, \qquad y' = \frac{dy}{dx} = \frac{y_1}{x_1}.$$

Since

$$\frac{\partial}{\partial x_1} = \frac{\partial y'}{\partial x_1} \frac{\partial}{\partial y'} = -\frac{y'}{x_1} \frac{\partial}{\partial y'}, \qquad \frac{\partial}{\partial y_1} = \frac{1}{x_1} \frac{\partial}{\partial y'},$$

we have from (26.7) and (26.4) in the case of two variables x and y

(26.16) $X_{(1)}f = \xi\dfrac{\partial f}{\partial x} + \eta\dfrac{\partial f}{\partial y} + \left[\dfrac{\partial \eta}{\partial x} + \left(\dfrac{\partial \eta}{\partial y} - \dfrac{\partial \xi}{\partial x}\right)y' - (y')^2\dfrac{\partial \xi}{\partial y}\right]\dfrac{\partial f}{\partial y'}.$

If the group is a G_r, we have r symbols $X_{(1)a}f$ of the form (26.16), for $a = 1, \cdots, r$. When the finite equations of G_r are written in the form

(26.17) $\bar{x} = f(x, y; a), \qquad \bar{y} = \varphi(x, y; a),$

we have

(26.18) $\bar{y}' = \dfrac{\dfrac{\partial \varphi}{\partial x} + \dfrac{\partial \varphi}{\partial y}y'}{\dfrac{\partial f}{\partial x} + \dfrac{\partial f}{\partial y}y'} \equiv \psi(x, y, y'; a).$

These are the equations in finite form of the extended group in the three variables x, y, y'. Under G_r a curve of the plane goes into a curve and the extended group gives also the relation between the slopes of the tangent at corresponding points. If a differential equation (26.15) admits the G_r, under a transformation of the group an integral curve is transformed into another integral curve.

If $f(x, y, y')$ is an absolute invariant of the extended group of a G_1, that is, if it is a solution of $X_{(1)}f = 0$ in which x, y and y' are treated as independent variables, we say that it is a *differential invariant* of the first order of the group G_1. For a given G_1, that is, for ξ and η given, the problem of finding its differential invariants reduces to the integration of

$$\frac{dx}{\xi} = \frac{dy}{\eta} = \frac{dy'}{\dfrac{\partial \eta}{\partial x} + \left(\dfrac{\partial \eta}{\partial y} - \dfrac{\partial \xi}{\partial x}\right)y' - (y')^2\dfrac{\partial \xi}{\partial y}}.$$

Let $u(x, y) = c$ be a solution of $\eta dx - \xi dy = 0$, or in other form $y = \varphi(x, c)$. Substituting this value of y in the equation

$$\frac{dy'}{dx} = \frac{1}{\xi}\left[\frac{\partial \eta}{\partial x} + \left(\frac{\partial \eta}{\partial y} - \frac{\partial \xi}{\partial x}\right)y' - (y')^2\frac{\partial \xi}{\partial y}\right],$$

we have a Riccati equation for the determination of y' as a function of x and c. It is readily shown that $y' = \eta/\xi$ is a particular solution of this equation, and in consequence of the known theory of Riccati equations we have that the complete solution of this equation reduces to quadratures. If we denote this solution by

$$w(x, y', c) = d,$$

where d is an arbitrary constant, then $v(x, y, y') \equiv w(x, y', u)$ is a solution of $X_{(1)}f = 0$ and the most general solution is any function of $u(x, y)$ and $v(x, y, y')$. The function $u(x, y)$ is an absolute invariant of G_1, that is, the curves $u = \text{const}$ are the trajectories of the group. Hence we have:

[26.5] *If the trajectories of a group G_1 in two variables are known, the differential invariants of the first order of G_1 can be found by quadratures.*

If the finite equations of G_1 are known, then the extended group is found by differentiation, and in accordance with §19 the determination of the differential invariants of the first order results from the elimination of the parameter a from the equations of the extended group.

27. Extensions of the second and higher orders. Ordinary differential equations of the second order. If we consider the extended group, which we now call the *first extension* of the given G_r, as a group in $2n$ variables and form its extended group, which we call the *second extension* of G_r, we obtain a group in x^i, x_1^i and the second differentials of the x's which we denote by x_2^i. Continuing this process we get extensions of all orders. Thus the p-th extension involves $(p + 1)n$ variables x^i, x_1^i, . . . , x_p^i.

From the manner in which (26.7) were obtained it follows that the symbols of the second extension are

$$(27.1) \qquad X_{(2)a}f = \xi_a^i \frac{\partial f}{\partial x^i} + \xi_{1a}{}^i \frac{\partial f}{\partial x_1^i} + \xi_{2a}{}^i \frac{\partial f}{\partial x_2^i},$$

where in consequence of (26.4)

$$(27.2) \qquad \xi_{2a}^i = d\xi_{1a}^i = \frac{\partial^2 \xi_a^i(x)}{\partial x^j \partial x^k} x_1^j x_1^k + \frac{\partial \xi_a^i(x)}{\partial x^k} x_2^k.$$

Generalizing this result, we have that the symbols of the p-th extension may be written in the form

$$(27.3) \qquad X_{(p)a}f = d^\alpha \xi_a^i \frac{\partial f}{\partial x_\alpha^i} \qquad (\alpha = 0, 1, \cdots, p),$$

with the understanding that $d^0\xi = \xi$ and $x_0 = x$.

As in the case of the first extension we have:

[27.1]   *The constants of structure and the first and second parameter groups of the p-th extension of a G_r are the same as for G_r.*

From the foregoing discussion it follows that the matrix of the first extension is obtained from $\|\xi_a^i\|$, the index i indicating the row and a the column, by augmenting the former by n rows containing the differentials of the ξ's. And in general the matrix of the p-th extension is that of the $(p-1)$-th extension augmented by n rows, the elements being the differentials of the preceding n rows. Since the p-th extension involves $(p+1)n$ variables, if $(p+1)n > r$ this extension is intransitive, and consequently there are absolute invariants of the p-th extension.

Consider for example the second extension of the group G_3 of motions of euclidean 2-space.   Its matrix is

$$\begin{Vmatrix} 1 & 0 & y \\ 0 & 1 & -x \\ 0 & 0 & y_1 \\ 0 & 0 & -x_1 \\ 0 & 0 & y_2 \\ 0 & 0 & -x_2 \end{Vmatrix}$$

Its absolute invariants are solutions of

$$(27.4) \quad \frac{\partial f}{\partial x} = 0, \quad \frac{\partial f}{\partial y} = 0, \quad y_1\frac{\partial f}{\partial x_1} - x_1\frac{\partial f}{\partial y_1} + y_2\frac{\partial f}{\partial x_2} - x_2\frac{\partial f}{\partial y_2} = 0.$$

From the first two of these equations it follows that these invariants are independent of x and y. By inspection we see that three independent solutions of the third equation are

$$(27.5) \quad \text{(i)} \ x_1^2 + y_1^2; \quad \text{(ii)} \ x_1y_2 - x_2y_1; \quad \text{(iii)} \ x_1x_2 + y_1y_2.$$

In accordance with the general theory of complete systems any absolute invariant of the second extension is a function of these three. The first and third of these are respectively the square of the linear element and one half of its differential.

For a curve $y = f(x)$, we have $y_1 = y'x_1$, $y_2 = y''x_1^2 + y'x_2$, consequently

$$(27.6) \qquad \frac{x_1y_2 - x_2y_1}{(x_1^2 + y_1^2)^{3/2}} = \frac{y''}{(1 + y'^2)^{3/2}},$$

which is the curvature of the curve.

We consider now the general case in two variables, x and y. We write (26.16) in the form

$$(27.7) \qquad X_{(1)}f = \xi\frac{\partial f}{\partial x} + \eta\frac{\partial f}{\partial y} + \eta_1\frac{\partial f}{\partial y'},$$

where

$$(27.8) \qquad \eta_1 = \frac{d\eta}{dx} - y'\frac{d\xi}{dx},$$

where $\dfrac{d\eta}{dx}$ is the total derivative, that is,

$$\frac{d\eta}{dx} = \frac{\partial \eta}{\partial x} + \frac{\partial \eta}{\partial y}y'.$$

Looking upon (27.7) as the symbol in three independent variables x, y and y', we have

$$(27.9) \qquad \delta x = \xi\delta t, \qquad \delta y = \eta\delta t, \qquad \delta y' = \eta_1\delta t,$$

and (27.8) follows from

$$\delta(dy - y'dx) = 0,$$

on interchanging the operators δ and d. In fact, this equation is in consequence of (27.9)

$$(d\eta - \eta_1 dx - y'd\xi)\delta t = 0.$$

In like manner if we consider the second extension but use y'' rather than x_2 and y_2 and put $\delta y'' = \eta_2\delta t$, then from

$$\delta(dy' - y''dx) = 0$$

we have

$$(27.10) \qquad \eta_2 = \frac{d\eta_1}{dx} - y''\frac{d\xi}{dx},$$

where $\dfrac{d\eta_1}{dx}$ denotes the total derivative with respect to x, that is,

$$\frac{d\eta_1}{dx} = \frac{\partial \eta_1}{\partial x} + \frac{\partial \eta_1}{\partial y}y' + \frac{\partial \eta_1}{\partial y'}y''.$$

And in general we have

$$(27.11) \qquad \eta_p = \frac{d\eta_{p-1}}{dx} - y^{(p)}\frac{d\xi}{dx},$$

and the operator in x, y, y', $\ldots$, $y^{(p)}$ is

$$(27.12) \quad X_{(p)}f = \xi\frac{\partial f}{\partial x} + \eta\frac{\partial f}{\partial y} + \eta_1\frac{\partial f}{\partial y'} + \cdots + \eta_p\frac{\partial f}{\partial y^{(p)}}.$$

A solution of the equation $X_{(p)}f = 0$ is an absolute invariant of the pth extension of G_1 with the symbol Xf, and is called a *differential invariant of the pth order of G_1*. Similar results hold for a G_r when we have a solution of the complete system

$$X_{(p)a}f = 0 \qquad\qquad (a = 1, \cdots, r).$$

If the finite equations of G_r are given by (26.17), the finite equations of the second extension in x, y, y' and y'' are given by (26.17), (26.18) and

$$(27.13) \qquad \bar{y}'' = \frac{d\bar{y}'}{d\bar{x}} = \frac{\dfrac{\partial\psi}{\partial x} + \dfrac{\partial\psi}{\partial y}y' + \dfrac{\partial\psi}{\partial y'}y''}{\dfrac{\partial f}{\partial x} + \dfrac{\partial f}{\partial y}y'}.$$

From the general theory it follows that:

[27.2]   *A necessary and sufficient condition that a differential equation $F(x, y, y', y'') = 0$ admit the G_1 with the symbol Xf is that F be a relative invariant of the second extension of G_1, that is, that $X_{(2)}F = 0$ when $F = 0$.*

If, however, $F = 0$ is given in the form $y'' = \omega(x, y, y')$, then the condition that this equation admit the G_1 is that ω satisfy the equation

$$(27.14) \qquad \eta_2 - \xi\frac{\partial\omega}{\partial x} - \eta\frac{\partial\omega}{\partial y} - \eta_1\frac{\partial\omega}{\partial y'} = 0,$$

it being understood that y'' in η_2 is replaced by ω.

We denote by $u(x, y)$ a solution of

$$(27.15) \qquad\qquad Xf = \xi p + yq = 0.$$

From §10 it follows that, when $u(x, y)$ is known, a function $v(x, y)$ can be formed by quadratures such that

$$\xi\frac{\partial v}{\partial x} + \eta\frac{\partial v}{\partial y} = 1.$$

If we put

$$\bar{x} = v(x, y), \qquad \bar{y} = u(x, y),$$

then in the coordinates $\bar{x}$ and $\bar{y}$

$$Xf = p$$

and the finite equations of the group are

$$\bar{x}' = \bar{x} + t, \qquad \bar{y}' = \bar{y}.$$

Consequently $\bar{y}$, $\dfrac{d\bar{y}}{d\bar{x}}, \ldots, \dfrac{d^p\bar{y}}{d\bar{x}^p}, \ldots$ are differential invariants.
The symbol of the pth extension in these coordinates is

$$X_{(p)}f = p.$$

Hence any differential invariant is a function of

$$\bar{y}, \bar{y}', \ldots, \bar{y}^{(p)}, \ldots$$

Consequently we have:

[27.3] *The determination of the differential invariants of a G_1 with the symbol Xf reduces to the determination of a solution of (27.15), a quadrature and differentiation.*

Since the property of invariance is independent of the coordinate system, it follows that the differential invariants of the pth order of Xf are of the form $F\left(u, \dfrac{du}{dv}, \ldots, \dfrac{d^pu}{dv^p}\right)$. In general the differential equations of order p which admit the G_1 with the symbol Xf are obtained by equating such functions to zero. However, there are exceptional cases. For example, if in the coordinate system of (27.15) the coefficient of $\partial f/\partial y^{(p)}$ in $X_{(p)}f$ involves $y^{(p)}$ as a factor, then $y^{(p)} = 0$ is invariant under (27.15) for this value of $y^{(p)}$.

28. Differential invariants. In §§ 26, 27 we have considered in particular the case where there are two variables and the effect of the transformations of a G_r on a curve $y = f(x)$ and the derivatives $y', \ldots, y^{(p)}, \ldots$. This is a particular case of the general problem in a V_n for a sub-space V_m of m dimensions defined by

$$(28.1) \qquad x^p = \varphi^p(x^1, \cdots, x^m) \quad (p = m + 1, \cdots, n).$$

If these values are substituted in the finite equations of the group

$$(28.2) \qquad x'^i = f^i(x; a),$$

we obtain for each set of values of the a's a V'_m into which V_m is transformed. In order that the equations of V'_m may be expressed in the form

$$(28.3) \qquad x'^p = \varphi'^p(x'^1, \cdots, x'^m; a) \quad (p = m + 1, \cdots, n),$$

it is necessary and sufficient that the first m of equations (28.2) after the substitution from (28.1) be solvable for $x^1, \ldots, x^m$ and the resulting expressions substituted in the last $n - m$ of (28.2). Hence the matrix

$$(28.4) \qquad \left\| \frac{\partial f^a}{\partial x^b} + \frac{\partial f^a}{\partial x^p} \frac{\partial \varphi^p}{\partial x^b} \right\| \qquad \binom{a,\, b = 1, \cdots, m;}{p = m + 1, \cdots, n},$$

must be of rank m. When the determinant of this matrix is developed in powers of $\dfrac{\partial \varphi^p}{\partial x^b}$, it is seen that the coefficients are minors of order m of the matrix $\left\| \dfrac{\partial f^a}{\partial x^i} \right\|$, for $a = 1, \cdots, m$; $i = 1, \cdots, n$, and every such minor is a coefficient. Since all of these minors cannot be zero in accordance with the requirement (4.2), it follows that the rank of (28.4) is m for general φ's. Accordingly on the assumption that the φ's are such that the rank of (28.4) is m, we have that each V_m is transformed into a V'_m defined by (28.3), in which the a's have particular values for each transformation of G_r.

From (28.1) and (28.3) we have

$$(28.5) \qquad dx^p - \frac{\partial \varphi^p}{\partial x^a} dx^a = 0, \qquad dx'^p - \frac{\partial \varphi'^p}{\partial x'^a} dx'^a = 0$$

$$(a = 1, \cdots, m).$$

From the first of these sets and (28.2) we have

$$(28.6) \qquad dx'^i = \left(\frac{\partial f^i}{\partial x^a} + \frac{\partial f^i}{\partial x^p} \frac{\partial \varphi^p}{\partial x^a} \right) dx^a$$

$$\binom{i = 1, \cdots, n; a = 1, \cdots, m;}{p = m + 1, \cdots, n},$$

by means of which the second set of (28.5) are expressible in the form

$$\left[\frac{\partial f^p}{\partial x^a} + \frac{\partial f^p}{\partial x^q}\frac{\partial \varphi^q}{\partial x^a} - \frac{\partial x'^p}{\partial x'^b}\left(\frac{\partial f^b}{\partial x^a} + \frac{\partial f^b}{\partial x^q}\frac{\partial \varphi^q}{\partial x^a}\right)\right]dx^a = 0$$
$$\begin{pmatrix} a, b = 1, \cdots, m; \\ p, q = m + 1, \cdots, n \end{pmatrix}.$$

Since the differentials $dx^1, \ldots, dx^m$ are arbitrary, the expressions in parenthesis are zero. The resulting equations can be solved for $\dfrac{\partial x'^p}{\partial x'^b}$ as functions of the x's, $\dfrac{\partial x^p}{\partial x^a}$ and the a's, because the rank of (28.4) is m. If we denote these partial derivatives by x'^p_b and x^p_a, the solution may be written as

$$(28.7) \qquad x'^p_b = f^p_b(x; x^q_a; a).$$

In order to find the second derivatives of x'^p with respect to $x'^1, \ldots, x'^m$, we make use of

$$dx'^p_b = \frac{\partial^2 x'^p}{\partial x'^b \partial x'^a}dx'^a.$$

By means of (28.7) and (28.5) the left-hand member is reducible to a linear function of $dx^1, \ldots, dx^m$, the coefficients being functions of the x's, the first and second derivatives of $x^{m+1}, \ldots, x^n$ with respect to $x^1, \ldots, x^m$ and the a's. By means of (28.6) the right hand member is

$$\frac{\partial^2 x'^p}{\partial x'^b \partial x'^a}\left(\frac{\partial f^a}{\partial x^c} + \frac{\partial f^a}{\partial x^q}\frac{\partial \varphi^q}{\partial x^c}\right)dx^c \qquad \begin{pmatrix} a, b, c = 1, \cdots, m; \\ p, q = m + 1, \cdots, n \end{pmatrix}.$$

Equating coefficients of each of the differentials dx^c on both sides of the equation, we obtain m equations with p and b fixed, which can be solved for $\dfrac{\partial^2 x'^p}{\partial x'^b \partial x'^a}$, because the rank of the matrix (28.4) is m. If we use the notation

$$x'^p_{\alpha_1 \cdots \alpha_m} = \frac{\partial^A x'^p}{\partial x'^{1\alpha_1} \cdots \partial x'^{m\alpha_m}} \qquad (A = \alpha_1 + \cdots + \alpha_m),$$

and similarly for x^p, these may be written in the form

$$(28.8) \qquad x'^p_{\alpha_1 \ldots \alpha_m} = f^p_{\alpha_1 \ldots \alpha_m}(x; x'^q_{\beta_1 \ldots \beta_m}; a)$$

$$\begin{pmatrix} p, q = m + 1, \cdots, n; \\ \alpha_1 + \cdots + \alpha_m = A; \\ \beta_1 + \cdots + \beta_m \leqslant A \end{pmatrix},$$

for $A = 2$; thus we have each second derivative of any x'^p equal to a function of the x's, first and second derivatives of $x^{m+1}, \ldots, x^n$ with respect to $x^1, \ldots, x^m$ and the a's.

If we apply to (28.8) for $A = 2$ the process which was applied to (28.7), we obtain a set of equations (28.8) for $A = 3$. Continuing this process step by step, we obtain equations of this sort for any integer N.

Consider now the set of equations (28.2) and (28.8) as A takes the values $1, \ldots, N$, where N is some positive integer. These equations define a group in the x's and the derivatives of $x^{m+1}, \ldots,$ x^n with respect to $x^1, \ldots, x^m$ of the first, second, $\ldots, N$th order. For, if we give the a's in these equations the values a_1^α, and apply the above processes to

$$(28.9) \qquad x''^i = f^i(x'; a_2),$$

we get in place of (28.8)

$$(28.10) \qquad x''^p_{\alpha_1 \ldots \alpha_m} = f^p_{\alpha_1 \ldots \alpha_m}(x'; x'^q_{\beta_1 \ldots \beta_m}; a_2).$$

Eliminating the x''s from (28.7) and (28.9), we have (cf. §4)

$$x''^i = f^i(x; a_3) \qquad a_3^\alpha = \varphi^\alpha(a_1; a_2).$$

Applying to these equations the processes by which (28.8) were obtained from (28.2), we get

$$x''^p_{\alpha_1 \ldots \alpha_m} = f^p_{\alpha_1 \ldots \alpha_m}(x; x^q_{\beta_1 \ldots \beta_m}; a_3).$$

Any such function on the right must result from the corresponding function in (28.10), when the x''s and $x'^q_{\beta_1 \ldots \beta_m}$ are replaced by their expressions from (28.2) and (28.8), otherwise we should have equations between the derivatives of $x^{m+1}, \ldots, x^n$ and a_1^α, a_2^α and a_3^α, which could not hold for general functions φ (28.1). Hence (28.2) and (28.8) is the expression of the Nth extension of the given group in another form. From the above discussion it follows that:

[28.1]   *Any extension of a group has the same structure and the same first and second parameter groups as the given group.*

Any function of the x's and derivatives of $x^{m+1}, \ldots, x^n$ with respect to $x^1, \ldots, x^m$ up to order N which is equal to the same function of the x''s and the same derivatives of $x'^{m+1}, \ldots, x'^n$ with respect to $x'^1, \ldots, x'^m$ for all the transformations of the Nth extension is called a *differential invariant of order N* of the given group G_r. From the results of §19 it follows that when the finite equations of a group G_r are known, the determination of the differential invariants of any order N involves differentiation in order to obtain the Nth extension and then the elimination of the parameters a^α from its equations. When the symbols of the group are given the problem of finding the differential invariants is that of finding solutions of the complete system obtained by equating to zero the symbols of the extension of given order. We proceed to the determination of these symbols.

We have seen in §10 that $\delta x^i = \xi^i \delta t$ and accordingly we put

$$\delta x^q_{\alpha_1 \ldots \alpha_m} = \xi^q_{\alpha_1 \ldots \alpha_m} \delta t.$$

We must have (cf. §27)

$$\delta\left(dx^p_{\alpha_1 \ldots \alpha_m} - \sum_a^{1 \cdots m} x^p_{\alpha_1 \ldots \alpha_{a-1}\alpha_a+1\alpha_{a+1} \cdots \alpha_m} dx^a \right) = 0.$$

Interchanging the operations δ and d, we have

$$(28.11) \quad d\xi^p_{\alpha_1 \ldots \alpha_m} - \sum_a^{1 \cdots m} \xi^p_{\alpha_1 \ldots \alpha_{a-1}\alpha_a+1\alpha_{a+1} \cdots \alpha_m} dx^a$$

$$- \sum_a^{1 \cdots m} x^p_{\alpha_1 \ldots \alpha_{a-1}\alpha_a+1\alpha_{a+1} \cdots \alpha_m} d\xi^a = 0.$$

Since $\xi^p_{\alpha_1 \ldots \alpha_m}$ involves the x's and derivatives of $x^{m+1}, \ldots, x^n$ to order $N(= \alpha_1 + \cdots + \alpha_m)$, we have

$$d\xi^p_{\alpha_1 \ldots \alpha_m} = \left[\frac{\partial \xi^p_{\alpha_1 \ldots \alpha_m}}{\partial x^a} + \frac{\partial \xi^p_{\alpha_1 \ldots \alpha_m}}{\partial x^q} \frac{\partial x^q}{\partial x^a} + \sum \frac{\partial \xi^p_{\alpha_1 \ldots \alpha_m}}{\partial x^q_{\beta_1 \ldots \beta_m}} \frac{\partial x^q_{\beta_1 \ldots \beta_m}}{\partial x^a} \right] dx^a,$$

where Σ indicates the sum of all the terms when

$$\beta_1 + \cdots + \beta_m \leqq N.$$

If we denote the expression in the parenthesis by $\dfrac{d\xi^p_{\alpha_1 \ldots \alpha_m}}{dx^a}$ and put

$$\frac{d\xi^p}{dx^a} = \frac{\partial \xi^p}{\partial x^a} + \frac{\partial \xi^p}{\partial x^q}\frac{dx^q}{\partial x^a},$$

on equating to zero the coefficients of each of $dx^1, \ldots, dx^m$ in (28.11), we have

$$(28.12) \quad \xi^p_{\alpha_1 \ldots \alpha_{a-1}\alpha_a+1\alpha_{a+1}\ldots \alpha_m} = \frac{d\xi^p_{\alpha_1 \ldots \alpha_m}}{dx^a}$$

$$- \sum_b^{1 \cdots m} x^p_{\alpha_1 \ldots \alpha_{b-1}\alpha_b+1\alpha_{b+1} \ldots \alpha_m}\frac{d\xi^b}{dx^a}.$$

By means of this formula we obtain the ξ's for any extension from those of the preceding one.   Equation (27.11) is a particular case of (28.12).

If we denote the symbol of the Nth extension by $X_{(N)}f$, we have

$$(28.13) \qquad X_{(N)}f = \xi^i\frac{\partial f}{\partial x^i} + \Sigma \xi^p_{\alpha_1 \ldots \alpha_m}\frac{\partial f}{\partial x^p_{\alpha_1 \ldots \alpha_m}},$$

where Σ indicates the sum of terms when $\alpha_1 + \cdots + \alpha_m$ takes the values 1 to N.   If we have a group G_r, for each symbol $X_a f$ we form a symbol $X_{(N)a}f$ by the above process and thus obtain the r symbols of the Nth extension.

Exercises

1. If an equation $Af = 0$ admits two operators X_1f and X_2f, such that $X_2f = \varphi X_1f$, then $A\varphi = 0$.

2. Show that the equation

$$Af \equiv x^1(x^2 - x^3)p_1 + x^2(x^3 - x^1)p_2 + x^3(x^1 - x^2)p_3 = 0$$

admits the operators

$$X_1f = x^i p_i, \qquad X_2f = \frac{1}{x^1 x^2 x^3}X_1f,$$

and thus $\varphi = x^1 x^2 x^3$ is a solution of the equation.   Show that $x^1 + x^2 + x^3$ is a second solution.

Lie-Scheffers, 1891, 3, p. 451.

3. Show that the equation

$$Af \equiv (\varphi + 2)p_1 - 2\varphi p_2 + \varphi p_3 = 0, \qquad \varphi = x^1 - x^2 - x^3$$

admits the two operators

$$X_1 f = p_1 + p_3, \qquad X_2 f = (x^2 + 2x^3 + 1)(p_1 + 2p_2 - p_3)$$

and that

$$(X_1, X_2)f = \frac{2}{x^2 + 2x^3 + 1} X_2 f,$$

and consequently $x^2 + 2x^3$ is a solution of $Af = 0$.

Lie-Scheffers, 1891, 3, p. 447.

4. If an equation $Af \equiv a^i p_i = 0$ $(i = 1, 2, 3)$ admits two operators $X_1 f$ and $X_2 f$, and the rank of the matrix $\|a^i, \xi_1^i, \xi_2^i\|$ is three, the integration of $Af = 0$ reduces to at most two quadratures.

Lie-Scheffers, 1891, 3, p. 444.

5. If $F(x)$ is an absolute invariant of a group G_r, $\frac{\partial F}{\partial x^i} x_1^i$ is an absolute invariant of the extended group.

6. If $F_1(x) = \cdots = F_p(x) = 0$ define a relative invariant of a group G_r, these equations and $\frac{\partial F_\alpha}{\partial x^i} x_1^i = 0$ $(\alpha = 1, \cdots, p)$ define a relative invariant of the first extended group.

7. Show that $y'' = \omega(x, y, y')$ is invariant under a G_1, if and only if the equation

$$\frac{\partial f}{\partial x} + y' \frac{\partial f}{\partial y} + \omega \frac{\partial f}{\partial y'} = 0$$

admits the first extension of the symbol Xf of G_1.

Lie-Scheffers, 1891, 3, p. 364.

8. Show that the following differential equations are invariant under the accompanying operator:

(1) $y'' = F(ax + by, y')$, $bp - aq$;

(2) $y'' = x^{n-2} F\left(\dfrac{y}{x^n}, \dfrac{y'}{x^{n-1}}\right)$, $xp + nyq$;

(3) $y'' x^{n+2} + (1 - n)x^{n+1} y' = F\left(\dfrac{y}{x^n}, xy' - ny\right)$, $x^n(xp + npq)$.

9. If $u(x, y)$ is an absolute invariant of a G_1 and $v(x, y, y')$ is a differential invariant of the first order, then $\dfrac{dv}{du}$ is a differential invariant of the second order.

Lie-Scheffers, 1891, 3, p. 377.

10. If $u(x, y)$ and $v(x, y, y')$ are invariants of a G_1, the most general differential equation invariant under G_1 is given by

$$\frac{dv}{du} = \varphi(u, v),$$

where φ is an arbitrary function of u and v.

Lie-Scheffers, 1891, 3, p. 377.

11. Show that because of Ex. 10 the solution of a differential equation of the second order admitting an operator requires at most the solution of two differential equations of the first order and quadratures.

Lie-Scheffers, 1891, 3, p. 378.

12. An ordinary differential equation of the second order in x and y is invariant under at most eight linearly independent (constant coefficients) operators.

Lie-Scheffers, 1891, 3, p. 405.

13. A differential equation of the second order which admits two operators can be integrated by quadratures.

Lie-Scheffers, 1891, 3, pp. 457–464.

14. When a group G_r has the equations (4.1), the equations

$$\text{(i)} \qquad x'^i_\alpha = f^i(x^1_\alpha, \cdots, x^n_\alpha; a^1, \cdots, a^r) \quad \begin{pmatrix} i = 1, \cdots, n; \\ \alpha = 1, \cdots, k \end{pmatrix}$$

define a group G_r in the nk variables x^i_α. If this group is transitive, any k ordinary points in V_n can be transformed into k other arbitrary ordinary points, and the given G_r is said to be *k-fold transitive*, when k is the maximum number for which this is true. The symbols of the group (i) are

$$X_{1a}f + \cdots + X_{ka}f,$$

where

$$X_{\alpha a} = \xi^i(x_\alpha)\frac{\partial f}{\partial x^i_\alpha} \qquad \begin{pmatrix} \alpha = 1, \cdots, k; \\ a = 1, \cdots, r \end{pmatrix}.$$

15. For k arbitrary points of coordinates x^i_α for $\alpha = 1, \cdots, k$, the ranks of the matrices

$$M_1 = \|\xi^i_a(x_1)\|, \qquad M_2 = \|\xi^i_a(x_1); \xi^i_a(x_2)\|, \cdots, M_k = \|\xi^i_a(x_1); \cdots; \xi^i_a(x_k)\|$$

form an increasing series of integers $\nu_1, \nu_2, \nu_3, \ldots$ of which the maximum is r. If $\nu_k < r$, there exists a sub-group $G_{r-\nu_k}$ which leaves the r points invariant. If k is the largest integer for which $\nu_k \geqq nk$, the given G_r is k-fold transitive.

16. If ν_k in Ex. 15 is less than nk, the equations

$$X_{1a}f + \cdots + X_{ka}f = 0$$

admit $nk - \nu_k$ independent solutions, called the *invariants of the k points* x^i_α. If k is such that $\nu_k = r$ for a G_r, the system of $k + 2$ points, or of a larger number, does not have any invariants which are not invariants of a lesser number of points.

Bianchi, 1918, 1, p. 336.

17. The projective group G_3 in one variable with the symbols p, xp, x^2p is 3-fold transitive. The invariant of four points is their cross-ratio

$$\frac{x_1 - x_3}{x_1 - x_4} \cdot \frac{x_2 - x_4}{x_2 - x_3}$$

and is the solution of

$$\sum_\alpha \frac{\partial f}{\partial x_\alpha} = 0, \qquad x_\alpha \frac{\partial f}{\partial x_\alpha} = 0, \qquad x^2_\alpha \frac{\partial f}{\partial x_\alpha} = 0 \quad (\alpha = 1, \cdots, 4).$$

CHAPTER III

INVARIANT SUB-GROUPS

29. Groups invariant under transformations. Consider a group G_r with symbols $X_a f$ and also a transformation S, namely

(29.1) $$\bar{x}^i = \varphi^i(x).$$

In order that the symbols

(29.2) $$\overline{X}_a f = \xi_a^i(\bar{x})\frac{\partial \bar{f}}{\partial \bar{x}^i}$$

generate in the $\bar{x}$'s the same group as the given group, it is necessary and sufficient that each of the operators (29.2) be a linear combination with constant coefficients (§11) of the symbols arising from $X_a f$ by the transformation (29.1), that is,

(29.3) $$\xi_a^i(\bar{x}) = c_a^b \xi_b^j(x)\frac{\partial \bar{x}^i}{\partial x^j}$$

must be satisfied identically in consequence of (29.1). From (29.3) we have

$$\overline{X}_a \bar{f} = c_a^b X_b f, \qquad \bar{f}(\bar{x}) = f(x).$$

If u^a are canonical parameters for G_r and we put

(29.4) $$\bar{u}^a = \bar{c}_b^a u^b, \qquad u^b = c_a^b \bar{u}^a,$$

where the constants c_b^a and $\bar{c}_a^b$ are in the relations (cf. §7)

(29.5) $$c_b^a \bar{c}_c^b = \delta_c^a,$$

then we have

(29.6) $$\bar{u}^a \overline{X}_a \bar{f} = u^b X_b f,$$

and consequently equations (11.11) are transformed into the same form in the $\bar{x}$'s by means of (29.1). Thus we have schematically

(29.7) $$ST_u S^{-1} = T_{\bar{u}},$$

or

(29.8) $$ST_u = T_{\bar{u}} S.$$

109

When the above conditions are satisfied, we say that the *group* G_r *admits the transformation* S, or is *invariant* under it.

When and only when we have $c_b^a = \delta_b^a$, equations (29.3) become

$$(29.9) \qquad \xi_a^i(\bar{x}) = \xi_a^j(x)\frac{\partial \bar{x}^i}{\partial x^j},$$

in which case (29.8) is

$$(29.10) \qquad ST_u = T_uS,$$

that is, S is *commutative* with each transformation of the group.

Consider next the case when the transformation (29.1) is replaced by a one-parameter group Γ_1 with the symbol $Xf = \xi^i p_i$. In this case the c's in (29.3) are functions of the parameter τ of the group such that, if $\tau = 0$ defines the identity, we have $(c_a^b)_{\tau=0} = \delta_a^b$. In order that (29.3) hold for the infinitesimal transformation

$$(29.11) \qquad \bar{x}^i = x^i + \xi^i \delta\tau,$$

we find on substitution that it is necessary that

$$(29.12) \qquad \xi^i\frac{\partial \xi_b^i}{\partial x^j} - \xi_b^j\frac{\partial \xi^i}{\partial x^j} + g_b^a\xi_a^i = 0,$$

where g_b^a are the constants defined by

$$(29.13) \qquad g_b^a = -\left(\frac{dc_b^a}{d\tau}\right)_{\tau=0} = \left(\frac{d\bar{c}_b^a}{d\tau}\right)_{\tau=0}.$$

From (29.12) we have that

$$(29.14) \qquad (X, X_b)f + g_b^a X_a f = 0$$

is a necessary condition that the group admits the infinitesimal transformation.

If the coordinates are chosen in accordance with theorem [10.2] so that

$$(29.15) \qquad \xi^i = \delta_n^i,$$

equations (29.12) become

$$(29.16) \qquad \frac{\partial \xi_b^i}{\partial x^n} + g_b^a\xi_a^i = 0.$$

In this coordinate system the finite equations of Γ_1 are

$$(29.17) \qquad \bar{x}^i = x^i + \delta_n^i\tau.$$

If in (29.6) we put $\bar{f} = \bar{x}^i$, then in consequence of (29.17) we have

$$(29.18) \qquad \bar{u}^a \xi_a^i(\bar{x}) = u^b \xi_b^i(x).$$

Differentiating with respect to τ, we have

$$(29.19) \qquad \frac{d\bar{u}^a}{d\tau}\xi_a^i(\bar{x}) + \bar{u}^a \frac{\partial \xi_a^i(\bar{x})}{\partial \bar{x}^n} = 0.$$

Since equations (29.16) are identities in the x's and consequently hold in the $\bar{x}$'s, the above equations are reducible to

$$\left(\frac{d\bar{u}^a}{d\tau} - \bar{u}^b g_b^a\right)\xi_a^i(\bar{x}) = 0,$$

from which it follows that $\bar{u}^a$ are the solutions of the equations

$$(29.20) \qquad \frac{d\bar{u}^a}{d\tau} = \bar{u}^b g_b^a$$

which reduce to u^a when $\tau = 0$.

Conversely, if equations (29.14) are satisfied, the coordinate system being any whatever, then in the particular coordinate system for which (29.17) are the finite equations of the group Γ_1 we have equations (29.15). If $\bar{u}_a$ is any set of solutions of (29.20), then $\bar{u}^a \xi_a^i(\bar{x})$ is independent of τ as is shown by differentiation with respect to τ, and we have (29.18) where $u^b = (\bar{u}^b)_{\tau=0}$, from which follow (29.6) in consequence of (29.17). We have thus established that equations (29.14) constitute a necessary and sufficient condition. Hence we have:

[29.1] *A necessary and sufficient condition that a group G_r generated by the infinitesimal transformations with the symbols $X_a f$ admit the transformations of a group Γ_1 with the symbol Xf is that equations (29.14) be satisfied, where the g's are constants.*

In order to obtain equations (29.3), we have made use of the fact that the transformations with the symbols $X_a f$ form a group. We shall show that the preceding theorem holds, even if the transformations $u^a X_a f$, where the u's are arbitrary parameters, do not form a group. In fact, if equations (29.6) are satisfied with the understanding that $(\bar{u}^a)_{\tau=0} = u^a$, we have equations (29.18) for the special coordinate system in which the equations of the group Γ_1 are given by (29.17). If in (29.19) we put $\tau = 0$, we get

$$\left(\frac{d\bar{u}^a}{d\tau}\right)_0 \xi_a^i(x) + u^a\frac{\partial\xi_a^i(x)}{\partial x^n} = 0.$$

Since these equations must hold for arbitrary values of the u's, if we take r different sets of u^a such that all but one is zero in a set and the other is unity, we obtain equations of the form (29.16) in which the g's are constants. Then the discussion proceeds as before. Hence we have:

[29.2] *If X_af are the symbols of r infinitesimal transformations, in order that the transformations of symbols u^aX_af, where the u's are arbitrary parameters, admit the group Γ_1 of transformations with the symbol Xf, it is necessary and sufficient that (29.14) be satisfied, the g's being constants. Then a given G_1 defined by u^aX_af is transformed by all the transformations of Γ_1 into the group G_1 defined by $\bar{u}^a\overline{X}_af$, where the $\bar{u}$'s are the solutions of (29.20) such that $\bar{u}^a = u^a$ when $\tau = 0$.* *

As a corollary to this theorem and because of the second fundamental theorem we have:

[29.3] *If X_af are the symbols of r transformations and all of them are invariant under each of them, the given transformations form a group of order r; and G_r is invariant under any sub-group G_1.*

From considerations analogous to those applied to the transformations (29.1) leading to (29.9), we have that, when and only when $\bar{u}^a = u^a$, each transformation of G_r is commutative with each transformation in Γ_1. Then from (29.20) we have $g_b^a = 0$. Hence we have:

[29.4] *A necessary and sufficient condition that each transformation of the group with the symbol Xf be commutative with each transformation of the group G_r with the symbols X_af is that*

$$(X, X_a)f = 0 \qquad (a = 1, \cdots, r).$$

As a corollary we have:

[29.5] *A necessary and sufficient condition that every two transformations of a group G_r with the symbols X_af be commutative is that*

$$(X_a, X_b)f = 0,$$

that is, the constants of composition are zero.

---

* Cf. *Bianchi*, 1918, 1, p. 198.

A group satisfying this condition is called *Abelian* (cf. §13).

When a sub-group G_1 of a group G_r is commutative with all the transformations of G_r, the G_1 is called *exceptional*. If its symbol is $e^a X_a f$, then we must have $(e^a X_a, X_b)f = 0$ and consequently

$$(29.21) \qquad e^a c_{ab}^d = 0.$$

These r^2 equations, as b and d take the values 1 to r, must be consistent. Hence the rank of the matrix

$$(29.22) \qquad ||c_{ab}^d||$$

where $a(= 1, \cdots, r)$ denotes the columns, and b and d the rows must be less than r. If the rank is s, there are $r - s$ independent sets of solutions of equations (29.21). If we put $r - s = m$, we may change the symbols of the given group G_r so that $X_1 f, \ldots, X_m f$ are the symbols of exceptional groups G_1. Hence we have that the corresponding constants of composition satisfy the conditions

$$(29.23) \qquad c_{ta}^b = 0 \qquad (a, b = 1, \cdots, r; t = 1, \cdots, m),$$

Since

$$(X_t, X_u)f = 0 \qquad (t, u = 1, \cdots, m),$$

it follows that:

[29.6] *The exceptional groups G_1 of a group G_r, if any, form an Abelian sub-group.*

30. Commutative groups. Reciprocal simply transitive groups.

If $X_a f$ are the symbols of a group G_r and $Zf = \zeta^i \dfrac{\partial f}{\partial x^i}$ is the symbol of a group G_1, then, as follows from theorem [29.4], a necessary and sufficient condition that each transformation of G_r be commutative with each transformation of G_1 is that

$$(30.1) \qquad \xi_a^i \frac{\partial \zeta^i}{\partial x^i} - \zeta^i \frac{\partial \xi_a^i}{\partial x^i} = 0.$$

We consider first the case when the group is simply transitive, that is, $r = n$ and the generic rank of the matrix $||\xi_a^i||$ is n. By means of (21.13), where in this case $\lambda, \mu, \beta, h = 1, \cdots, n$, equations (30.1) are equivalent to

$$(30.2) \qquad \frac{\partial \zeta^i}{\partial x^i} + \zeta^k \Lambda_{ki}^j = 0.$$

Expressing the conditions of integrability of these equations, we find that they are satisfied identically because of (21.19). Hence equations (30.2) admit n sets of solutions ζ_b^i, for $b = 1, \cdots, n$, such that the generic rank of the ζ's is n. Moreover, any solution of (30.2) is a linear combination (constant coefficients) of these n sets of solutions.

If we put $Z_a f \equiv \zeta_a^i p_i$, since

$$(Z_a, X_c)f = 0, \qquad (Z_b, X_c)f = 0 \quad (a, b, c = 1, \cdots, n),$$

it follows from the Jacobi identity (2.4) that

$$((Z_a, Z_b), X_c)f = 0.$$

Hence $\zeta_a^i \dfrac{\partial \zeta_b^i}{\partial x^i} - \zeta_b^i \dfrac{\partial \zeta_a^i}{\partial x^i}$ is a solution of (30.2) and in consequence of the above observation we have

$$(30.3) \qquad \zeta_a^i \frac{\partial \zeta_b^i}{\partial x^i} - \zeta_b^i \frac{\partial \zeta_a^i}{\partial x^i} = \bar{c}_{ab}^e \zeta_e^i,$$

where the $\bar{c}$'s are constants. Consequently $Z_a f$ are the symbols of a continuous group $\bar{G}_r$. Clearly the relation between the groups G_r and $\bar{G}_r$ is reciprocal; they are said to be *reciprocal* groups. From (8.5) and (30.1) it is seen that the first and second parameter groups of any group G_r are reciprocal groups.

From the results of §29 it follows that if $x'^i = \varphi^i(x)$ is any transformation of the group $\bar{G}_r$ then

$$\xi_a^i(x') = \xi_a^j \frac{\partial \varphi^i}{\partial x^j},$$

that is, each transformation of $\bar{G}_r$ transforms the group G_r of symbols $\xi_a^i(x)\dfrac{\partial f}{\partial x^i}$ into the group G_r' of the symbols $\xi_a^i(x')\dfrac{\partial f}{\partial x'^i}$. Consequently G_r and G_r' are equivalent groups, and from theorem [22.3] it follows that the equations of transformation, which are in fact the equations of the group $\bar{G}_r$, involve n parameters. From (22.7) it follows that they are the integrals of the completely integrable system of equations

$$(30.4) \qquad \frac{\partial x'^i}{\partial x^j} = \xi_a^i(x')\xi_j^a(x) \quad (a, i, j = 1, \cdots, n).$$

Thus we have:

[30.1] *If $\xi_a^i(x)$ are the vectors of a simply transitive group, the equations (30.4) are completely integrable, and their solutions, involving n*

arbitrary constants, are the transformations of another simply transitive group, the reciprocal of the given group.

Returning to the consideration of the solutions of equations (30.2), we may write

$$(30.5) \qquad \zeta_a^i = \psi_a^b \xi_b^i.$$

Substititing in (30.2) and making use of (21.13), we have

$$\frac{\partial \psi_a^b}{\partial x^i} \xi_b^j + \psi_a^b \xi_b^k (\Lambda_{ki}^j - \Lambda_{ik}^j) = 0.$$

For the present case equations (21.15) are

$$(30.6) \qquad \Lambda_{ki}^j - \Lambda_{ik}^j = c_{hl}^m \xi_i^h \xi_k^l \xi_m^j \quad (h, i, j, k, l, m = 1, \cdots, n).$$

From these two sets of equations it follows that

$$(30.7) \qquad \frac{\partial \psi_a^b}{\partial x^i} + \psi_a^l c_{hl}^b \xi_i^h = 0.$$

If we substitute for the derivatives of the ζ's in (30.3) their expressions from (30.2), we obtain

$$\zeta_a^i \zeta_b^k (\Lambda_{ik}^j - \Lambda_{ki}^j) = \bar{c}_{ab}^e \zeta_e^j.$$

From these equations, (30.5) and (30.6) we get, since the determinant of the ξ's is not zero,

$$\bar{c}_{ab}^e \psi_e^m = -c_{hl}^m \psi_a^h \psi_b^l.$$

From (30.7) it follows that the ψ's are not constants, and they are not all zero when the x's are zero, otherwise they would be zero identically, as follows from the theory of differential equations of the form (30.7). Since the above equations must hold identically, we have when the x's are zero

$$\bar{c}_{ab}^e c_e^m = -c_{hl}^m c_a^h c_b^l,$$

where the c's are constants. Hence the groups G_r and $\bar{G}_r$ have the same structure and consequently we have (§22):

[30.2] *Two reciprocal simply transitive groups are equivalent.*

If $r < n$ and the generic rank q of the matrix $\|\xi_a^i\|$ is equal to r, we choose a coordinate system such that (21.8) is satisfied, and proceed as in the case of simply transitive groups. Then equations (30.2) hold with $i, j, k = 1, \cdots, r$, and in addition we have from (30.1)

$$(30.8) \qquad \frac{\partial \zeta^\sigma}{\partial x^i} = 0 \qquad \begin{pmatrix} \sigma = r + 1, \; \cdots \;, n; \\ i = 1, \; \cdots \;, r \end{pmatrix}.$$

The system of equations (30.2) is completely integrable because of (21.19), and consequently a solution is determined by taking for initial values arbitrary functions of $x^{r+1}, \ldots, x^n$. Thus each solution of (30.2) and (30.8) involves n arbitrary functions of these coordinates. For any two of these solutions we have equations of the form (30.3), where now the $\bar{c}$'s are in general functions of the coordinates $x^{r+1}, \ldots, x^n$. Hence although each solution yields a symbol Zf which generates a group $\bar{G}_1$ each of whose transformations is commutative with every transformation of G_r, these $\bar{G}_1$'s are not sub-groups of a group of higher order.

However, if it is possible to choose the coordinates so that (21.8) hold and the other ξ's are independent of $x^{r+1}, \ldots, x^n$, then solutions of (30.2) can be found not involving these variables, and the r operators $Z_a f$ are the symbols of a group $\bar{G}_r$ reciprocal to G_r. In this case we have in fact two simply transitive groups in r variables and the above result is in keeping with theorem [30.1]. Hence we have:

[30.3] *If the rank of the matrix* $\|\xi_a^i\|$ *of an intransitive group* G_r *is* r, *there exist groups* $\bar{G}_1$, *whose vectors involve arbitrary functions, such that each transformation of the latter is commutative with each transformation of* G_r, *but in general these* $\bar{G}_1$*'s are not sub-groups of a group of higher order.*

If the group G_r is transitive but not simply transitive, we may in all generality assume that the matrix $\|\xi_a^i\|$ for $a = 1, \; \cdots \;, n$ is of rank n, and that for $p = n + 1, \; \cdots \;, r$ we have (21.2). When a takes the values 1 to n in (30.1), we have the equivalent system (30.2) and for $a = n + 1, \; \cdots \;, r$ equations (30.1) reduce by means of (30.1) for $a = 1, \; \cdots \;, n$ to

$$(30.9) \qquad \zeta^i \Phi_{lp}^h \xi_i^l = 0,$$

in consequence of (21.11). Hence the problem reduces to the solution of the mixed system (30.2) and (30.9). The conditions of integrability of (30.2) assume the form (cf. (21.17))

$$(30.10) \qquad \zeta^k \Lambda_{kli}^j = 0.$$

From (21.18) it follows that in this case equations (30.10) are not satisfied identically.

Equations (30.9) are in this case the set F_0 referred to in §1. Equations (30.10) and those resulting from (30.9) by differentiation and substitution from (30.2) are the system F_1. We observe that F_1 is linear and homogeneous in the ζ's and the same is true of the succeeding systems $F_2, \ldots$. The number of independent equations, say s, in this set is at most $n - 1$. From algebraic considerations it follows, in accordance with theorem [1.1], that the conditions of the problem are that there exist a positive integer N such that the rank of the matrix of the sets $F_0, F_1, \ldots, F_N$ is $n - s$ ($s \geqslant 1$) and that this is also the rank of the matrix of $F_0, \ldots,$ F_{N+1}. If these conditions are satisfied, the solution of the problem reduces to the integration of a completely integrable set of equations of the form (1.11) in $n - s$ functions θ^v and the $\overline{\psi}$'s are linear and homogeneous in these functions. Consequently any solution is expressible as a linear function with constant coefficients of $n - s$ particular solutions, and such an expression is a solution. If then there are $n - s$ (>1) solutions, we have (30.3) holding for any two of them, where the $\bar{c}$'s are constants. Hence we have:

[30.4] *If a group G_r is multiply transitive, there is not in general one, or more groups, $\overline{G}_1$ each of whose transformations is commutative with every transformation of G_r; if there are s (>1) of such groups $\overline{G}_1$, they are sub-groups of a group $\overline{G}_s$.*

If G_r is intransitive and the generic rank q of the matrix $\|\xi_a^i\|$ is less than r, we choose the coordinates so that (21.8) hold. We make use of (21.13) to obtain (30.2) for $i, j, k = 1, \cdots, q$. In this case we have (30.8) for $i = 1, \cdots, q$ and $\sigma = q + 1, \cdots, n$, and also (30.9) and (30.10) for $h, i, j, k, l = 1, \cdots, q$ and $p = q + 1, \cdots, r$. The analytical processes are the same as in the case of multiply transitive groups, except that in the equations corresponding to (1.11) the independent variables are $x^1, \ldots, x^q$, and the variables $x^{q+1}, \ldots, x^n$ appear as parameters. Hence arbitrary functions of the latter enter in the solutions, and ζ^σ for $\sigma = q + 1, \cdots, n$ are arbitrary functions of these parameters. Consequently we have a theorem similar to theorem [30.4], but in general the groups $\overline{G}_1$ are not sub-groups of a group of higher order.

In consequence of the preceding results we are enabled to establish the theorem:

[30.5] *A simply transitive group is imprimitive, the systems of imprimitivity being the invariant varieties of the subgroups of the reciprocal group.*

If $\overline{G}_m$ for $m \geqslant 1$ is a sub-group of the reciprocal group $\overline{G}_n$ of a simply transitive group G_n and its symbols are $Z_p f$ for $p = 1, \cdots, m$, then evidently $Z_p f = 0$ is a complete system and since $(X_a, Z_p) f = 0$, for $a = 1, \cdots, n$ we have that the invariant varieties of $\overline{G}_m$ form a system of imprimitivity of G_n.

In order to show that all the systems of imprimitivity are of this kind, we remark that since the matrix of the reciprocal group is of rank n, the vectors b^i_α of equations (23.3) are expressible linearly in terms of the ζ's. Since the former may be replaced by any linear combination of these, we may in all generality put

$$B_\alpha f = Z_\alpha f + \mu^\sigma_\alpha Z_\sigma f \qquad (\alpha = 1, \cdots, p; \sigma = p + 1, \cdots, n).$$

Then equations (23.6) become in consequence of (30.1)

$$X_a \mu^\sigma_\alpha Z_\sigma f = \lambda^\beta_{a\alpha}(Z_\beta f + \mu^\sigma_\beta Z_\sigma f).$$

Since the operators Zf are independent, we have $\lambda^\beta_{a\alpha} = 0$ and $X_a \mu^\sigma_\alpha = 0$ from which it follows that the μ's are constants, and from the requirement that the equations $B_\alpha f = 0$ form a complete system, it follows that $B_a f$ are the symbols of a sub-group of $\overline{G}_n$.

31. Invariant sub-groups. If $X_1 f, \cdots, X_m f$ for $m < r$ are the symbols of a sub-group G_m of a group G_r, then (§16)

$$(31.1) \qquad (X_t, X_u) f = c^v_{tu} X_v f \qquad (t, u, v = 1, \cdots, m).$$

A necessary and sufficient condition (cf. (29.14)) that every transformation of G_r transforms a member of the sub-group G_m into a member of the latter is that

$$(31.2) \qquad (X_a, X_t) f = c^u_{at} X_u f \qquad \begin{pmatrix} a = 1, \cdots, r; \\ t, u = 1, \cdots, m \end{pmatrix},$$

that is,

$$(31.3) \qquad c^p_{at} = 0 \qquad \begin{pmatrix} a = 1, \cdots, r; \\ t = 1, \cdots, m; \\ p = m + 1, \cdots, r \end{pmatrix}.$$

When these conditions are satisfied, the G_m is called an *invariant sub-group*, or a *self-conjugate* sub-group. The group itself and the identity are trivial examples of such sub-groups possessed by any G_r. When a group G_r does not have any invariant sub-groups other than the group itself and the identity, the group is said to be *simple*.

From §29 it follows that, if a group G_r admits a sub-group G_m ($m \geqslant 1$) of exceptional G_1's, then G_m is an invariant sub-group of G_r.

Consequently, if the matrix $\|c_{ab}^e\|$ in which a indicates the columns and b and e the rows is of rank less than r, the group is not simple (cf. §35).

Let V_p be an invariant variety for an invariant sub-group G_m of a G_r and S any transformation of G_m, and denote by V_p' the transform of V_p by a transformation T of G_r which is not of G_m, then

$$TST^{-1}V_p' = V_p'.$$

But by (29.7) TST^{-1} is a transformation of G_m, since the latter is an invariant sub-group of G_r. Hence V_p' is an invariant variety of G_m. If a sub-group of G_m leaves V_p point-wise invariant, the transform by T of the members of this sub-group leaves V_p' point-wise invariant. Hence the groups induced by G_m in V_p and V_p' are of the same order. If T is infinitesimal, V_p' is infinitely near V_p, unless it coincides with it. The latter case arises, if V_p is *isolated* in the sense that it is not one of a continuous array of invariant varieties in which the induced group is of the same order for all of these varieties. Hence we have:

[31.1] *If V_p is an invariant variety for an invariant sub-group of a G_r, the transform of V_p by any transformation of G_r, not of the sub-group, is an invariant variety of the sub-group and the group induced by the sub-group in the two varieties is of the same order; if V_p is isolated, the transform coincides with V_p.*[*]

If there are two invariant sub-groups G_m and $G_{m'}$ of a G_r and there are p independent transformations common to the two sub-groups, these form an invariant sub-group of G_r. For, the transform of any of the latter is contained in G_m and $G_{m'}$, and consequently is common to the two.

Suppose that two invariant sub-groups G_m and $G_{m'}$ of a given G_r do not have a common sub-group, and we take as their symbols $X_1f, \ldots, X_mf$ and $X_{m+1}f, \ldots, X_{m+m'}f$. Then

(31.4) $$(X_p, X_u)f = c_{pu}^t X_t f = c_{pu}^q X_q f \quad \begin{pmatrix} p, q = 1, \cdots, m; \\ t, u = m+1, \cdots, m+m' \end{pmatrix}.$$

Since the operators of the two sub-groups are independent, we must have in this case

---

[*] *Lie-Scheffers*, 1893, 1, pp. 529, 531.

$$(31.5) \qquad c_{pu}^{\,t} = c_{pu}^{\,q} = 0 \quad \begin{pmatrix} p, q = 1, \cdots, m; \\ t, u = m + 1, \cdots, m + m' \end{pmatrix},$$

and consequently the transformations of the two sub-groups are commutative. It is evident that the symbols of the two groups generate a group $G_{m+m'}$, which is G_r itself when $m + m' = r$, and which is an invariant sub-group of G_r when $m + m' < r$. In the latter case because of (31.5) the sub-groups G_m and $G_{m'}$ are invariant sub-groups of $G_{m+m'}$. We say that $G_{m+m'}$ is the *direct product* of G_m and $G_{m'}$. Hence we have:

[31.2] *If a group G_r has two invariant sub-groups not containing a common sub-group, the transformations of the two sub-groups are commutative, and the direct product of the two sub-groups is either G_r itself, or an invariant sub-group of G_r of which G_m and $G_{m'}$ are invariant sub-groups.*

Suppose that G_m and $G_{m'}$ are two invariant sub-groups of a G_r and they have a sub-group G_p in common. In this case it follows that the second and third terms in (31.4) being respectively symbols of $G_{m'}$ and G_m must be a symbol of G_p. Consequently if the symbols are chosen, so that the symbols of G_m and $G_{m'}$ are

$$(31.6) \quad \begin{aligned} G_m &: X_1 f, \ldots, X_p f, X_{p+1} f, \ldots, X_m f; \\ G_{m'} &: X_1 f, \ldots, X_p f, X_{m+1} f, \ldots, X_{m+m'-p} f, \end{aligned}$$

then

$$(31.7) \qquad (X_k, X_u) f = c_{ku}^{\,h} X_h f$$
$$\begin{pmatrix} k = 1, \cdots, m; \, h = 1, \cdots, p; \\ u = 1, \cdots, p, m + 1, \cdots, m + m' - p \end{pmatrix}.$$

From this result and the fact that G_m and $G_{m'}$ are sub-groups it follows that the $m + m' - p$ independent symbols in (31.6) generate a group $G_{m+m'-p}$ which is called the *product* of G_m and $G_{m'}$; we use the term product in this case as distinguished from direct product when there is no common sub-group. In consequence of (31.7) G_m and $G_{m'}$ are invariant sub-groups of $G_{m+m'-p}$ and G_p is an invariant sub-group of $G_m, G_{m'}, G_{m+m'-p}$ and G_r. Clearly $m + m' - p \le r$; if it is less than r, $G_{m+m'-p}$ is an invariant sub-group of G_r. Hence we have:

[31.3] *If two invariant sub-groups of G_r, namely G_m and $G_{m'}$, have a sub-group G_p in common, their product $G_{m+m'-p}$ is either G_r or an* *invariant sub-group of it; in the latter case* G_m, $G_{m'}$ *and* G_p *are invariant sub-groups of* $G_{m+m'-p}$ *and of* G_r.

Suppose that we have two groups G_m and $G_{m'}$ such that all the transformations of each group are invariant under all the transformations of the other but not commutative. Then we have equations of the form (31.4) in which the c's are constants (§29), from which it follows that there are certain independent groups G_1 common to the two, say p. Combining the two sets of symbols we have $m + m' - p$ linearly independent ones (constant coefficients). Because of (31.4) and similar equations for each of the given groups it follows from the second fundamental theorem that these symbols generate a group $G_{m+m'-p}$, and from these equations it follows that G_m and $G_{m'}$ are invariant sub-groups of the enlarged group, as is also the sub-group of order p common to G_m and $G_{m'}$.

Let T, T_1 and T', T_1 be any two transformations each of G_m and $G_{m'}$ respectively; it follows that $T'T$ is a transformation of the group $G_{m+m'-p}$. We have also that $T_1T' = T'_2T_1$, where T'_2 is another transformation of $G_{m'}$; this is due to the hypothesis that the transformations of $G_{m'}$ are invariant under those of G_m. Consequently we have

$$T'_1T_1T'T = T'_1T'_2T_1T = T'_3T_2,$$

and thus the products $T'T$ form a group Γ. As we take T' or T equal to the identity, these transformations are those of G_m or $G_{m'}$, and therefore Γ is not a sub-group of $G_{m+m'-p}$, but is in fact this group, which we call the *product* of the two given groups. Hence we have:

[31.4] *If two groups G_m and $G_{m'}$ are invariant with respect to each other and are not commutative, they possess a common sub-group G_p; the two groups and also G_p are invariant sub-groups of the group $G_{m+m'-p}$ generated by the independent (constant coefficients) symbols of the two groups; and each transformation of the enlarged group consists of the product of a transformation of G_m and one of $G_{m'}$.*

By similar reasoning we have:

[31.5] *If two groups G_m and $G_{m'}$ are commutative, their symbols generate a group of order $m + m'$, the direct product of the two, of which each is an invariant sub-group.*

---

* *Bianchi*, 1918, 1, p. 204.

An invariant sub-group is said to be a *maximum* one, if it is not a sub-group of an invariant sub-group of higher order. We shall prove the theorem:

[31.6] *If G_{r_1} and G_{ρ_1} are maximum invariant sub-groups of a G_r and they have a sub-group G_p of all common transformations, their product is G_r, that is, $r_1 + \rho_1 - p = r$, and G_p is a maximum invariant sub-group of G_{r_1} and G_{ρ_1}.* *

The first part of this theorem follows from theorem [31.3]. For, if $r_1 + \rho_1 - p < r$, the product of G_{r_1} and G_{ρ_1} is an invariant sub-group of G_r containing G_{r_1} and G_{ρ_1}, contrary to hypothesis. If G_p is not a maximum invariant sub-group of G_{r_1}, there is an invariant sub-group G_{p+k} of it containing G_p. Consider the product of G_{p+k} and G_{ρ_1}. In consequence of (31.7) the commutator of any symbol of this product and a symbol of G_{r_1} is a symbol of G_{p+k}. The commutator of a symbol of this product and one of G_{ρ_1} is a combination of symbols of G_p and G_{ρ_1}, that is, of G_{ρ_1} itself. Hence the product is an invariant sub-group of G_r and it contains G_{ρ_1}, contrary to the hypothesis that the latter is a maximum. Consequently G_p is a maximum invariant sub-group of G_{r_1} and similarly of G_{ρ_1}.

If G_{r_1} and G_{ρ_1} are maximum invariant sub-groups of G_r and they have no sub-group in common, by the above reasoning it follows that $r_1 + \rho_1 = r$. Moreover, by theorem [31.2] it follows that the transformations of the two groups are commutative. Furthermore, each of these sub-groups is simple. For, if G_{r_1} contained an invariant sub-group G_k, then by reasoning similar to the above the symbols of G_k and G_{ρ_1} generate an invariant sub-group of G_r which contains G_{ρ_1}, contrary to hypothesis. Hence we have:

[31.7] *If G_{r_1} and G_{ρ_1} are maximum invariant sub-groups of a G_r and they do not contain a common sub-group, then $r_1 + \rho_1 = r$ and G_{r_1} and G_{ρ_1} are simple groups commutative with one another; consequently G_r is the direct product of these sub-groups.*

Exercises

1. If a series of infinitesimal transformations $u^a X_a f$ admit separately two G_1's of symbols $Y_1 f$ and $Y_2 f$, they admit the group G_1 of symbol $a Y_1 f + b Y_2 f$,

---

* *Lie-Engel*, 1893, 2, p. 706.

where a and b are any constants; also they admit the G_1 whose symbol is $(Y_1, Y_2)f$.

Bianchi, 1918, 1, p. 197.

2. If a set of transformations $x'^i = f^i(x; a)$ involving r essential parameters is such that a combination of any two, that is, $T_{a_2}T_{a_1}$, involves only r essential parameters, then $T_a = S_b T_{a_0}$, where S_b are the transformations of a group G_r and T_{a_0} is a transformation of the set; moreover, G_r is invariant under T_{a_0}.

Bianchi, 1918, 1, p. 118.

3. Show that, in consequence of theorems [24.1] and [30.5], a systatic group is imprimitive, and a primitive group is asystatic.

Bianchi, 1918, 1, p. 190.

4. A G_3 with the structure

$$(X_1, X_2)f = X_1 f, \qquad (X_1, X_3)f = 2X_2 f, \qquad (X_2, X_3)f = X_3 f$$

is simple (cf. §35); show also that the basis can be chosen so that

$$(X_1, X_2)f = X_3 f, \qquad (X_2, X_3)f = X_1 f, \qquad (X_3, X_1)f = X_2 f.$$

5. If a G_r admits a non-invariant sub-group G_{r-1}, the basis can be chosen so that $X_1 f, \ldots, X_{r-1} f$ are the symbols of G_{r-1} and

$$(X_a, X_r)f = c_{ar}^b X_b f, \qquad (X_{r-1}, X_r)f = c_{r-1}^b X_b f + X_f \begin{pmatrix} a = 1, \cdots, r-2; \\ b = 1, \cdots, r-1 \end{pmatrix}.$$

Then show by means of the Jacobi identity (2.4) that $X_1 f, \ldots, X_{r-2} f$ are the symbols of a G_{r-2}, which is an invariant sub-group of G_{r-1}.

Lie-Scheffers, 1893, 1, p. 544.

6. If a G_r contains a simple G_{r-1}, the latter is an invariant sub-group of G_r (cf. Ex. 5).

Lie-Scheffers, 1893, 1, p. 572.

7. If a G_4 has an invariant sub-group G_3 with the structure

$$(X_1, X_2)f = X_1 f, \qquad (X_1, X_3)f = 2X_2 f, \qquad (X_2, X_3)f = X_3 f,$$

then $(X_i, X_4)f = a_i^j X_j f$ $(i, j = 1, 2, 3)$. Because of the Jacobi identity the a's satisfy the conditions

$$2a_2^1 = a_3^2, \qquad a_3^3 = -a_1^1, \qquad a_1^2 = 2a_2^3, \qquad a_1^3 = a_3^1 = a_2^2 = 0.$$

If $X_4 f$ is replaced by $X_4' f - a_2^1 X_1 f + a_1^1 X_2 f + a_2^3 X_3 f$, it follows that

$$(X_i, X_4')f = 0 \qquad\qquad (i = 1, 2, 3).$$

8. Establish theorem [29.6] by means of the Jacobi identity.

9. Show that the projective group G_6 with the symbols

$$\begin{aligned} &X_1 f = p + yr, \qquad X_2 f = xp + zr \\ &X_3 f = x(xp + yq + zr) - zq, \qquad X_4 f = q + xr \\ &X_5 f = yq + zr, \qquad X_6 f = y(xp + yq + zr) - zp \end{aligned} \qquad \left(r = \frac{\partial f}{\partial z} \right)$$

leaves invariant the quadric $z - xy = 0$, and that the induced group Γ_6 on the quadric has the symbols

$$p, \qquad xp, \qquad x^2 p, \qquad q, \qquad yq, \qquad y^2 q.$$

Lie-Engel, 1893, 2, p. 199.

10. Show for the induced group Γ_6 in Ex. 9 the two sub-groups with the symbols p, xp, x^2p and q, yq, y^2q are invariant sub-groups of Γ_6; derive therefrom the two invariant sub-groups of G_6.

32. Isomorphisms of groups. Let G_r and H_p be two continuous groups whose finite equations are

$$(32.1) \qquad x'^i = f^i(x; a), \qquad y'^i = h^i(y; b) \quad \begin{pmatrix} i = 1, \cdots, n; \\ j = 1, \cdots, m \end{pmatrix}.$$

If $p = r$ and $\varphi^\alpha(a^1, \cdots, a^r)$ for $\alpha = 1, \cdots, r$ are r functionally independent functions, the equations

$$(32.2) \qquad\qquad b^\alpha = \varphi^\alpha(a)$$

establish a one-to-one correspondence between the transformations of the two groups. The correspondence is said to be *isomorphic*, and the two groups are said to be isomorphic, if T_{a_1}, S_{b_1} and T_{a_2}, S_{b_2} being any two pairs of corresponding transformations, the transformations $T_{a_2}T_{a_1}$ and $S_{b_2}S_{b_1}$ correspond. Later we shall have occasion to consider the case when $p < r$ in which case the correspondence is said to be *multiply* isomorphic, and when $p = r$ *simply* isomorphic.

When the isomorphism is simple, if T_{a_1} is the identity, then T_{a_2} corresponds to S_{b_2} and also to $S_{b_2}S_{b_1}$, and consequently S_{b_1} is the identity, since the correspondence is one-to-one. Hence if T_{a_1} and S_{b_1} correspond so also do $T_{a_1}^{-1}$ and $S_{b_1}^{-1}$.

If the quantities b in (32.1) are replaced by their expressions (32.2), we obtain a new set of equations for H_r in terms of the a's, say

$$(32.3) \qquad\qquad y'^i = k^i(y; a),$$

and then corresponding transformations have the same values of the a's. If the correspondence is to be isomorphic, equations (4.6), namely $a_3^\alpha = \varphi^\alpha(a_1; a_2)$, must apply to both groups and conversely. Consequently the two groups must have the same first and second parameter-groups. Since this condition is evidently sufficient, we have:

[32.1] *A necessary and sufficient condition that two groups be simply isomorphic is that they have the same parameter-groups.*

If this condition is satisfied it follows from (14.19) that the vectors $U_a^\alpha(u)$ are the same for the two groups, and then from the corresponding equations (6.3) that the constants of structure of the two groups are the same. Conversely, when the latter condition is satisfied, it follows from (14.6) and (14.8) that the two groups have the same vectors $U_\alpha^a(u)$ and $\bar{U}_\alpha^a(u)$, and from (14.18) and (14.17) that equations (14.16) are the same for both groups. Hence we have:

[32.2] *A necessary and sufficient condition that two groups be simply isomorphic is that their bases can be chosen so that the constants of structure of the two groups are the same for both.*

Thus from the results of §9 it follows that the parameter-groups of any group are simply isomorphic with it. Also from §§26, 27 it follows that any extension of a group is isomorphic with it.

We consider next the case of multiple isomorphism, that is, when $p < r$, and we understand that in (32.2) where now $\alpha = 1, \cdots, p$ the jacobian matrix $\left\| \dfrac{\partial \varphi}{\partial a} \right\|$ is of rank p; in accordance with the theory of implicit functions this insures that there are values of the a's in (32.2) for any values of the b's. Then there are ∞^{r-p} transformations T of G_r corresponding to a transformation S of H_p. In particular, to the identity of H_p corresponds a sub-group G_{r-p} of G_r. Since to any T there corresponds only one S, it follows by the argument applied to the case $p = r$ that the identity of G_r is a member of this sub-group, and that if T_{a_1} corresponds to S_{b_1} then $T_{a_1}^{-1}$ corresponds to $S_{b_1}^{-1}$. Since then $T_{a_2} T_{a_1} T_{a_2}^{-1}$ corresponds to $S_{b_2} S_{b_1} S_{b_2}^{-1}$, if S_{b_1} is the identity, then $T_{a_2} T_{a_1} T_{a_2}^{-1}$ corresponds to the identity whatever be T_{a_2}. Since this is true for every T_{a_1} corresponding to the identity, we have:

[32.3] *If G_r and H_p ($p < r$) are two groups multiply isomorphic, the transformations of G_r corresponding to the identity in H_p form an invariant sub-group of G_r of order $r - p$.*

If we substitute from (32.2) in the second of (32.1) and obtain (32.3), these are finite equations of H_p but all the parameters are not essential. If we take two transformations S_{b_1} and S_{b_2}, and a pair T_{a_1} and T_{a_2} corresponding to them and apply (4.6) to get $T_{a_2} T_{a_1}$, then when (4.6) is applied to the transformations S in the form (32.3), we get the transformation $S_{b_2} S_{b_1}$. In this sense the two groups have the same constants of structure, for suitable choice of the bases, following the procedure in the case of simple isomorphism. This means that, if we consider the infinitesimal generators of the two groups and denote their respective bases by $X_1f, \ldots, X_rf$ and $Y_1'f, \ldots, Y_p'f$, we can put

(32.4) $$Y_af = a_a^\alpha Y_\alpha'f \qquad \left(\begin{matrix} a = 1, \cdots, r; \\ \alpha = 1, \cdots, p \end{matrix}\right),$$

where the a's are constants so chosen that

(32.5) $$(Y_a, Y_b)f = c_{ab}^e Y_ef \quad (a, b, e = 1, \cdots, r),$$

the constants c_{ab}^e being given by

(32.6) $$(X_a, X_b)f = c_{ab}^e X_ef.$$

If now we put

(32.7) $$(Y_\alpha', Y_\beta')f = c_{\alpha\beta}'^\gamma Y_\gamma'f \quad (\alpha, \beta, \gamma = 1, \cdots, p),$$

it follows from (32.4) and (32.5) that

(32.8) $$a_a^\alpha a_b^\beta c_{\alpha\beta}'^\gamma = c_{ab}^e a_e^\gamma \qquad \left(\begin{matrix} a, b, e = 1, \cdots, r; \\ \alpha, \beta, \gamma = 1, \cdots, p \end{matrix}\right).$$

Conversely, if c_{ab}^e and $c_{\alpha\beta}'^\gamma$ are the constants of structure of two groups G_r and H_p respectively, and equations (32.8) are consistent, then from (32.4) we have (32.5). If we change the basis of G_r and put as in §7

$$c_{ab}^e = \bar{c}_{a_1 b_1}^{e_1} \bar{c}_a^{a_1} \bar{c}_b^{b_1} c_{e_1}^e,$$

then (32.8) becomes

$$\bar{a}_{a_1}^\alpha \bar{a}_{b_1}^\beta c_{\alpha\beta}'^\gamma = \bar{c}_{a_1 b_1}^{e_1} \bar{a}_{e_1}^\gamma,$$

where $\bar{a}_{a_1}^\alpha = a_a^\alpha c_{a_1}^a$. Hence we have:

[32.4] *If G_r and H_p are two groups with the respective constants of composition c_{ab}^e and $c_{\alpha\beta}'^\gamma$, a necessary and sufficient condition that they admit multiple isomorphism is that equations (32.8) be compatible.*

Thus the problem of determining whether two groups are isomorphic is an algebraic one.

Suppose that the bases for H_p and G_r are chosen so that

$$Y_1f, \ldots, Y_pf$$

in (32.5) are independent, and that

(32.9) $$Y_\alpha f = b_\alpha^t Y_t f \qquad \left(\begin{matrix} \alpha = p + 1, \cdots, r; \\ t = 1, \cdots, p \end{matrix}\right),$$

where the b's are constants. Then to each of the $r - p$ independent infinitesimal transformations

(32.10) $$X_\alpha f - b_\alpha^t X_t f$$

of G_r corresponds the identity in H_p. Evidently these transformations determine a sub-group G_{r-p} of G_r. Conversely, if λ^a are constants such that

$$Yf \equiv \lambda^a Y_a f = 0,$$

then $\lambda^t + \lambda^\alpha b_\alpha^t = 0$, and $Xf = \lambda^a X_a f = \lambda^\alpha (X_\alpha - b_\alpha^t X_t)f$ is the symbol of a G_1 of transformations of G_{r-p}.

Since $(Y_i, (Y_\alpha - b_\alpha^t Y_t))f = 0$, we have

$$c_{i\alpha}^s - b_\alpha^t c_{it}^s + (c_{i\alpha}^\beta - b_\alpha^t c_{it}^\beta)b_\beta^s = 0 \quad \begin{pmatrix} i = 1, \cdots, r; \\ s, t = 1, \cdots, p; \\ \alpha, \beta = p + 1, \cdots, r \end{pmatrix},$$

by means of which it can be shown that G_{r-p} is an invariant sub-group, as previously established from the finite equations of the group. This result is a special case of the theorem:

[32.5] *If G_r and H_p $(p \leqslant r)$ are isomorphic, to a sub-group of order m of G_r corresponds a sub-group of order $k(\leq m)$ of H_p; to a sub-group of order m of H_p corresponds a sub-group of order $m + r - p$ of G_r; if either sub-group is invariant, so is the other.*

This theorem follows immediately from the definition of isomorphic correspondence at the beginning of this section. Also if infinitesimal transformations are used, we have at once that, if $\lambda_\sigma^i X_i f$ for $\sigma = 1$, $\cdots$, m define a sub-group of G_r, then $\lambda_\sigma^i Y_i f$ generate a sub-group of H_p of order $m - s$, where s is the number of linearly independent (constant coefficients) generators of this set yielding the identity. Conversely if $\lambda_\sigma^i Y_i f$ generate a sub-group of order m of H_p, then $\lambda_\sigma^i X_i f$ and (32.10) are the bases of a sub-group of G_r of order $m + r - p$.

33. Series of composition of a group. Theorem of Lie-Jordan. If a group G_r admits a maximum invariant sub-group G_{r_1}, and the latter in turn a maximum sub-group G_{r_2} and so on, the series

(33.1) $$G_r, G_{r_1}, \ldots, G_{r_q}, 1$$

(where necessarily G_{r_q} is a simple group) is called a *series of composition* of the group, and the differences

(33.2) $$r - r_1, \cdots, r_{q-1} - r_q, r_q$$

are called the *indices of composition* of the group; by some writers (33.1) is called a *normal series*. It may be that a group G_r admits more than one series of composition. For example, if a group admits two maximum invariant sub-groups G_{r_1}, G_{ρ_1} which do not contain a common sub-group, then as shown in §31 these sub-groups are simple, and we have the two series of composition

$$G_r, \quad G_{r_1}, \quad 1; \quad G_r, \quad G_{\rho_1}, \quad 1.$$

Moreover, since $r_1 + \rho_1 = r$ (§31), we have that the respective indices of composition are equal to

$$\rho_1, r_1; \quad r_1, \rho_1.$$

We shall establish the following theorem due to Lie and Jordan:

[33.1] *If a group G_r admits two series of composition, (33.1) and*

$$(33.3) \qquad G_r, G_{\rho_1}, \ldots, G_{\rho_p}, 1,$$

the indices of composition of the two series are the same set of integers, but not necessarily in the same order; consequently the number of indices is the same.

We have just seen that this theorem holds when the group admits two maximum invariant sub-groups which do not have a sub-group in common. If $r = 2$, this is the only case which can arise, and the theorem holds for $r = 2$. We shall establish the theorem by induction for the case when G_{r_1} and G_{ρ_1} have a common sub-group Γ_p, which by theorem [31.6] is a maximum invariant sub-group of both G_{r_1} and G_{ρ_1}. We take a series of composition for Γ_p, say $\Gamma_p, \Gamma_{p_1}, \Gamma_{p_2}, \ldots, \Gamma_{p_s}, 1$, and consider the series (33.1) and

$$(33.4) \qquad G_r, G_{r_1}, \Gamma_p, \Gamma_{p_1}, \ldots, \Gamma_{p_s}, 1.$$

In §31 it was seen that $r = r_1 + \rho_1 - p$. Assuming the theorem to hold for the two series for G_{r_1}, we have that (33.2) and

$$r - r_1, r_1 - p, p - p_1, \cdots, p_{s-1} - p_s, p_s$$

are the same numbers, possibly in different orders. Similarly the indices of (33.3) are the same numbers as

$$r - \rho_1, \rho_1 - p, p - p_1, \cdots, p_{s-1} - p_s, p_s.$$

But

$$r - r_1 = \rho_1 - p, \qquad r - \rho_1 = r_1 - p,$$

and consequently the theorem is proved for G_r.

34. Factor groups. Suppose that a group G_r admits an invariant sub-group G_m and that the basis of G_r is chosen so that $X_t f$ for $t = 1, \cdots, m$ is the basis of the sub-group; then we have

$$(34.1) \qquad c_{ta}^{\,p} = 0$$

$$(a = 1, \cdots, r; t = 1, \cdots, m; p = m + 1, \cdots, r).$$

Consider the Jacobi relations

$$c_{pq}^{\,i} c_{is}^{\,t} + c_{qs}^{\,i} c_{ip}^{\,t} + c_{sp}^{\,i} c_{iq}^{\,t} = 0 \quad \begin{pmatrix} p, \ q, \ s, \ t = m + 1, \cdots, r; \\ i = 1, \cdots, r \end{pmatrix}$$

In consequence of the preceding equations these relations become

$$(34.2) \qquad c_{pq}^{\,u} c_{us}^{\,t} + c_{qs}^{\,u} c_{up}^{\,t} + c_{sp}^{\,u} c_{uq}^{\,t} = 0$$

$$(p, \ q, \ s, \ t, \ u = m + 1, \cdots, r).$$

By the third fundamental theorem there exists a group $\overline{G}_{r-m}$ whose constants of composition are $c_{pq}^{\,u}$. All such groups being simply isomorphic, we call any one of them the *factor group* of G_r determined by the invariant sub-group G_m, and write symbolically

$$\overline{G}_{r-m} = \frac{G_r}{G_m}.$$

If we denote its basis by $Y_p f$ for $p = m + 1, \cdots, r$ and adjoin the operators $Y_a f \equiv 0$ for $a = 1, \cdots, m$ (cf. §32), we have that G_r and $\overline{G}_{r-m}$ are multiply isomorphic, and G_m corresponds to the identity in $\overline{G}_{r-m}$. Hence we have:

[34.1] *If a group G_r admits an invariant sub-group G_m, there exists a group $\overline{G}_{r-m}$ multiply isomorphic to G_r, and G_m corresponds to the identity in $\overline{G}_{r-m}$.*

Conversely, if a group $\overline{G}_{r-m}$ is multiply isomorphic to a group G_r, the former is the factor group of G_r determined by the invariant sub-group of G_r corresponding to the identity in $\overline{G}_{r-m}$.

If $\overline{G}_s$ is a maximum invariant sub-group of $\overline{G}_{r-m}$, then the corresponding invariant sub-group G_{s+m} of G_r is maximum, and conversely. Consequently if $\overline{G}_{r-m}$ is simple, then G_m which determines it as a factor group of G_r is a maximum invariant sub-group, and conversely. Hence we have:

[34.2] *A necessary and sufficient condition that a factor group $\dfrac{G_r}{G_m}$ be simple is that G_m be a maximum invariant sub-group of G_r.*

If $\bar{G}_{r-m}$ is not simple, it admits a series of composition (§33), say

$$\bar{G}_{r-m}, \; \bar{G}_{r-m_1}, \; \cdots, \; \bar{G}_{r-m_q}, \; 1,$$

each of the sub-groups being a maximum invariant sub-group of its predecessor; then the corresponding sub-groups of G_r, namely

$$G_{r-m_1+m}, \; \cdots, \; G_{r-m_q+m}, \; G_m, \; 1,$$

form a series of composition. As a consequence we have:

[34.3] *Any invariant sub-group G_m of a G_r is a member of a series of composition of G_r.*

Suppose that G_r admits a series of composition

$$(34.3) \qquad\qquad G_r, \; G_{r_1}, \; \cdots, \; G_{r_p}, \; 1.$$

From theorem [34.2] it follows that each of the factor groups such as

$$\bar{G}_{r_a-r_{a+1}} = \frac{G_{r_a}}{G_{r_{a+1}}}$$

is simple. In this way we obtain a series of simple factor groups

$$(34.4) \qquad\qquad \bar{G}_{r-r_1}, \; \bar{G}_{r_1-r_2}, \; \cdots, \; \bar{G}_{r_{p-1}-r_p}, \; G_{r_p}.$$

Suppose now that G_r admits a second series of composition

$$(34.5) \qquad\qquad G_r, \; G_{\rho_1}, \; \cdots, \; G_{\rho_p}, \; 1;$$

as shown in §33 this must have the same number of elements as (34.3). Let Γ_m (which may be the identity) be the sub-group common to G_{r_1} and G_{ρ_1}; we recall that it is a maximum invariant sub-group of G_{r_1} and of G_{ρ_1} and that $r = r_1 + \rho_1 - m$. We choose the basis of G_r so that $X_a f$ for $a = 1, \; \cdots, \; m$ is the basis of Γ_m; $X_a f$ and $X_h f$ for $h = m + 1, \; \cdots, \; r_1$ the basis for G_{r_1}; and $X_a f$ and $X_l f$ for $l = r_1 + 1, \; \cdots, \; r$ the basis for G_{ρ_1}. Then $X_a f$, $X_h f$ and $X_l f$ is the basis for G_r. From the considerations at the beginning of this section and in particular equations (34.2), it follows that the factor groups $\dfrac{G_r}{G_{r_1}}$ and $\dfrac{G_{r_1}}{\Gamma_m}$ have the constants of structure of G_r for which the indices take the respective values $r_1 + 1, \; \cdots, \; r$ and $m + 1, \; \cdots, \; r_1$, and that the factor groups

$\dfrac{G_{\rho_1}}{\Gamma_m}$ and $\dfrac{G_r}{G_{\rho_1}}$ have these respective constants of structure. Hence we have the theorem:

[34.4] *If G_{r_1} and G_{ρ_1} are maximum invariant sub-groups of G_r with a common sub-group Γ_m, the factor groups $\dfrac{G_r}{G_{r_1}}$ and $\dfrac{G_{\rho_1}}{\Gamma_m}$ are simply isomorphic, as are also $\dfrac{G_r}{G_{\rho_1}}$ and $\dfrac{G_{r_1}}{\Gamma_m}$.*

In particular, if Γ_m is the identity, G_{ρ_1} is the factor group $\dfrac{G_r}{G_{r_1}}$ so that the factor groups of the series G_r, G_{r_1}, 1 are G_{ρ_1} and G_{r_1}; in like manner the factor groups of the series G_r, G_{ρ_1}, 1 are G_{r_1} and G_{ρ_1}.

We are now in position to establish the following theorem of Hölder-Engel:[*]

[34.5] *If a group G_r admits two series of composition, the simple factor groups of the two series, to within their ordering, are simply isomorphic.*

We assume that this holds for order less than r and show that it holds for r. If Γ_m admits the series Γ_m, Γ_{m_1}, . . . , 1, then the theorem holds for the series (34.3) and

$$(34.6) \qquad G_r, \ G_{r_1}, \ \Gamma_m, \ \Gamma_{m_1}, \ . \ . \ . \ , \ 1$$

since by hypothesis it holds for G_{r_1}. Similarly the theorem holds for (34.5) and

$$(34.7) \qquad G_r, \ G_{\rho_1}, \ \Gamma_m, \ \Gamma_{m_1}, \ . \ . \ . \ , \ 1.$$

Then it holds for (34.3) and (34.5), because it holds for (34.6) and (34.7) in consequence of the above discussion. When $r = 2$, we have the particular case discussed above, and hence the theorem is proved by induction.

35. Derived groups. Consider the symbols $X_{ab}f$ defined by

$$(35.1) \qquad X_{ab}f = (X_a, X_b)f = c_{ab}^e X_e f.$$

Each of them determines a sub-group G_1 of G_r. Moreover, since

$$(X_{ab}, X_{cd})f = (c_{ab}^e X_e, c_{cd}^h X_h)f = c_{ab}^e c_{cd}^h X_{eh}f,$$

_____

[*]*Lie-Engel*, 1893, 2, p. 765.

we have that these symbols determine a group, which is either G_r itself, or a sub-group of G_r. Since also

$$(X_d, X_{ab})f = c_{ab}^e(X_d, X_e)f = c_{ab}^e X_{de} f,$$

we have that, if this group is a sub-group, it is an invariant sub-group. It is called the *derived* group of G_r.

Consider the matrix

$$(35.2) \qquad\qquad C = \|c_{ab}^e\|,$$

where e indicates the columns, and a and b the rows. If the rank of the matrix is p, then any one of the symbols $X_{ab}f$ is expressible linearly (constant coefficients) in terms of p of them and the derived group is of order p. If the rank is zero, that is, if the group is Abelian, the derived group consists of the identity. If the rank is r, the derived group coincides with the given group. As a corollary we have:

[35.1] *If G_r is a simple group, the rank of C is r; if the rank is less than r, G_r is not simple.*

Suppose that the rank of C is $r_1 < r$, then as seen above the derived group G'_{r_1} is an invariant sub-group. If we denote by $X_1 f, \ldots, X_{r_1} f$ the symbols of this group, these and $r - r_1$ others $X_{r_1+1} f, \ldots, X_r f$ may be taken as the symbols of G_r. From the definition of G'_{r_1} it follows that

$$(35.3) \qquad (X_a, X_b)f = c_{ab}^p X_p f \quad \begin{pmatrix} a, b = 1, \cdots, r; \\ p = 1, \cdots, r_1 \end{pmatrix}.$$

If in these equations we let a and b take the values $1, \ldots, r_1$, $r_1 + 1, \cdots, r_1 + h$ $(h < r - r_1)$, we see that the symbols with these indices form a sub-group G_{r_1+h}, and then from (35.3), when a takes the values $1, \ldots, r$ and b from 1 to $r_1 + h$, it follows that G_{r_1+h} is an invariant sub-group of G_r. Moreover, if a takes the values 1 to $r_1 + h$ and b the values 1 to $r_1 + h - 1$, we have from (35.3) that G_{r_1+h-1} is an invariant sub-group of G_{r_1+h}. Hence as h takes the values $r - r_1 - 1, r - r_1 - 2, \cdots, 2, 1$, we have a set of invariant sub-groups of G_r

$$(35.4) \qquad\qquad G_{r-1}, G_{r-2}, \ldots, G_{r_1+1}, G'_{r_1},$$

each of which is an invariant sub-group of those of higher order. Hence if G'_{r_1} is a simple group, then G_r and the set (35.4) form a series of composition of indices $1, \ldots, 1, r_1$.

As defined G_{r-1} has the symbols $X_1f, \ldots, X_{r-1}f$, but if any of the symbols $X_{r_1+1}f, \ldots, X_rf$ other than X_rf had been omitted, we should have obtained another invariant sub-group of order $r - 1$. Hence there are $r - r_1$ such sub-groups. Each of these in turn admits $r - r_1 - 1$ invariant sub-groups of order $r - 2$, and so on. Hence we have:

[35.2] *If the derived group of a G_r is of order $r_1 < r$, then G_r admits $r - r_1$ invariant sub-groups of order $r - 1$, each of the latter $r - r_1 - 1$ invariant sub-groups of order $r - 2$, and so on; and all these sub-groups contain the derived group.*

Conversely, we have:*

[35.3] *If a group G_r admits an invariant sub-group of order $r - 1$, the derived group of G_r is this sub-group or is a sub-group of it; and consequently its order is less than r.*

For, if $X_1f, \ldots, X_{r-1}f$ are the symbols of the sub-group G_{r-1}, then

$$(X_a, X_h)f = c_{ah}^k X_k f \quad \begin{pmatrix} a = 1, \cdots, r; \\ h, k = 1, \cdots, r - 1 \end{pmatrix}$$

and consequently the derived group either coincides with G_{r-1} or is contained in it.

From theorem [35.3] it follows that, if the rank of the matrix C (35.2) is r, the group does not admit an invariant sub-group of order $r - 1$. In particular, if $r = 3$ and the rank of C is 3, the group is simple. For, if it had an invariant sub-group, it would be of order one; assume X_1f to be its symbol, then $c_{a1}^\alpha = 0$ for $a = 1, 2, 3$ and $\alpha = 2, 3$, but in this case the rank of C is less than 3.

From the definition of a derived group it is evident that:

[35.4] *The derived group of a sub-group G_m of a G_r is contained in the derived group of G_r.*

If G_m is an invariant sub-group of G_r and the basis of the latter is chosen so that $X_\alpha f$ for $\alpha = 1, \cdots, m$ is the basis of the sub-group, in the identities

$$((X_\alpha, X_\beta), X_a) + ((X_\beta, X_a), X_\alpha) + ((X_a, X_\alpha), X_\beta) = 0$$
$$\begin{pmatrix} \alpha, \beta = 1, \cdots, m; \\ a = 1, \cdots, r \end{pmatrix}$$

---

* *Bianchi*, 1918, 1, p. 212.

the second and third terms are symbols of transformations of the derived group of G_m, and consequently it follows from these identities that the derived group of G_m is an invariant sub-group of G_r. The same reasoning may be applied to this derived group and we have:

[35.5] *The derived groups of an invariant sub-group of a group are invariant sub-groups of the group.*

If G'_{r_1} in (35.4) is not a simple group, its derived group, called the *second derived group of* G_r, may be of order less than r_1, say r_2. Then as above we show the existence of sub-groups G_{r_1-1}, $\cdots$, G''_{r_2}, such that each of them is an invariant sub-group of the one of one higher order, and all of them are invariant sub-groups of G'_{r_1}. This process may be continued as long as each derived group is of lower order than the one preceding it. Consequently we get a sequence such as

(35.5) $G_r, G_{r-1}, \ldots, G_1, 1$

or

(35.6) $G_r, G_{r-1}, \ldots, G_{r_s}, 1,$

the latter case arising when the $(s + 1)$th derived group is the same as the sth derived group. In either case each group G_h of the sequence for $h = 1, \cdots, r - 1$ is an invariant sub-group of the group G_{h+1}.

36. Integrable groups. A group is said to be *integrable* when there exists an array of sub-groups as (35.5) such that each sub-group G_h for $h = 1, \cdots, r - 1$ is an invariant sub-group of the group G_{h+1}; the significance of the term integrable will appear later. If we denote by X_1f the generator of G_1, by X_1f and X_2f the generators of G_2 and so on, we have

$(X_2, X_1)f = c_{21}^1 X_1 f,$
$(X_3, X_1)f = c_{31}^1 X_1 f + c_{31}^2 X_2 f,$ $(X_3, X_2)f = c_{32}^1 X_1 f + c_{32}^2 X_2 f,$

and in general

(36.1) $(X_{h+k}, X_h)f = c_{h+k\,h}^l X_l f$
$$\left(\begin{matrix} h = 1, \cdots, r - 1; k = 1, \cdots, r - h; \\ l = 1, \cdots, h + k - 1 \end{matrix} \right).$$

Hence each c_{ab}^l for which l is equal to or greater than the larger of a and b is equal to zero.

From the method of §35 yielding (35.5) it follows that, if one of the successive derived groups is the identity, then the given G_r is integrable. In particular, the first derived group of an Abelian group is the identity. Hence we have:

[36.1] *An Abelian group is integrable.*

Also from theorem [16.1] we have:

[36.2] *Any G_2 is integrable.*

Conversely, if a group G_r is integrable, its first derived group is contained in G_{r-1} of a sequence (35.5) by theorem [35.3]. And by theorem [35.4] the second derived group is contained in the derived group of G_{r-1}, and consequently is contained in G_{r-2}. Similarly the third derived group is in G_{r-3}, and so on. Consequently we have:

[36.3] *A necessary and sufficient condition that a G_r be integrable is that there be an integer p ($\leq r$) such that the pth derived group be the identity.*

If G_r is integrable and G_p is a sub-group of G_r, then the first derived group of G_p is contained in the first derived group of G_r by theorem [35.4]. If these two derived groups are equivalent, then G_p is integrable; if not, then by the same theorem the second derived group of G_p is contained in the second of G_r, and so on. Hence in consequence of the preceding theorem we have:

[36.4] *Any sub-group of an integrable group is integrable.*

The significance of the designation *integrable* as applied to a group is set forth in the theorem:

[36.5] *If a linear homogeneous partial differential equation of the first order in n variables $Af = 0$ admits an integrable group G_{n-1} for which the symbols and Af are independent, the integration of the equation reduces to quadratures.*

Let X_1f, X_2f, . . . , $X_{n-1}f$ be the symbols of the group, so chosen that the first b symbols for any b from 1 to $n-2$ generate a sub-group which is invariant under the sub-group of the first $b+1$ symbols. If $Af = 0$ is the equation, then

$$(36.2) \qquad Af = 0, \qquad X_1f = 0, \cdots, X_{n-2}f = 0$$

form a complete system, and in consequence of the above statement this system admits the operator $X_{n-1}f$. Accordingly, if u is a solution of (36.2) other than a constant, so also is $X_{n-1}u$ a solution by theorem [25.1]. If $X_{n-1}u$ were zero, the system of equations (36.2) and $X_{n-1}f = 0$ would not be independent. Hence $X_{n-1}u = \varphi(u)$, since any two solutions of (36.2) are functions of one another. If we put $u_1 = \int \dfrac{du}{\varphi(u)}$, we have that u_1 is a solution of (36.2) and that $X_{n-1}u_1 = 1$. This equation and (36.2) in which f is replaced by u_1 can be solved for its derivatives, say

$$\frac{\partial u_1}{\partial x^i} = u_{1i}.$$

Since $X_1f, \ldots, X_{n-2}f$ are the symbols of an invariant sub-group G_{n-2}, and Af admits the group, we have

$$(X_a, X_b)u_1 = 0, \qquad (X_a, A)u_1 = 0 \qquad (a, b = 1, \cdots, n-1),$$

and consequently $u_{1i}\,dx^i$ is an exact differential and therefore u_1 is obtained by a quadrature.

If we change the coordinates putting $x'^1 = u_1, x'^2 = x^2, \cdots,$ $x'^n = x^n$, then in the new coordinate system the equations (dropping primes)

$$(36.3) \qquad Af = 0, \qquad X_1f = 0, \cdots, X_{n-3}f = 0$$

do not involve derivatives with respect to x^1. Hence we may treat these equations as a set in $n-1$ variables, with x^1 entering possibly as a parameter. Since $X_1f, \ldots, X_{n-3}f$ are the symbols of an invariant sub-group of G_{n-2}, defined above, we may apply the same process, and find by a quadrature an integral u_2, which is independent of u_1, previously found, since it involves variables other than x^1, which is u_1 in the present coordinate system. By repeating the process we obtain by $n-1$ quadratures $n-1$ independent solutions of $Af = 0$, that is, the complete solution. The transformations of coordinates leading to the equations in the forms stated above involve quadratures, differentiation and direct processes.

Exercises

1. Show that the G_4 with the symbols

$$p, \qquad xp, \qquad x^2p, \qquad q$$

and the sub-group with the first three symbols are multiply isomorphic.

2. The transformations of a transitive imprimitive group which leave invariant a variety of imprimitivity form are invariant sub-group.

Lie-Engel, 1888, 1, vol. 1, p. 307.

3. If $u^\mu(x^1, \cdots, x^n) = c^\mu$ for $\mu = 1, \cdots, n - q$ are the equations of the systems of imprimitivity of a transitive imprimitive group G_r, then

$$X_a(u^\mu) = \varphi_a^\mu(u^1, \cdots, u^{n-q})$$

(cf. §23), and $U_a f = \varphi_a^\mu \dfrac{\partial f}{\partial u^\mu}$ are the symbols of the group induced in each variety V_q of imprimitivity. Show also that the induced group is isomorphic with the given group.

Bianchi, 1918, 1, pp. 294, 295.

4. A necessary and sufficient condition that the factor group of an invariant sub-group G_m of a G_r be Abelian is that the first derived group of G_r be contained in G_m.

Bianchi, 1918, 1, p. 289.

5. The first derived group is the invariant sub-group of lowest order determining an Abelian factor group.

Bianchi, 1918, 1, p. 289.

6. Show that the G_4 with the symbols

$$G_4: \quad p, \quad q, \quad xq, \quad x^2q$$

is integrable and that the sub-groups

$$G_3: q, \quad xq, \quad ap + bx^2q; \qquad G_2: q, \quad xq; \qquad G_1: q$$

are invariant sub-groups of G_4 and each one is an invariant sub-group of every one of higher order.

Lie-Scheffers, 1893, 1, p. 538.

7. If ρ is a non-zero root of the determinant equation $|c_{ab}^e u^a - \rho \delta_b^e| = 0$, the quantities v^a defined by $(c_{ab}^e u^a - \rho \delta_b^e)v^b = 0$ are such that $u^a X_a f$ and $v^b X_b f$ determine a sub-group G_2 of G_r, and $v^b X_b f$ is the symbol of the derived group of G_2 (cf. §16).

8. If $u^a X_a f$ and $v^a X_a f$ determine a sub-group G_2 of a G_r, its derived group has the symbol $u^a X_a f$ if $c_{ab}^e u^a v^b = u^e$, and this is possible only if the matrix $||c_{ab}^e u^a||$ and the augmented matrix $||c_{ab}^e u^a, u^e||$ are of the same rank (cf. §16).

9. The basis of a G_3 can be chosen so as to have one of the following structures:

(1) $(X_1, X_2)f = X_1 f,$ $(X_1, X_3)f = 2X_2 f,$ $(X_2, X_3)f = X_3 f;$
(2) $(X_1, X_2)f = 0,$ $(X_1, X_3)f = X_1 f,$ $(X_2, X_3)f = cX_2 f\ (c \neq 0, 1);$
(3) $(X_1, X_2)f = 0,$ $(X_1, X_3)f = X_1 f,$ $(X_2, X_3)f = X_2 f;$
(4) $(X_1, X_2)f = 0,$ $(X_1, X_3)f = X_1 f,$ $(X_2, X_3)f = X_1 f + X_2 f;$
(5) $(X_1, X_2)f = 0,$ $(X_1, X_3)f = X_1 f,$ $(X_2, X_3)f = 0;$
(6) $(X_1, X_2)f = 0,$ $(X_1, X_3)f = 0,$ $(X_2, X_3)f = X_1 f;$
(7) $(X_1, X_2)f = 0,$ $(X_1, X_3)f = 0,$ $(X_2, X_3)f = 0.$

All of these groups are integrable except the first.

Lie-Scheffers, 1893, 1, p. 571.

10. The basis of any non-integrable G_3 can be chosen so as to have the structure (1) of Ex. 9; show that this G_3 is isomorphic to the projective group on the line (Ex. 8, p. 66).

11. In 4-space of coordinates x^i the linear homogeneous transformations $x'^i = a^i_j x^j$ leave invariant the hypersurface $\Sigma x^{i2} = \dfrac{1}{K}$, where K is a constant, if $a^i_j a^i_k = \delta_{jk}$ and thus form a G_6. There are two sub-groups, each of the third order of this group for which $x'^i x^i = \dfrac{\alpha}{K}$, where α is a constant. The values for the a's in these cases are

(i) $\qquad\qquad a^i_i = \alpha$ (i not summed), $\qquad a^i_i = -a^j_i \qquad$ ($i, j = 1, \cdots, 4$)

and either

(ii) $\qquad a^2_1 = a^4_3 \equiv \beta, \qquad a^3_1 = a^2_4 \equiv \gamma, \qquad a^3_2 = a^4_1 \equiv \delta,$

or

(iii) $\qquad a^2_1 = a^3_4 \equiv \beta, \qquad a^3_1 = a^4_2 \equiv \gamma, \qquad a^3_2 = a^1_4 \equiv \delta,$

where in each case

(iv) $\qquad \alpha^2 + \beta^2 + \gamma^2 + \delta^2 = 1.$

Bianchi, 1902, 1, p. 445.

12. Show that each of the two sub-groups of G_6 of Ex. 11 are invariant sub-groups of G_6, and the only ones.

13. If the symbols of the G_6 in Ex. 11 are denoted by (cf. Ex. 12, p. 57)

$$X_{ij}f = x^i p_j - x^j p_i,$$

the symbols of the two invariant sub-groups of G_6 are

$$(X_{12} \pm X_{34})f, \qquad (X_{31} \pm X_{24})f, \qquad (X_{23} \pm X_{14})f.$$

14. If we put

$$x^\alpha = \frac{y^\alpha}{1 + \dfrac{K}{4}r^2}, \qquad x^4 = \frac{1}{\sqrt{K}} \frac{1 - \dfrac{K}{4}r^2}{1 + \dfrac{K}{4}r^2}, \qquad r^2 = y^\alpha y^\alpha \qquad (\alpha = 1, 2, 3),$$

these equations define the hypersurface $\Sigma x^{i2} = \dfrac{1}{K}$. The symbols of the induced group Γ_6 on this hypersurface (cf. §20) of G_6 of Ex. 11 are given by

$$X_{ab}f = Y_{ab}f = y^a \frac{\partial f}{\partial y^b} - y^b \frac{\partial f}{\partial y^a}, \qquad X_{4a}f = Y_a f \qquad (a, b = 1, 2, 3),$$

were the vectors η^α_a of $Y_a f$ are given by

$$\eta^\alpha_a = \frac{1}{\sqrt{K}} \left[\delta^\alpha_a \left(1 - \frac{K}{4} r^2 \right) + \frac{K}{2} y^a y^\alpha \right].$$

Robertson, 1932, 3, p. 512.

15. Show that the vectors of the invariant sub-groups of Γ_6, which are induced by the sub-groups of Ex. 13, can be written in the form

$$\zeta_a^\alpha = \sqrt{K}(\epsilon_{\alpha\alpha\beta}y^\beta + \eta_a^\alpha), \qquad \bar{\zeta}_a^\alpha = \sqrt{K}(\epsilon_{\alpha\alpha\beta}y^\beta - \eta_a^\alpha) \quad (a, \alpha, \beta = 1, 2, 3),$$

where $\epsilon_{\alpha\alpha\beta}$ is $+$ or -1 as $\alpha\alpha\beta$ is an even or odd permutation of 1, 2, 3, and zero otherwise; show also that

$$\sum_a \zeta_a^\alpha \zeta_a^\beta = \sum_a \bar{\zeta}_a^\alpha \bar{\zeta}_a^\beta = \delta^{\alpha\beta}\left(1 + \frac{K}{4}r^2\right)^2.$$

Robertson, 1932, 3, p. 512.

CHAPTER IV

THE ADJOINT GROUP

37. Linear homogeneous groups and vector-spaces. Consider the group G_1 for which the functions ξ^i in a particular coordinate system are

$$(37.1) \qquad \xi^i = a^i_j x^j,$$

where the a's are constants. If we interpret the x's as cartesian coordinates of a euclidean space, the condition that a point $P(x)$ be transformed into a point collinear with P and the origin is that

$$(37.2) \qquad Xx^i = a^i_j x^j = \rho x^i.$$

Thus the vector OP of components x^i is transformed into the vector with the same direction. Consequently each solution of the equations

$$(37.3) \qquad (a^i_j - \rho \delta^i_j) x^j = 0$$

determines an invariant direction through the origin. Accordingly the problem of finding invariant directions reduces to the determination of the roots of the determinant equation

$$(37.4) \qquad \Delta = |a^i_j - \delta^i_j \rho| = 0,$$

where the superscript indicates the row and the subscript the column. Since the solutions of (37.3) are determined only to within a multiplier, we see that any vector in an invariant direction is transformed into a vector in that direction. Equation (37.4) is called the *characteristic equation* of the matrix $\|a^i_j\|$.

In the foregoing we have considered all the vectors through the origin as constituting the space, which we call a vector-space E_n, and we have called x^i, for particular values of these quantities, the components of a vector. In particular, the n vectors of components

$$(37.5) \qquad e^i_a = \delta^i_a \qquad\qquad (a = 1, \cdots, n)$$

are the unit vectors in the direction of the coordinate axes. Any linear combination of the base vectors is a vector of the space, so that

$$(37.6) \qquad \bar{x}^i = x^a e_a^i$$

is a vector when the base e_a is general. The quantities $\bar{x}^i$ are called the absolute components and x^a the components relative to the base e_a. When the base is chosen as in (37.5), then $\bar{x}^i = x^i$. Hence, when the base is taken as the axes of a coordinate system, the absolute components are equal to the *relative*, or *coordinate*, components in this coordinate system.

If we change the base in accordance with the equations

$$(37.7) \qquad e'^i_a = e^i_b c^b_a,$$

where the c's are any constants such that the determinant $|c^b_a|$ is different from zero, then, since $\bar{x}^i$ are to be absolute in (37.6), for any vector we have

$$(37.8) \qquad x^a = c^a_b x'^b, \qquad x'^b = \bar{c}^b_a x^a,$$

where

$$(37.9) \qquad c^b_a \bar{c}^a_d = \delta^b_d.$$

Thus to a change in base (37.7) there corresponds a change in coordinates (37.8), that is, a change in relative components.

If we have n independent vectors with respect to a base e_a, their components x^a_h must be such that the determinant $|x^a_h|$ is not zero. If then we assign arbitrarily values x'^a_h such that their determinant is not zero, the equations

$$\bar{c}^b_a x^a_h = x'^b_h$$

can be solved for the $\bar{c}$'s and we have:

[37.1] *The base of a vector-space can be chosen so that the components of n independent vectors take arbitrary values, subject to the condition that their determinant is different from zero.*

For the change of coordinates (37.8) we have

$$\xi'^k = \xi^i \frac{\partial x'^k}{\partial x^i} = a^i_j x^j \bar{c}^k_i = a'^k_l x'^l,$$

where

$$(37.10) \qquad a'^k_l = \bar{c}^k_i a^i_j c^j_l,$$

and consequently

$$a''^k_l x'^l = \bar{c}^k_i a^i_j x^j = \rho x^i \bar{c}^k_i = \rho x'^k.$$

Hence we have:

[37.2] *The roots of the characteristic equation are invariant under change of base.*

If x denotes the matrix of one column whose elements are x^i and C denotes the matrix $\|c^b_a\|$, then in accordance with the law of multiplication of matrices, the first of (37.8) may be written

(37.11) $x = Cx'.$

Similarly, if A denotes the matrix $\|a^i_j\|$ and C^{-1} the inverse of C, we have from (37.10)

(37.12) $A' = C^{-1}AC,$

and consequently

(37.13) $(A' - I\rho) = C^{-1}(A - I\rho)C,$

where I denotes the unit matrix $\|\delta^j_i\|$.

We return to the consideration of equations (37.3) and denote by ρ_1 one of the roots of (37.4). We have observed that there is at least one invariant direction for ρ_1, that is, a vector-space E_1, and in accordance with theorem [37.1] the base may be chosen so that the components of a vector x_1 satisfying (37.3) are $(1, 0, \ldots, 0)$. Then from the equations (37.3) for this base we have

(37.14) $a^1_1 = \rho_1,$ $a^{i_1}_1 = 0$ $(i_1 = 2, \cdots, n).$

If we adjoin to E_1 any other vector x^i of E_n not in E_1, the components of any vector in the pencil of the two are of the form $\lambda + \mu x^1, \mu x^2, \cdots, \mu x^n$. When the vectors of this pencil are subjected to a transformation of G_1, ordinarily they go into vectors of another pencil through E_1. However, if $x^2, \ldots, x^n$ are chosen so that

(37.15) $(a^{i_1}_{j_1} - \delta^{i_1}_{j_1}\rho)x^{j_1} = 0$ $(i_1, j_1 = 2, \cdots, n),$

the E_2 determined by E_1 and the vector, say x_2, of these components and x^1_2 arbitrary is invariant under G_1, although an individual direction in E_2 may not be invariant. We remark that because of (37.14) the quantity ρ in (37.15) is a root of the characteristic equation (37.4). If we choose the basis so that the vectors x_1 and x_2 have the respective components $(1, 0, \ldots, 0)$ and $(0, 1, 0, \ldots, 0)$, then it follows from the corresponding equations (37.15) that

$$(37.16) \qquad a_2^{i_2} = 0 \qquad (i_2 = 3, \cdots, n)$$

and (37.14) obtain.

In like manner if we take E_2 and a vector x_3 not in E_2, such that $x_3^3, \ldots, x_3^n$ satisfy the equations

$$(37.17) \qquad (a_{j_2}^{i_2} - \delta_{j_2}^{i_2}\rho)x^{j_2} = 0 \qquad (i_2, j_2 = 3, \cdots, n),$$

where now as before ρ is necessarily a root of (37.4), then the E_3 determined by E_2 and x_3, whatever be x_3^1 and x_3^2, is invariant under transformations of G_1. Furthermore, by a suitable choice of basis we have

$$(37.18) \qquad a_3^{i_3} = 0 \qquad (i_3 = 4, \cdots, n),$$

as well as (37.14) and (37.16). Evidently this process may be continued, using the roots of (37.4), and we have:

[37.3] *The transformations of a linear homogeneous group G_1 leave invariant at least one sequence of vector-spaces $E_1, E_2, \ldots, E_{n-1}$ in E_n such that any E_k is contained in E_{k+1}.*

Also we have shown that:

[37.4] *A coordinate system can be chosen so that for the matrix of a linear homogeneous group we have*

$$(37.19) \qquad a_j^t = 0 \qquad (j = 1, \cdots, n-1; t = j+1, \cdots, n).$$

We have seen that a solution of (37.15) determines with E_1 an E_2 such that any vector in E_2 is transformed into a vector in E_2. If x_2 is such a vector, we say that it is invariant modulus x_1, or in matrix form

$$Ax_2 = \rho x_2 \ (\text{mod } x_1).$$

In like manner we say that E_2 and a solution of (37.17) determine an E_3, and for a vector x_3 in E_3 we have

$$Ax_3 = \rho x_3 \ (\text{mod } x_1, x_2),$$

where ρ in both of these equations indicates a root of (37.4).

If the root ρ_1 is a multiple root of order ν_1 and it is used for the determination of the vectors $x_1,\ \ldots\ ,\ x_{\nu_1}$, then the above results may be written in the form

$$(37.20)\quad (A - I\rho_1)x_1 = 0,\ (A - I\rho_1)^2 x_2 = 0,\ \cdots,$$
$$(A - I\rho_1)^{\nu_1}x_{\nu_1} = 0.$$

We observe that in (37.15) the principal minor $a_{j_1}^{i_1}$ of the matrix A operates only on the coordinates $x^2,\ \ldots\ ,\ x^n$, say on $\bar{E}_{n-1}$ which in the terminology of Weyl is the *projection* of E_n with respect to E_1, and ρ is a root of (37.4). Similarly in (37.17) the principal minor $a_{j_2}^{i_2}$ operates on $x^3,\ \ldots\ ,\ x^n$, which are the components of a vector in $\bar{E}_{n-2}$, the projection of E_n with respect to E_2. Thus in obtaining a vector-space E_{p+1} in the above theorem, we operate with the minor $a_{j_p}^{i_p}\ (i_p, j_p = p + 1,\ \cdots\ ,\ n)$ on the projection of E_n with respect to E_p.

Suppose that G_r is an integrable group of linear homogeneous transformations; then we have a sequence

$$G_r,\ G_{r-1},\ \ldots\ ,\ G_1,\ 1$$

such that for any $k < r$ the sub-group G_k is an invariant sub-group of G_{k+1}. We choose the basis of G_r so that $X_1 f$ is the generator of G_1; $X_1 f$ and $X_2 f$ of G_2, and so on as in §36. We denote by A_i the matrix of $X_i f$. For each root of the characteristic equation of A_1, we have one or more invariant directions given by

$$(37.21)\qquad\qquad\qquad A_1 x = \rho x.$$

If there are p of these invariant directions for a given root, any linear combination of vectors in these directions is a solution of (37.21). Hence if we denote by $\rho_1,\ \ldots\ ,\ \rho_s$ the different roots of the characteristic equation of A_1, we have vector-spaces $E_{p_1},\ \ldots\ ,$ E_{p_s}, of dimensions $p_1,\ \ldots\ ,\ p_s$ respectively. Moreover, each direction in any of these vector-spaces is invariant under G_1. Furthermore, no two of these vector-spaces have a vector in common, since (37.21) and $A_1 x = \rho' x\ (\rho' \neq \rho)$ are incompatible; and consequently these vector-spaces are discrete.

Since G_1 is an invariant sub-group of G_2 and if T_1 is any transformation of G_1 and T_2 a transformation of G_2 but not of G_1, then

$$T_2 T_1 T_2^{-1} = T_1',$$

where T_1' is a transformation of G_1, or $T_2 T_1 = T_1' T_2$. Hence if E is any one of the invariant spaces referred to above, and we put $T_2 E = E'$, we have from $T_2 T_1 E = T_1' T_2 E$ that $E' = T_1' E'$, and consequently E' is an invariant space of G_1. But since the invariant vector-spaces are discrete and we are dealing with continuous groups, it follows that E' is E. Hence each of the vector-spaces is invariant under G_2, but each direction in such a space is not necessarily invariant. If then we apply to any one of these vector-spaces the group generated by $X_2 f$, that is, the matrix A_2, there is in it at least one invariant direction, and there may be sub-sets of invariant directions as in the case of A_1, all of which being invariant for G_1 also are therefore invariant for G_2. Hence there exist for G_2 an array of discrete vector-spaces, each vector of which is invariant for G_2. We may then proceed in like manner with G_3, and so on and have that there is at least one invariant direction for G_r, say E_1. Then we project E_n with respect to E_1 and obtain an $\bar{E}_{n-1}$. We can then carry on the same process for $\bar{E}_{n-1}$ as was done for E_n and obtain in it a direction invariant for G_r. Hence the E_2 in E_n represented by this direction is invariant with respect to G_r. Then project E_n with respect to E_2, and apply the process to the resulting $\bar{E}_{n-2}$, and so on. Accordingly we have the theorem of Lie:*

[37.5] *If a group G_r of linear homogeneous transformations is integrable, there is at least one sequence of invariant vector-spaces $E_1, \ldots, E_{n-1}$ such that any E_k is contained in E_{k+1}.*

By choosing the coordinates as was done in the case of a one-parameter group, we have:

[37.6] *When a linear homogeneous group is integrable, the basis can be chosen so that*

$$(37.22) \qquad a_{\alpha j}^{t} = 0 \quad \left(\begin{matrix} \alpha = 1, \cdots, r; \\ j = 1, \cdots, n-1; t = j + 1, \cdots, n \end{matrix} \right).$$

38. Canonical form of linear homogeneous transformations. Fundamental vectors. The reduction by means of homogeneous linear transformations of the transformation (37.1) to the form for which (37.19) obtain is only a partial step in the reduction which can be made. Equation (37.12) sets forth the transformation of

---

*1893, 1 p., 533.

the matrix A under a linear homogeneous transformation. Hence we may take over the known results of the reduction of such forms. The reduction to normal form of a matrix A is based upon the elementary divisors of the matrix Δ (37.4), which are defined as follows: Let F_i be the highest common factor of the i-rowed minors of Δ and l_i the highest power of $\rho - \rho_1$ in F_i, where ρ_1 is a root of (37.4). As i takes the values $1, \ldots, n$, we get numbers $\nu_i = l_i - l_{i-1}$ which are equal to or greater than zero; then for such values of ν_i which are not equal to zero the expressions $(\rho - \rho_1)^{\nu_i}$ are the elementary divisors of Δ for the root ρ_1. The fundamental theorem is as follows:

If the elementary divisors of Δ are

$$(38.1) \qquad (\rho - \rho_1)^{\nu_1}, \ (\rho - \rho_2)^{\nu_2}, \ \cdots , \ (\rho - \rho_k)^{\nu_k},$$
$$(\nu_1 + \cdots + \nu_k = n),$$

(the ρ_a's in these factors are not necessarily different, nor are the ν's), there exists a real homogeneous transformation of the variables such that Δ assumes the form

$$(38.2) \qquad \Delta = \begin{vmatrix} M_1 & & & 0 \\ & M_2 & & \\ & & \ddots & \\ 0 & & & M_k \end{vmatrix},$$

where M_i represents the matrix of order ν_i

$$\begin{vmatrix} \rho_i - \rho & 1 & & 0 & \cdot & \cdot & 0 \\ 0 & & \rho_i - \rho & 1 & 0 & \cdot & 0 \\ \cdot & \cdot & \cdot & \cdot & \cdot & \cdot & \cdot & \cdot & \cdot \\ \cdot & \cdot & \cdot & \cdot & \cdot & \cdot & \cdot & \cdot & \cdot \\ 0 & \cdot & \cdot & \cdot & \cdot & 0 & \rho_i - \rho & 1 \\ 0 & \cdot & \cdot & \cdot & \cdot & \cdot & \cdot & 0 & \rho_i - \rho \end{vmatrix},$$

and all the terms of Δ not in any M_i are zero.[*]

We have remarked that several of the ρ_a's in (38.1) may be the same. If we gather into one principal minor all terms involving the same ρ, and denote by ν_a the order of the root ρ_a, we see that the minor corresponding to the root ρ_a is of the form

---

[*] *Bôcher*, 1907, 1, pp. 270, 271, 288, 289.

$$(38.3) \qquad M_a = \begin{vmatrix} \rho_a - \rho & e_{a1} & 0 & \cdot & \cdot & \cdot & \cdot & 0 \\ 0 & \rho_a - \rho & e_{a2} & 0 & \cdot & \cdot & \cdot & 0 \\ \cdot & \cdot & \cdot & \cdot & \cdot & \cdot & \cdot & \cdot \\ \cdot & \cdot & \cdot & \cdot & \cdot & \cdot & \cdot & \cdot \\ 0 & \cdot & \cdot & \cdot & \cdot & 0 & \rho_a - \rho & e_{a[\nu_a - 1]} \\ 0 & \cdot & \cdot & \cdot & \cdot & \cdot & 0 & \rho_a - \rho \end{vmatrix},$$

where any e is one or zero, as the case may be, and that all the other terms in $\|a_j^i\|$ in the same rows or columns as terms of M_a are zero.

We denote by x_1 a vector for the root ρ_1 satisfying (37.3), that is,

$$(38.4) \qquad Ax_1 = \rho_1 x_1.$$

For the above normal form of A it follows that in this coordinate system one solution is

$$(38.5) \qquad x_1^i = \delta_1^i,$$

and also that

$$a_{b_1}^{b_1} = \rho_1, \qquad a_{b_1+1}^{b_1} = e_{1b_1}, \qquad a_c^{b_1} = 0$$
$$(b_1 = 1, \cdots, \nu_1; c \neq b_1 \neq (b_1 + 1)).$$

Hence each of the vectors

$$(38.6) \qquad x_{b_1}^i = \delta_{b_1}^i$$

satisfies the corresponding equation

$$(38.7) \qquad Ax_{b_1} = \rho_1 x_{b_1} + e_{1[b_1-1]}x_{b_1-1}.$$

In particular, we have

$$(a_j^i - \delta_j^i \rho_1)x_2^j = e_{11}x_1^i,$$

and consequently

$$(a_i^k - \delta_i^k \rho_1)(a_j^i - \delta_j^i \rho_1)x_2^j = e_{11}(a_i^k - \delta_i^k \rho_1)x_1^i = 0,$$

or, in matrix notation

$$(A - I\rho_1)^2 x_2 = 0.$$

And in general we have

$$(38.8) \qquad (A - I\rho_1)^{b_1}x_{b_1} = 0 \qquad (b_1 = 1, \cdots, \nu_1).$$

We say that the vectors (38.6) determine a vector-space E_{ρ_1} of dimension ν_1 associated with the root ρ_1, called the *root-space* for the root ρ_1. Any vector of this space being a linear combination (constant coefficients) of the vectors (38.6), it follows that any vector of the space E_{ρ_1}, say $\bar{x}_1$, satisfies the condition

$$(A - I\rho_1)^b \bar{x}_1 = 0,$$

where b depends on $\bar{x}_1$ and is $\leqslant \nu_1$. We shall refer to the vectors (38.6) as the *fundamental* vectors of E_{ρ_1} corresponding to the first, . . . , ν_1th rows of the matrix M_1, and any two in the set for consecutive values of b_1 we call *consecutive*.

In like manner the vectors

(38.9) $$x_{b_2}^i = \delta_{b_2}^i \quad (b_2 = \nu_1 + 1, \cdots, \nu_1 + \nu_2)$$

determine a space E_{ρ_2} corresponding to the root ρ_2. We have

(38.10) $$A x_{b_2} = \rho_2 x_{b_2} + e_{2[c_2-1]} x_{b_2-1}, \qquad (A - I\rho_2)^{c_2} x_{b_2} = 0$$
$$(c_2 = b_2 - \nu_1),$$

$$(A - I\rho_2)^c \bar{x}_2 = 0 \qquad\qquad (c \leqslant \nu_2).$$

In this way we proceed with the other roots $\rho_3, \ldots,$ and we have:

[38.1] *If x_α is any vector of the root-space for ρ_α, then*

(38.11) $$(A - I\rho_\alpha)^c x_\alpha = 0 \qquad\qquad (c \leqslant \nu_\alpha).$$

If we write equation (37.4) in the form

$$\varphi(\rho) = (\rho - \rho_1)^{\nu_1} \cdots (\rho - \rho_p)^{\nu_p} = 0,$$

there exist functions $\psi_a(\rho)$ such that

$$\frac{1}{\varphi(\rho)} = \sum_a^{1, \ldots, p} \frac{\psi_a(\rho)}{(\rho - \rho_a)^{\nu_a}},$$

from which we obtain

$$1 = \sum_a f_a(\rho),$$

where

$$f_a(\rho) = \psi_a(\rho) \Pi_{b \neq a} (\rho - \rho_b)^{\nu_b} \quad (a, b = 1, \cdots, p).$$

Hence we have the matrix equation

(38.12) $$I = \sum_a f_a(A), \qquad f_a(A) = \psi_a(A) \Pi_{b \neq a} (A - \rho_b I)^{\nu_b}.$$

If x is a vector of the space, we have

$$(38.13) \qquad x = x_1 + \cdots + x_p,$$

where x_a is a certain vector of the root-space for ρ_a. In consequence of the above theorem and the definition of $f_a(A)$ we have

$$f_a(A)x_b = 0 \qquad\qquad (b \neq a),$$

and then from the first of (38.12)

$$(38.14) \qquad x_b = Ix_b = f_b(A)x_b.$$

Hence we have

$$f_a(A)x = f_a(A)x_a = x_a.$$

Now if $(A - \rho_1 I)^h x = 0$ for $h \leqslant \nu_1$, then

$$(A - \rho_1 I)^{\nu_1} x = 0.$$

Hence if we write x in the form (38.13), we have for $a \neq 1$

$$f_a(A)x = 0.$$

As a takes the values $2, \ldots, p$, it follows from (38.14) that $x_2 = \cdots = x_p = 0$. Consequently we have:

[38.2] *If x is any non-zero vector such that*

$$(A - I\rho_a)^c x = 0 \qquad\qquad (c \leqslant \nu_a),$$

then x is a vector of the root-space of ρ_a.

Suppose now that σ is not a root and that

$$(A - I\sigma)^c x = 0.$$

Since σ is not a root, there exist polynomials $M(\rho)$ and $N(\rho)$ such that

$$M(\rho)(\rho - \sigma)^c + N(\rho)\varphi(\rho) = 1.$$

Since this is an integral identity in one variable and is a direct consequence of the rules of addition and multiplication, it will hold when a matrix is substituted for ρ, so that we have

$$M(A)(A - I\sigma)^c + N(A)\varphi(A) = I.$$

Since $\varphi(A) = (A - I\rho_1)^{r_1} \cdots (A - I\rho_p)^{r_p}$, we have $\varphi(A)x = 0$. Hence we have

$$M(A)(A - I\sigma)^c x = Ix = x.$$

But by hypothesis the left-hand member is zero and hence $x = 0$. Accordingly we have:

[38.3] *If*

$$(A - I\sigma)^c x = 0$$

either σ is not a root and x is a zero vector, or σ is a root and x is a vector of its root-space.

39. Adjoint group of a given group.

From equations (29.14) we have that a group G_r is invariant under transformations of any G_1 of the group. If we denote by $\lambda^a X_a f$, where the λ's are constants, the symbol of a particular G_1, and put

(39.1) $$v^a = \lambda^a \tau,$$

then the v's are canonical parameters of this G_1. If u^a are canonical parameters of a given G_1 and $\bar{u}^a$ those for the G_1 into which it is transformed by the above G_1, it follows from (29.4) that

(39.2) $$\bar{u}^a = \sigma_b^a(v)u^b,$$

where the σ's are functions of the v's, such that $(\sigma_b^a)_{\tau=0} = \delta_b^a$. These are equations of the transformations of the u's into the $\bar{u}$'s, involving r parameters v^a. We shall show that they form a group. In fact, the equations analogous to (29.7) are

(39.3) $$T_v T_u T_v^{-1} = T_{\bar{u}}.$$

If we put similarly

$$T_{v_1} T_{\bar{u}} T_{v_1}^{-1} = T_{\bar{u}_1},$$

then

$$T_{\bar{u}_1} = T_{v_1} T_v T_u T_v^{-1} T_{v_1}^{-1} = T_{v_2} T_u T_{v_2}^{-1},$$

where

(39.4) $$v_2^a = \psi^a(v; v_1),$$

these being the equations in canonical parameters analogous to (4.6) in general parameters. Hence we have:

[39.1] *Equations (39.2) define a group and the equations connecting the parameters are the same as for the given group G_r.*

This group Γ is called the *adjoint* of the given group. It is not necessarily of order r, as will be shown later; that is, all of the parameters v^a are not necessarily essential.

From the equations

$$T_v T_{u_1} T_v^{-1} = T_{\bar{u}_1}, \qquad T_v T_{u_2} T_v^{-1} = T_{\bar{u}_2},$$

we have

$$T_{\bar{u}_2} T_{\bar{u}_1} = T_v T_{u_2} T_v^{-1} T_v T_{u_1} T_v^{-1} = T_v T_{u_2} T_{u_1} T_v^{-1}.$$

Hence each transformation of the adjoint group establishes an *automorphism* of the given group, that is, an isomorphism of the group to itself in the sense of §32, and consequently we have:

[39.2] *The constants of structure of a group are invariant, when the canonical parameters undergo a transformation of the adjoint group.*

If in (29.14) we replace the symbol Xf by $\lambda^a X_a f$, we have in consequence of (7.2) that $g_b^a = \lambda^e c_{be}^a$, and consequently in place of equations (29.20), we have

$$(39.5) \qquad \frac{d\bar{u}^a}{d\tau} = \bar{u}^b \lambda^e c_{be}^a.$$

When the expressions (39.2) are substituted in (39.5), we obtain

$$(39.6) \qquad \frac{d\sigma_b^a(v)}{d\tau} = \sigma_b^d \lambda^e c_{de}^a.$$

Comparing this result with (14.12), we see that the functions σ_b^a are those given by (14.13). Furthermore, these functions σ_b^a are invariants expressed in terms of canonical parameters, and are equal to the functions $\rho_b^a(a)$ in general parameters as defined by (8.14). Thus the equations of the adjoint group may be written

$$(39.7) \qquad \bar{u}^a = \rho_b^a(a) u^b.$$

From this point of view the transformations of the adjoint group are isomorphic to the transformations connecting each set of the vectors $\bar{A}_a^\alpha$ and A_a^α in the group-space.

If we put

$$(39.8) \qquad \eta_a^\alpha(u) = c_{ba}^\alpha u^b, \qquad E_a\varphi = \eta_a^\alpha \frac{\partial \varphi}{\partial u^\alpha},$$

equations (39.5) become

$$(39.9) \qquad \frac{d\bar{u}^\alpha}{d\tau} = \lambda^a \eta_a^\alpha(\bar{u}),$$

from which it follows that $E_a\varphi$ are the symbols of the adjoint group (cf. (11.8)).

From (39.8) it follows that, in consequence of (7.4),

$$(39.10) \quad (E_a, E_b)\varphi = (c_{ea}^f c_{fb}^e + c_{be}^f c_{fa}^e) u^e \frac{\partial \varphi}{\partial u^c} = c_{ab}^f c_{ef}^e u^e \frac{\partial \varphi}{\partial u^c} = c_{ab}^e E_e \varphi.$$

Consequently we have:

[39.3] *The symbols $E_a f$ have the same constants of structure as the group G_r.*

In order that the adjoint group be of order r, it is necessary and sufficient that there do not exist any relations of the form

$$(39.11) \qquad g^a c_{ba}^\alpha = 0,$$

where the g's are constants, as follows from (39.8). If we denote by C the matrix of r columns and r^2 rows

$$(39.12) \qquad C = \|c_{ba}^\alpha\|,$$

the indices b and α indicating rows and a the columns, we have that if the matrix C is of rank $s(\leqslant r)$, equations (39.11) admit $r - s$ independent sets of solutions, and the order of the adjoint group is s. In this case there are $r - s$ linearly independent (constant coefficients) exceptional sub-groups G_1 (§29).

From (39.8) and (7.3) it follows that $u^a \eta_a^\alpha = 0$, and consequently the generic rank of the matrix $\|\eta_a^\alpha\|$ is less than r. Hence we have:

[39.4] *The adjoint group is always intransitive.*

In §16 it was seen that if $X_1 f, \ldots, X_m f$ generate a sub-group G_m of G_r, then $c_{tu}^z = 0$ for $t, u = 1, \cdots, m; z = m + 1, \cdots, r$. Then from (39.10) we have:

[39.5]. *If $X_1f, \ldots, X_mf$ generate a sub-group G_m of G_r, then $E_1f, \ldots, E_mf$ generate a sub-group of the adjoint group, and conversely.*

Furthermore, if this G_m is an invariant sub-group of G_r, then $c_{at}^{\ z} = 0$ for $a = 1, \cdots, r; t = 1, \cdots, m; z = m + 1, \cdots, r$, and we have:

[39.6] *If $X_1f, \ldots, X_mf$ generate an invariant sub-group of G_r, then $E_1f, \ldots, E_mf$ generate an invariant sub-group of the adjoint group, and conversely.*

Herein lies the significance and importance of the adjoint group, in that the determination of sub-groups and invariant sub-groups of a given group reduces to such determinations for the adjoint group, which is linear homogeneous in the canonical parameters.

To a sub-group G_1 of G_r with the generator $v^a X_a f$ corresponds the sub-group $\mathcal{G}_1$ of the adjoint group with the generator $v^a E_a f$. As in §37 the quantities v^a are the components of a vector in a vector-space E_r,* such that each sub-space E_1 represents a G_1 of G_r and also a $\mathcal{G}_1$ of the adjoint group, and each transformation in G_1 and $\mathcal{G}_1$ is represented by a vector in E_1. To each sub-group G_m of G_r there corresponds a sub-group of the adjoint group represented by the vectors of a sub-space E_m. When the vectors of this E_m are subjected to any transformation of the adjoint group, the E_m goes into a new E_m' unless the transformation is represented by a vector of E_m, as follows from the first of the above theorems, or unless the sub-group G_m is invariant under G_r. Hence we have:

[39.7] *A necessary and sufficient condition that a sub-space E_m of E_r represent a sub-group G_m of G_r is that it be an invariant variety of all the transformations of the adjoint group represented by vectors of E_m. A necessary and sufficient condition that G_m be an invariant sub-group is that E_m be an invariant variety of the adjoint group.*

If u_1 and u_2 are any two vectors of E_r, then

$$(u_1^a X_a, u_2^b X_b)f = u_{12}^e X_e f \quad (a, b, e = 1, \cdots, r),$$

where

$$u_{12}^e = c_{ab}^e u_1^a u_2^b.$$

---

* E_r, E_1 and E_m are not to be confused with the symbols $E_a f$ of the adjoint group.

Hence u_{12}, denoted by (u_1, u_2), is a vector of the derived group of G_r. Also if u_1 and u_2 are vectors of a sub-group of G_r, then u_{12} is a vector of this sub-group. Furthermore, if u_1 is any vector of an invariant sub-group of G_r and u_2 is any vector of G_r, then u_{12} is a vector of the sub-group.

In consequence of the Jacobi relations (7.4) for any three vectors u_1, u_2, u_3 we have

$$((u_1, u_2), u_3) + ((u_2, u_3), u_1) + ((u_3, u_1), u_2) = 0.$$

As an application of this identity we consider the case when u_1 and u_2 are vectors of an invariant sub-group G_m and u_3 is any vector of G_r. Since the sub-group is invariant (u_2, u_3) and (u_3, u_1) are vectors of G_m, and consequently the second and third terms of the above identity are vectors of the derived group of G_m, and hence their sum is a vector of the derived group. Consequently the first term, which is the transform of a vector in the derived group of G_m by a general transformation of G_r, is a vector of the derived group, and consequently the latter is an invariant sub-group of G_r. Since the second derived group of G_m bears to its first derived sub-group, if it is not equal to it, the same relation that the latter bears to G_m, it is an invariant sub-group of G_r, and so on. Hence we have:

[39.8] *All of the derived groups of an invariant sub-group of a G_r are invariant sub-groups of G_r.*[*]

In particular we have:

[39.9] *All of the derived groups of a G_r are invariant sub-groups of G_r.*

Exercises

1. If in the general linear homogeneous operator $a^i_j x^i p_i$ non-homogeneous coordinates $y^\alpha = \dfrac{x^\alpha}{x^n}$ $(\alpha = 1, \cdots, n-1)$ are introduced, it becomes

$$(a^\alpha_n + a^\alpha_\beta y^\beta - a^n_n y^\alpha - a^n_\beta y^\beta y^\alpha)\frac{\partial f}{\partial y^\alpha} \quad (\alpha, \beta = 1, \cdots, n-1).$$

2. Show that theorem [39.2] follows from (8.22) and (39.7).
3. The adjoint group of the simple group G_3 with basis such that

$$(X_1, X_2)f = X_1 f, \qquad (X_1, X_3)f = 2X_2 f, \qquad (X_2, X_3)f = X_3 f$$

---

[*] *Lie-Engel*, 1893, 2, p. 679.

has the symbols

$$E_1f = -x^2p_1 - 2x^3p_2, \qquad E_2f = x^1p_1 - x^3p_3, \qquad E_3f = 2x^1p_2 + x^2p_3,$$

and this group leaves invariant the conic $4x^1x^3 - (x^2)^2 = 0$.

4. The adjoint group of an Abelian group is the identity.

5. Find the adjoint groups of the various types of G_3 of Ex. 9, p. 137.

6. A sub-group G_2 of a G_r belongs to at least one sub-group G_3 of G_r.

Lie-Scheffers, 1893, 1, p. 564.

7. If $u^1, \ldots, u^r$ are considered to be the homogeneous coordinates of a point in a flat space R_{r-1}, each infinitesimal transformation is represented by a point in R_{r-1}, and the operator $(u^aX_a, v^bX_b)f$ is represented by the point $c_{ab}^e u^a v^b$. What does theorem [39.7] become under this interpretation?

8. Show with the aid of Ex. 11, p. 44 that, if the G_1 with the symbol u^aX_af is transformed by the transformations of the G_1 with the symbol v^bX_bf, the point of coordinates u^a in R_{r-1} describes a curve whose tangent at the point u^a is the line joining the latter to the point $c_{ab}^e u^a v^b$.

Lie-Scheffers, 1893, 1, p. 471.

9. For the group G_4 of Ex. 7, p. 123 the invariant sub-group is represented in R_3 by the plane $u^4 = 0$, and the adjoint group of G_4 leaves a conic invariant by Ex. 3 and theorem [31.1]. Show that X_4f can be chosen so that E_4f leaves the plane $u^4 = 0$ point-wise invariant and deduce therefrom the result of Ex. 7, p. 123.

Lie-Scheffers, 1893, 1, p. 573.

10. In the R_2 of a G_3 the equations $w^e = c_{ab}^e u^a v^b$ define a correlation between the point w^a and the line joining u^a and v^a, whose Plücker coordinates are $p^{ab} = u^a v^b - u^b v^a$. This correlation is singular unless G_3 is simple.

11. If a sub-group G_3 of a G_4 is not invariant, its representative plane R_3 is transformed by the adjoint group of G_4 into ∞^1 planes, which envelope a cone, a general developable surface or all pass through a line. In each case there is an invariant configuration and consequently there is no simple G_4.

Lie-Scheffers, 1893, 1, p. 576.

40. The characteristic equation of a group. The rank of a group. The equation

$$(40.1) \qquad \Delta(u, \rho) \equiv |\eta_b^\alpha(u) - \rho\delta_b^\alpha| = 0,$$

where η_b^α is defined by (39.8), is called the *characteristic equation of the group*, and $\Delta(u, \rho)$ the *characteristic matrix*. From §38 it follows that the latter is unaltered when the u's undergo a homogeneous linear transformation. The classification of groups is based upon the properties of the characteristic matrix (cf. §46).

Since the rank of $||\eta_b^e||$ is less than r (§39), we have

$$(40.2) \quad (-1)^r\Delta(u, \rho) = \rho^r - \psi_1(u)\rho^{r-1} + \psi_2(u)\rho^{r-2} - \psi_3(u)\rho^{r-3} + \cdots + (-1)^{r-1}\psi_{r-1}(u)\rho.$$

To within sign at most $\psi_q(u)$ is the sum of the principal minors of $||\eta_b^e||$ of order q. Hence if the rank of the matrix is q, the functions

ψ_s for $s > q$ are zero and the equation (40.1) admits zero as a root of order $r - q$ at least.

From (40.2) we have

$$\psi_1 = \eta_a^a,$$
$$\psi_2 = \tfrac{1}{2}(\eta_a^a \eta_b^b - \eta_a^b \eta_b^a),$$
$$\psi_3 = \eta_{[a}^a \eta_b^b \eta_{c]}^c \equiv \frac{1}{3!}(\eta_a^a \eta_b^b \eta_c^c + \eta_b^a \eta_c^b \eta_a^c + \eta_c^a \eta_a^b \eta_b^c)$$
$$- \frac{1}{3!}(\eta_a^a \eta_c^b \eta_b^c + \eta_c^a \eta_b^b \eta_a^c + \eta_b^a \eta_a^b \eta_c^c)$$

and so on. If we put

(40.3) $$\qquad \varphi_1 = \eta_a^a, \qquad \varphi_2 = \eta_a^b \eta_b^a, \qquad \varphi_3 = \eta_a^b \eta_b^c \eta_c^a,$$

and so on, we have

$$\psi_1 = \varphi_1, \qquad \psi_2 = \tfrac{1}{2}(\varphi_1^2 - \varphi_2), \qquad \psi_3 = \tfrac{1}{6}\varphi_1^3 - \tfrac{1}{2}\varphi_1\varphi_2 + \tfrac{1}{3}\varphi_3,$$

and consequently

(40.4) $$\quad \varphi_1 = \psi_1, \qquad \varphi_2 = \psi_1^2 - 2\psi_2, \qquad \varphi_3 = \psi_1^3 - 3\psi_1\psi_2 + 3\psi_3.$$

Hence φ_1 is the sum of the roots, φ_2 the sum of their squares and φ_3 the sum of their cubes. The sum of the terms of the principal diagonal of a matrix, in this case φ_1, is called the *trace* of the matrix.

From the definition of φ_2 it follows that

(40.5) $$\qquad\qquad \varphi_2 = g_{ij}u^i u^j, \qquad g_{ij} = c_{ia}^b c_{jb}^a.$$

Each set of values of u^a determines a G_1 of a given G_r. We call u^a the components of the vector u for G_1 in the general vector-field E_r of all the vectors of the group. The roots of the characteristic equation (40.2) are evidently functions of the u's. For a general set of values of the u's equation (40.2) may be written

$$(-1)^r \Delta(u, \rho) = \rho^{\nu_0}(\rho - \rho_1)^{\nu_1} \cdots (\rho - \rho_p)^{\nu_p},$$
$$\nu_0 + \nu_1 + \cdots + \nu_p = r,$$

in which case we say that the ρ_a's are the *generic roots* of the equation. There may be particular vectors u for which two or more of these roots are equal. In order to determine these we observe that, if the highest common factor of Δ and $\dfrac{\partial \Delta}{\partial \rho}$ is denoted by D, in the general case the equation $\Delta/D = 0$ has $0, \rho_1, \cdots, \rho_p$ for simple roots. Accordingly if a vector u is such that the discriminant of Δ/D vanishes, two or more of the roots are equal. We call such a vector *special*, and the others *regular*.

Suppose we have two vectors u^a and v^a, then

$$(40.6) \qquad (u^a X_a,\, v^b X_b)f = w^h X_h f,$$

where

$$(40.7) \qquad w^h = c_{ab}{}^h u^a v^b.$$

Now

$$(40.8) \qquad \eta_j^i(w) = c_h{}_j^i w^h = c_h{}_j^i c_{ab}{}^h u^a v^b = (c_b{}_j^h c_{ah}{}^i - c_{aj}{}^h c_{bh}{}^i) u^a v^b$$

$$= \eta_h^i(u)\eta_j^h(v) - \eta_h^i(v)\eta_j^h(u).$$

From this it follows that the trace of the matrix $\eta(w)$, that is, $\eta_i^i(w)$, is equal to zero.

If t^a is a vector, so also is $\eta_b^a(u)t^b$ a vector, which we may call the transform of t^a by the matrix $\eta(u)$. Suppose that we have a sub-space E_m of E_r such that the transform of any vector in E_m is a vector in E_m, then we say that E_m is invariant under $\eta(u)$. If the canonical parameters are chosen so that the m vectors determining E_m have the components δ_p^a for $p = 1, \cdots, m$, then

$$\eta_b^a(u)\delta_p^b = \lambda_p^a \delta_q^a \qquad (p,\, q = 1, \cdots, m),$$

where the λ's are constants. Consequently $\eta_p^a(u) = 0$ for $a > m$. If the vector-space E_m is invariant also for $\eta(v)$, then it is invariant for $\eta(w)$, as follows from (40.8). Then from (40.8) and

$$\eta_p^a(u) = \eta_p^a(v) = 0$$

for $a > m$ we have

$$\eta_p^p(w) = \eta_q^p(u)\eta_p^q(v) - \eta_q^p(v)\eta_p^q(u) = 0 \quad (p,\, q = 1, \cdots, m).$$

But $\eta_p^p(w)$ is the trace of the minor of the matrix $\eta(w)$ operating on the sub-space E_m. Since the trace is independent of the choice of canonical parameters, we have:

[40.1] *If $\eta(u)$ and $\eta(v)$ leave a sub-space invariant, the trace of the minor of the matrix $\eta(w)$, where $w^e = c_{ab}{}^e u^a v^b$, operating on the sub-space is zero.*

If a G_r admits a sub-group G_m, and $u^a X_a f$ is a generator of a G_1 of G_m, then u^a is called a vector of G_m. If the basis of G_r is chosen so that $X_1 f, \ldots, X_m f$ is the basis of G_m, then for any vector of G_m we have $u^z = 0\,(z = m + 1, \cdots, r)$. From this result and (16.5) it follows that $\eta_t^z(u) = 0$ for $t = 1, \cdots, m$, and the characteristic matrix Δ for u^a is of the form

$$
\left\|
\begin{array}{c|c}
\eta_v^t - \rho \delta_v^t & \eta_z^t \\
\hline
0 & \eta_w^z - \rho \delta_w^z
\end{array}
\right\|
\quad
\left(
\begin{array}{l}
t, v = 1, \cdots, m; \\
w, z = m + 1, \cdots, r
\end{array}
\right).
$$

Since $\eta_v^t(u) = c_{sv}^t u^s$, where $t, v, s = 1, \cdots, m$, it follows that $\|\eta_v^t(u)\|$ is the matrix of the sub-group with respect to the vector u. Hence we have

$$
\Delta = \Delta_m \, |\eta_w^z - \rho \delta_w^z|,
$$

where Δ_m is the characteristic matrix of the sub-group G_m. We may write

$$
(-1)^m \Delta_m = \rho^m - \psi_{m,1}(u)\rho^{m-1} + \cdots + (-1)^{m-1}\psi_{m,m-1}(u)\rho.
$$

If G_m is an invariant sub-group, then as follows from (31.3) $\eta_w^z(u) = 0$, and we have

$$
\Delta = (-1)^{r-m}\rho^{r-m}\Delta_m.
$$

In this case $\psi_{m,a} = \psi_a$, and zero is a root of the characteristic equation of order $r - m + 1$ at least. Hence we have:

[40.2]   *If a G_r admits an invariant sub-group of order m, the characteristic equation of G_r for a vector of the sub-group has a zero root of order $r - m + 1$ at least.*

Since in particular ψ_1 and ψ_2 for Δ and Δ_m are equal in this case, we have:

[40.3]   *If G_m is an invariant sub-group of a G_r, the sum of the roots and the sum of the squares of the roots of the characteristic equation of G_m with respect to a vector of G_m are equal to the same quantities for the characteristic equation of G_r with respect to this vector.*

We shall prove the following theorem of Killing:[*]

---

[*] 1888, 3, p. 260.

[40.4] *The coefficients $\psi_\kappa(u)$ in the characteristic equation of a group are invariants of the adjoint group, that is, they satisfy the system of equations*

$$(40.9) \qquad\qquad E_a f = 0 \qquad\qquad (a = 1, \cdots, r).$$

If we put

$$\gamma_b^e = \eta_b^e - \rho\delta_b^e,$$

then

$$E_a\Delta(u, \rho) = \eta_a^\kappa\frac{\partial\Delta}{\partial u^\kappa} = \eta_a^\kappa\frac{\partial\Delta}{\partial\gamma_b^e}\frac{\partial\eta_b^e}{\partial u^\kappa} = \eta_a^\kappa c_{\kappa b}^{\;\;e}\frac{\partial\Delta}{\partial\gamma_b^e}$$

$$= u^d c_{da}^{\;\;\kappa} c_{\kappa b}^{\;\;e}\frac{\partial\Delta}{\partial\gamma_b^e} = u^d(c_{ab}^{\;\;\kappa} c_{d\kappa}^{\;\;e} + c_{\kappa a}^{\;\;e} c_{db}^{\;\;\kappa})\frac{\partial\Delta}{\partial\gamma_b^e}$$

$$= (c_{ab}^{\;\;\kappa}\eta_\kappa^e + c_{\kappa a}^{\;\;e}\eta_b^\kappa)\frac{\partial\Delta}{\partial\gamma_b^e}$$

$$= (c_{ab}^{\;\;\kappa}\gamma_\kappa^e + c_{\kappa a}^{\;\;e}\gamma_b^\kappa)\frac{\partial\Delta}{\partial\gamma_b^e}.$$

Since

$$\gamma_\kappa^e\frac{\partial\Delta}{\partial\gamma_b^e} = \delta_\kappa^b\Delta, \qquad \gamma_b^\kappa\frac{\partial\Delta}{\partial\gamma_b^e} = \delta_e^\kappa\Delta,$$

the last member of the above equation is zero. Hence $E_a\Delta(u, \rho) = 0$ whatever be ρ and consequently the theorem is proved.

We have seen that the functions $\psi_1, \ldots, \psi_{r-1}$ are homogeneous functions in the u's of orders $1, \cdots, r-1$ respectively. Among them there may be relations so that only p of them are functionally independent; p is called the *rank of the group*. If the rank of $\|\eta_b^a\|$ is q, the equations (40.9) admit at most $r - q$ independent solutions; consequently

$$(40.10) \qquad\qquad p \leqslant r - q.$$

But, as we have remarked, if the rank of $\|\eta_b^a\|$ is q, the equation $\Delta(u, \rho) = 0$ has a zero root of order $r - q$ at least. Hence as a corollary to (40.10) we have:

[40.5] *The rank of a group is equal to, or less than, the order of the zero root of the characteristic equation.*

In particular, when all the ψ's vanish and consequently zero is a root of order r, the rank of the group is zero.

Since the roots are functions of the ψ's, we have:

[40.6] *The non-zero generic roots of the characteristic equation as functions of the u's are invariants of the adjoint group.*

Since the ψ's are functions of the roots, the number of independent ψ's is equal to the number of independent non-zero generic roots. Hence we have:

[40.7] *The rank of a group is equal to the number of functionally independent non-zero generic roots.*

If $u^a X_a f$ and $v^a X_a f$ are the generators of a sub-group G_2 of a given G_r, then from (40.6) and (40.7) we have

$$c_{ab}^{h} u^a v^b = \sigma_2 u^h + \sigma_1 v^h.$$

If $\sigma_1 \neq 0$, we may take $\bar{v}^a = v^a + \dfrac{\sigma_2}{\sigma_1} u^a$ as the vector of the second generator, in which case the above equation becomes

$$(c_{ab}^{h} u^a - \delta_b^h \sigma_1) \bar{v}^b = 0,$$

that is, σ_1 is a non-zero root of the characteristic equation for $\eta_a^b(u)$ and $\bar{v}$ is an invariant vector for the matrix $\|\eta_a^b(u)\|$. Similarly if $\sigma_2 \neq 0$, it is a non-zero root of the characteristic equation for $\eta_a^b(v)$. As a consequence of this result we have:

[40.8] *When the rank of the group is zero, every sub-group G_2 is Abelian.*

41. Determination of invariant sub-groups. That the coefficients ψ_a in the characteristic equation are invariants for the adjoint group may be established by another method, which has important consequences.

When the canonical parameters undergo any linear homogeneous transformation (12.10), the constants c_{ij}^{k} are components of a tensor of the third order, as follows from (7.13). Consequently such quantities as $c_{ij}^{j} u^i$ and $g_{ij} u^i u^j$ are scalars under such transformations. However, when the linear homogeneous transformation is a member of the adjoint group, the c's are the same in the two sets of parameters u^a and $\bar{u}^a$ by theorem [39.2]. Consequently we have

$$c_{ij}^{j} \bar{u}^i = c_{ij}^{j} u^i, \quad g_{ij} \bar{u}^i \bar{u}^j = g_{ij} u^i u^j.$$

Since any one of the functions ψ_a in (40.2) is a homogeneous function of the u's and the coefficients are functions of the c's such that ψ_a is a scalar, it follows that all of these functions are invariant under the adjoint group.

When a combination of the quantities c_{ij}^{k} is made by addition, subtraction, multiplication and contraction, the resulting quantity is called a *comitant* of the adjoint group. For example, the coefficients of the u's in any of the quantities $\psi_a(u)$ of the characteristic equation of a group are comitants. From theorem [39.2] it follows that:

[41.1] *A comitant is invariant under any transformation of the adjoint group.*

Suppose that a set of equations

$$(41.1) \qquad C_j^h \colon\colon\colon{}_{ki}^{l}\, u^i = 0,$$

where the C's are comitants, are consistent. If v^i is any vector of the group and we put

$$(41.2) \qquad \bar{u}^i = c_{jk}^{i} v^j u^\kappa,$$

then $u^i + \bar{u}^i \delta t$ is the transform of u^i by the infinitesimal transformation of the adjoint group determined by v. In view of the preceding theorem the equation obtained on replacing u^i by $u^i + \bar{u}^i \delta t$ holds and consequently

$$(41.3) \qquad C_j^h \colon\colon\colon{}_{ki}^{l}\, \bar{u}^i = 0.$$

From (40.7) it follows that $\bar{u}^i$ arises from the commutator of $u^a X_a f$ and $v^b X_b f$. Hence if equation (41.1) admits exactly $(m-1)$ solutions other than u^i, and v^i is any vector of E_r, then since $\bar{u}^i$ satisfies (41.3), it follows that the m sets of solutions of (41.1) determine an invariant sub-group G_m. Hence we have:

[41.2] *If equations of the form* (41.1) *admit m independent sets of solutions, they determine an invariant sub-group G_m.*

Consider the equations

$$(41.4) \qquad c_{ij}^{k} u^i = 0$$

which admit solutions, if the rank of the matrix $\lVert c_{ij}^{k} \rVert$, where i indicates the column and j and k the row is of rank less than r. If this rank is $r - m$, then the group G_r admits m exceptional sub-groups G_1 (§29), and these determine an invariant sub-group G_m. Moreover, G_m is Abelian and consequently integrable.

Again if $c_{ij}^j \neq 0$, the equation

$$(41.5) \qquad\qquad c_{ij}^j u^i = 0$$

admits $r - 1$ sets of independent solutions. Hence we have:

[41.3] *If $c_{ij}^j \neq 0$, the group G_r admits an invariant sub-group of order $r - 1$.*

As a corollary we have:

[41.4] *For a simple group $c_{ij}^j = 0$.*

Another important invariant sub-group is defined by

$$(41.6) \qquad\qquad g_{ij} u^j \equiv c_{ih}^l c_{jl}^h u^j = 0.$$

Evidently it admits as a sub-group the sub-group of G_r defined by (41.4), if the latter exists. Since the latter is an invariant sub-group of G_r, it is an invariant sub-group of the sub-group defined by (41.6).

An invariant sub-group is defined also by

$$(41.7) \qquad\qquad c_{ijk} u^j = 0,$$

where

$$(41.8) \qquad\qquad c_{ijk} = c_{ij}^h g_{hk} = c_{ij}^h c_{hl}^m c_{km}^l.$$

From the definition it follows that c_{ijk} is skew-symmetric in i and j. That it is skew-symmetric in j and k follows from the following expression which is equivalent to the right-hand member of (41.8) in consequence of the Jacobi relations (7.4):

$$c_{km}^l (c_{li}^h c_{jh}^m + c_{jl}^h c_{ih}^m) = c_{li}^h (c_{jh}^m c_{km}^l - c_{kh}^m c_{jm}^l).$$

Consequently c_{ijk} is skew-symmetric in all its indices.

42. Integrable groups. If a G_r is integrable and the adjoint group is of order r, the adjoint group is integrable, since we have equations of the form (36.1) in the generators $E_a f$. If the order of the adjoint group is $s(<r)$, that is, if G_r admits $r - s$ independent exceptional sub-groups (cf. §29), we choose as the generators of the adjoint group the first s of the $E_a f$ which are independent (constant coefficients), and we denote them by $E_t f$. If we take any two of this limited set, say E_t and E_u $(u > t)$, then $(E_t, E_u)f = c_{tu}^l E_l f$, where in consequence of (36.1) $l < u$. If l is a value not taken on by t, then $E_l f = a_l^v E_v f$, where the a's are constants and v takes on the values of the set taken on by t which are less than l. Hence we have

$$(E_t, E_u)f = d_{tu}^v E_v f,$$

where $u > t$, $v < u$ and the d's are at most linear combinations (constant coefficients) of c_{tu}^b where $b \leqslant v$. Consequently conditions analogous to (36.1) are satisfied, and we have:

[42.1] *The adjoint group of an integrable group is integrable.**

Since the adjoint group is linear homogeneous, we have in consequence of theorem [37.5] the following theorem of Lie:†

[42.2] *If a group G_r is integrable, in the representative vector-space of the adjoint group there is at least one sequence of invariant sub-spaces $E_1, \ldots, E_{r-1}$ such that any E_κ is contained in $E_{\kappa+1}$.*

Consider any one of these invariant sub-spaces, say E_h. To the vectors of E_h and $E_{h+\kappa}$ for $\kappa = 1, \cdots, r - h$ there correspond sub-groups G_h and $G_{h+\kappa}$ of G_r, and since $E_{h+\kappa}$ contains E_h, G_h is a sub-group of $G_{h+\kappa}$. Moreover, by theorem [39.7], since E_h is invariant under all the transformations of the adjoint group, G_h is an invariant sub-group of G_r and consequently of $G_{h+\kappa}$. Hence we have the theorem of Lie:‡

[42.3] *If a group G_r is integrable, it admits a sequence of sub-groups $G_1, G_2, \ldots, G_{r-1}$ such that G_h for any h is an invariant sub-group of $G_{h+\kappa}$, for $\kappa = 1, \cdots, r - h$.*

Hence the generators of an integrable group can be chosen so that

$$(42.1) \qquad (X_h, X_{h+\kappa})f = c_{h\,h+\kappa}^l X_l f$$

$$\begin{pmatrix} h = 1, \cdots, r - 1; l = 1, \cdots, h; \\ \kappa = 1, \cdots, r - h \end{pmatrix},$$

from which it follows that

$$(42.2) \qquad c_{ab}^i = 0 \text{ for } i > a \text{ or } b.$$

---

* *Lie-Scheffers*, 1893, 1, p. 536.
† 1893, 1, p. 536.
‡ 1893, 1, p. 537.

When these conditions are satisfied, we have

$$\eta_i^k(u) = c_{ij}^k u^i = 0 \text{ for } k > j.$$

Hence the terms of the matrix $\|\eta_j^k\|$ below the main diagonal are zero, and accordingly the roots of the characteristic equation are $u^i c_{i1}^1, \ldots, u^i c_{ir}^r$. Since these roots are invariant under the adjoint group, we have

$$c_{ij}^a u^i \frac{\partial}{\partial u^a}(c_{\kappa e}^e u^\kappa) = c_{ij}^a u^i c_{ae}^e = 0 \qquad (e \text{ not summed}).$$

Consequently

$$c_{ij}^a u^i v^j c_{ae}^e = 0 \qquad\qquad (e \text{ not summed}),$$

and since $c_{ij}^a u^i v^j$ is a vector of the derived group, and any vector of the derived group is a linear expression of such vectors, we have:

[42.4] *For an integrable group the roots of the characteristic equation with respect to any vector of the derived group are all zero.*

Since $g_{ij}u^i u^j$ is equal to the sum of the squares of the roots of the characteristic equation of $\eta(u)$, we have the necessity of the following theorem of Cartan:*

[42.5] *A necessary and sufficient condition that a group be integrable is that $g_{ij}u^i u^j$ be zero for each vector of its derived group.*

That this condition is sufficient will be shown in §44.

We shall prove by induction the following theorem of Engel:†

[42.6] *If all the roots of the characteristic equation for an arbitrary vector of a group are zero, the group is integrable.*

We assume that the given group admits an integrable sub-group G_m, and choose the generators of G_r so that $X_1 f, X_2 f, \ldots, X_m f$ are the generators of G_m. We denote by E_m the vector-space of G_m and denote by $\bar{E}_{r-m}$ the projection of E_r with respect to E_m. Since G_m is integrable, the transformations of the adjoint group for vectors of E_m leave E_m invariant and at least one direction of $\bar{E}_{r-m}$ by the results of §37. If we adjoin this direction to E_m and denote by E_{m+1} the resulting sub-space, we have that E_{m+1} is invariant for the

---

* 1894, 1, p. 47.

† *Lie-Engel*, 2, p. 774; cf. *Killing*, 1888, 3, p. 289.

transformations of the adjoint group for vectors of E_{m+1}, and consequently these vectors represent a sub-group G_{m+1} of G_r. In accordance with the results of §37, when a particular vector u of E_m is used and v is a vector in E_{m+1} but not in E_m we have

$$c_{ij}^k u^i v^j = \rho(u) v^k \qquad (\text{mod } E_m),$$

where $\rho(u)$ is a root of the characteristic equation for G_r with respect to u, and by hypothesis is zero. Hence for all the vectors u of E_m we have

$$c_{ij}^k u^i v^j = 0 \qquad (\text{mod } E_m)$$

and consequently G_m is an invariant sub-group of G_{m+1}. Since any G_1 is integrable, we can by the repeated application of the above result establish the existence of a sequence of sub-groups as in (35.5), each invariant with respect to the sub-group of one higher order, and hence the theorem is proved.

43. The root-spaces of the matrix $\eta(u_0)$ for a regular vector u_0. The classification of continuous groups is based upon the canonical form to which the constants of structure of the group are reducible by a suitable choice of the basis. This is accomplished by a study of the matrix $\|\eta_b^a(u_0)\| = \|c_{ib}^a u_0^i\|$, where u_0 is a regular vector of the group, that is, one for which the generic roots $0, \rho_1, \ldots, \rho_p$ of (40.1) are different. From §38 it follows that the parameters u^a may be subjected to a linear homogeneous transformation such that in the new coordinate system the matrix $\|\eta_b^a(u_0)\|$ assumes the form (here the superscript indicates the row and the subscript the column):

(43.1)

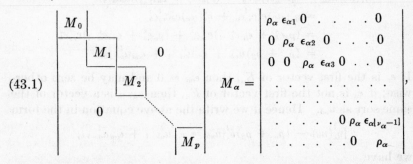

where $\rho_0 = 0$, and the ϵ's are one or zero, as the case may be. To each root there corresponds a vector sub-space of E_r of the same order as the root which we call the *root-space* for this root. If in any system of parameters we denote by $e_1, \ldots, e_{\nu_0}$ the fundamental vectors of E_0 of the root zero and by $e_{\nu_0+1}, \ldots, e_{\nu_0+\nu_1}$ those of E_1 for the root ρ_1, and so on, then in the particular system of parameters for which (43.1) holds these vectors have the components δ_a^i, where a takes the values for the given minors. From this we have that, if e_a and e_{a+1} are consecutive fundamental vectors of the root-space of the root ρ_α, where a takes the range of values for the matrix M_α, then (§38)

$$\eta(u_0)e_{a+1} = \rho_\alpha e_{a+1} + \epsilon_{\alpha\beta}e_a,$$

where $\epsilon_{\alpha\beta}$ is the ϵ for the row of the vector e_a. This vector equation holds in any parameter system. Since $\eta_b^a(u_0)u_0^b = 0$, it follows that u_0 may be taken as the first fundamental vector of the root-space E_0.

Suppose that e_a^i and e_b^i are fundamental vectors of the root-spaces E_α and E_β respectively, which may be the same or different root-spaces; this means that a and b take values appropriate to these respective root-spaces. Then $e_a^i X_i f$ and $e_b^i X_i f$ are the generators of two sub-groups G_1 of G_r. Now we have

$$(e_a^i X_i, e_b^j X_j)f = u_{ab}^k X_k f,$$

where

$$u_{ab}^k = c_{ij}^k e_a^i e_b^j,$$

and consequently are the components of a vector of E_r, unless they are all zero. We seek the sub-space of this vector. We have

$$
\begin{aligned}
\eta_\kappa^l(u_0)u_{ab}^\kappa &= c_{h\kappa}^l u_0^h c_{ij}^\kappa e_a^i e_b^j = (c_{jh}^\kappa c_{\kappa i}^l + c_{h i}^\kappa c_{\kappa j}^l)u_0^h e_a^i e_b^j \\
&= \eta_j^\kappa(u_0)e_b^j c_{i\kappa}^l e_a^i + \eta_i^\kappa(u_0)e_a^i c_{\kappa j}^l e_b^j \\
&= (\rho_\beta e_b^\kappa + \epsilon_{\beta c}e_{b-1}^\kappa)c_{i\kappa}^l e_a^i + (\rho_\alpha e_a^\kappa + \epsilon_{\alpha d}e_{a-1}^\kappa)c_{\kappa j}^l e_b^j \\
&= (\rho_\alpha + \rho_\beta)u_{ab}^l + \epsilon_{\beta c}u_{ab-1}^l + \epsilon_{\alpha d}u_{a-1b}^l.
\end{aligned}
$$

If e_a is the first vector of E_α, then $\epsilon_{\alpha d} = 0$ and may be zero otherwise; if e_a is not the first vector of E_α, then u_{a-1b} is a vector of the same sort as u_{ab}. Hence if we write the above equation in the form

$$[\eta_\kappa^l(u_0) - (\rho_\alpha + \rho_\beta)\delta_\kappa^l]u_{ab}^\kappa = \epsilon_{\beta c}u_{ab-1}^l + \epsilon_{\alpha d}u_{a-1b}^l,$$

we have

$$[\eta_\kappa^l(u_0) - (\rho_\alpha + \rho_\beta)\delta_\kappa^l]^2 u_{ab}^\kappa = \epsilon_{\beta c}\epsilon_{\beta c_1}u_{ab-2}^l + (\epsilon_{\beta c}\epsilon_{\alpha d_1} + \epsilon_{\beta c_2}\epsilon_{\alpha d})u_{a-1b-1}^l + \epsilon_{\alpha d}\epsilon_{\alpha d_2}u_{a-2b}^l.$$

Consequently we have

$$[\eta_\kappa^l(u_0) - (\rho_\alpha + \rho_\beta)\delta_\kappa^l]^p u_{ab}^\kappa = 0,$$

where $p \leqslant \nu_\alpha + \nu_\beta$, since the ultimate effect of repeating the operation is to operate on the first vectors of the root-space and get ϵ's which are zero. A similar result will follow, if in place of fundamental vectors e_a and e_b we take any vectors of E_α and E_β respectively, since any such vector is a linear combination (constant coefficients) of the fundamental vectors of the root-space. From this result and theorems [38.2] and [38.3] we have:*

[43.1] *If u_α and u_β are vectors of the root-spaces E_α and E_β respectively, which may be different root-spaces or the same one, the quantities*

$$(43.2) \qquad u_{\alpha\beta}^k = c_{ij}^k u_\alpha^i u_\beta^j, \qquad\qquad u_{\alpha\beta} \equiv (u_\alpha, u_\beta)$$

are zero, if $\rho_\alpha + \rho_\beta$ is not a root, and, if they are not zero, then $\rho_\alpha + \rho_\beta$ is a root and they are the components of a vector of the root-space of this root; moreover, if $\rho_\alpha + \rho_\beta$ is a root, this vector, which may be a zero vector, is a vector of the root-space of this root.

Another way of stating this result is as follows:

[43.2] *If the matrix $\eta(u_\alpha)$ for a vector u_α of the root-space E_α operates on a vector of E_β, the resulting vector, if it is not zero, is a vector of the root-space for $\rho_\alpha + \rho_\beta$.*

When in particular u_α and u_β are vectors of E_0, so also is (u_α, u_β). Hence we have:

[43.3] *The generators of G_r for the vectors of the root-space E_0 for the matrix $\eta(u_0)$ generate a sub-group of G_r.*

We call this the sub-group U.

Again if u is a vector of E_0 and u_α of any other root-space E_α, then (u, u_α) is a vector of E_α, that is, the root-spaces for $\eta(u)$ are the same as for $\eta(u_0)$; consequently $\eta(u)$ operating on a vector of E_α yields a vector of this root-space. In the special parameter system yielding (43.1) the fundamental vectors of E_α have the components δ_a^i, where a takes the values appropriate to E_α. Then $\eta_j^i(u)\delta_a^j = \lambda_a^b\delta_b^i$, where the λ's are constants and b takes the values for E_α. From these equations it follows that $\eta_a^i(u) = \lambda_a^b\delta_b^i$ and

---

* *Cartan*, 1894, 1, p. 41; *Weyl*, 1926, 2, p. 357.

consequently $\eta_a^i(u) = 0$, unless i takes the values on the same range as a. Hence the matrix $\eta(u)$ consists only of principal minors, and each minor is of the same order as for $\eta(u_0)$, but none of the terms of one of these minors is necessarily zero. Thus the matrix $\eta(u)$ is of the form (43.1), where now

$$(43.3) \qquad M_\alpha = \begin{vmatrix} \eta_{a+1}^{a+1} & \eta_{a+2}^{a+1} & \cdots & \eta_{a+\nu_\alpha}^{a+1} \\ \cdots & \cdots & \cdots & \cdots \\ \cdots & \cdots & \cdots & \cdots \\ \cdots & \cdots & \cdots & \cdots \\ \eta_{a+1}^{a+\nu_\alpha} & & & \eta_{a+\nu_\alpha}^{a+\nu_\alpha} \end{vmatrix} \qquad \begin{array}{l} (a = \nu_0 + \nu_1 + \cdots \\ \qquad + \nu_{\alpha-1}). \end{array}$$

Accordingly the characteristic determinant of $\eta(u)$ is equal to the product of the characteristic determinants of these principal minors, and the order of any determinant is equal to the order of the corresponding root-space. Hence the roots of the equation obtained by equating to zero the characteristic determinant of the matrix M_α are generic roots of the equation (40.2), and in fact all the roots are equal. For suppose that two of them were different so that the equation were of the form $(\rho - \rho_\alpha(u))^\mu(\rho - \bar\rho_\alpha(u))^\nu$. But when u is replaced by u_0, the determinant is $(\rho - \rho_\alpha(u_0))^{\nu_\alpha}$, and consequently $\rho_\alpha(u_0) = \bar\rho_\alpha(u_0)$, which is contrary to the hypothesis that u_0 is a regular vector. Hence for any vector u of the sub-group U, the characteristic determinants of the respective minors of the matrix $\eta(u)$ are

$$(43.4) \qquad \rho^{\nu_0}, \ (\rho - \rho_1(u))^{\nu_1}, \ \cdots, \ (\rho - \rho_p(u))^{\nu_p}.$$

It should be remarked that if u is a special vector two or more of the roots are equal. However, the vectors corresponding to each minor are the same in all cases. But only when u is regular will we speak of them as the root-spaces of $\eta(u)$. Accordingly we have:

[43.4] *The root-spaces for the matrix* $\eta(u)$, *where* u *is a regular vector of the sub-group* U, *are the same as for the matrix* $\eta(u_0)$.

Since the trace of the matrix M_α is equal to $\nu_\alpha\rho_\alpha$, we have

$$(43.5) \qquad \nu_\alpha\rho_\alpha(u) = \eta_a^a(u) = c_{ba}^a u^b \qquad (\alpha \text{ not summed}),$$

where a takes the values appropriate to the root-space E_α, and b takes the values $1, \ldots, \nu_0$, since the vector u is a linear combination of the vectors of components δ_b^i for $b = 1, \cdots, \nu_0$. Hence we have:

[43.5]   *Each non-zero root of the characteristic equation for* $\eta(u)$ *is a linear homogeneous function of the components of* u.*

Incidentally we have that if (u, u_a) is zero, where u_a is a vector of the root-space E_α and u is a vector of the sub-group U, then u is not a regular vector. For, if u were a regular vector, then u_a is a vector of its zero root-space and the latter would be of order greater than ν_0. Hence we have:

[43.6]   *If* $\eta(u)$, *where* u *is a regular vector of the sub-group* U, *operates on a vector of any root-space, the resulting vector is in this root-space.*

Hence theorem [43.1] holds for the matrix $\eta(u)$, where u is any regular vector of the sub-group U.

Since the roots of the characteristic equation for the matrix M_0 are zero, we have in consequence of theorem [42.6]:

[43.7]   *The sub-group* U *as determined by any regular vector of a group* G_r *is integrable.*

Since U is integrable, its derived group is either the identity, or an invariant sub-group of U. The vectors of this sub-group, if it exists, are linear combinations of vectors of the form $\bar{u} = (u, u')$, where u and u' are vectors of E_0. Since each root-space is invariant under $\eta(u)$ and $\eta(u')$, the trace of the minor of $\eta(\bar{u})$ for the root-space of any root is zero by theorem [40.1], that is, $\nu_\alpha \rho_\alpha(\bar{u}) = 0$ (α not summed), and consequently all the roots are zero. Hence we have:

[43.8]   *The characteristic equation of a group with respect to a vector of the derived group of* U *is* $\rho^r = 0$.†

44. Canonical form of the matrix $\eta(u)$. **Cartan's criterion for integrable groups.** Since the sub-group U is integrable, the corresponding sub-group of the adjoint group is integrable, and in consequence of theorem [37.5] when this sub-group is applied to any one of the root-spaces, say the one of root ρ_α, there is an invariant

---

* *Cartan*, 1894, 1, p. 38; *Weyl*, 1926, 2, p. 358.
† *Cartan*, 1894, 1, p. 42.

direction, E_1, an invariant E_2 through it and so on.   If we take the first of these as the first fundamental direction of the root-space, a vector determining with the latter the invariant E_2 as the second fundamental vector, and so on, then the principal minor of the matrix $\eta(u)$ for the root-space E_α is of the form (cf. §37)

$$
(44.1) \qquad
\begin{vmatrix}
\rho_\alpha & & & & \\
0 & \rho_\alpha & & & \\
0 & 0 & \rho_\alpha & & \\
\cdot & \cdot & \cdot & \cdot & \cdot & \cdot & \cdot \\
\cdot & \cdot & \cdot & \cdot & \cdot & \cdot & \cdot \\
0 & \cdot & \cdot & \cdot & 0 & \rho_\alpha
\end{vmatrix},
$$

where the terms above the main diagonal do not enter into the discussion.

Since the vector-spaces E_1, E_2, . . . in E_α, which led to the above result are invariant for the sub-group of the adjoint group corresponding to the sub-group U, they are invariant for every one-parameter group of this sub-group of the adjoint group, whether the vector of this one-parameter group be regular or special.   We have seen (§43) that if u_α is any vector of the root-space E_α and $(\bar{u}, u_\alpha) = 0$, then $\bar{u}$ is a special vector of the sub-group U.   Suppose this condition holds for all vectors u_α of this root-space.   Then in particular it holds for the first fundamental vector e_a.   But for a regular vector u of U, we have

$$(u, e_a) = \rho_\alpha(u)e_a.$$

Since by hypothesis $(\bar{u}, e_a) = 0$ and the vector e_a is independent of u, it follows that $\rho_\alpha(\bar{u}) = 0$.   Hence we have:

[44.1]   *If $\bar{u}$ is a vector of the sub-group U such that $(\bar{u}, u_\alpha) = 0$ for all the vectors u_α of the root-space E_α, then $\rho_\alpha(\bar{u}) = 0$, that is, $\bar{u}$ is a special vector of U.*

If e', e'', . . . , $e^{(\nu_\alpha)}$ are the fundamental vectors of the root-space of the root ρ_α, we have (cf. §37)

$$(\eta(u) - I\rho_\alpha)e' = 0, \qquad (\eta(u) - I\rho_\alpha)^2 e'' = 0, \cdots,$$
$$(\eta(u) - I\rho_\alpha)^{\nu_\alpha} e^{(\nu_\alpha)} = 0.$$

These equations are equivalent to

$$e' = \eta(u)\rho_\alpha^{-1}e',$$
$$e'' = \eta(u)\rho_\alpha^{-2}(2\rho_\alpha - \eta(u))e'' = \eta(u)u_\alpha'',$$
$$e''' = \eta(u)\rho_\alpha^{-3}(3\rho_\alpha^2 - 3\rho_\alpha\eta(u) + \eta^2(u))e''' = \eta(u)u_\alpha''',$$

. . . .

. . . .

where u_α'', u_α''', . . . are vectors of the root-space E_α. Hence if we take any vector of the root-space, that is, a linear combination (constant coefficients) of e', ... , $e^{(\nu_\alpha)}$, we find from the right-hand members of the above equations the vector of the root-space upon which $\eta(u)$ operates to yield the given vector. Hence we have the following converse of theorem [43.6]:

[44.2] *If u_α is a vector of the root-space E_α, there is a unique vector $\bar{u}_\alpha$ of this root-space such that $u_\alpha = \eta(u)\bar{u}_\alpha$, where u is any regular vector of the sub-group U.*

The vectors defined by (43.2) are evidently vectors of the derived group of G_r. From theorem [43.2] we have that, if such a vector belongs to the root-space E_0, it arises either from two vectors of E_0, or from the vectors of two root-spaces whose roots differ only in sign. Suppose we denote two such roots by $\rho_\alpha(u)$ and $\rho_{\alpha'}(u)$, and vectors of these spaces by u_α and $u_{\alpha'}$, respectively, then $\bar{u} = (u_\alpha, u_{\alpha'})$ is a vector of E_0. The characteristic equation for $\eta(\bar{u})$ has the same root-spaces as for $\eta(u)$, and the roots for the spaces of u_α and $u_{\alpha'}$ will differ in sign, since they are $\rho_\alpha(\bar{u})$ and $\rho_{\alpha'}(\bar{u})$, respectively.

If $\rho_\beta(u)$ is any root of the characteristic equation of $\eta(u)$ other than ρ_α or $\rho_{\alpha'}$, and we operate with $\eta(u_\alpha)$ on any vector u_β of the root-space E_β of root ρ_β, we have that the resulting vector is zero, or that it is a vector of the $\rho_\alpha + \rho_\beta$ root-space and that $\rho_\alpha + \rho_\beta$ is a root in accordance with theorem [43.1]. Similarly, if we operate with $\eta(u_{\alpha'})$, we obtain a zero vector, or $\rho_\beta - \rho_\alpha$ is a root. Since

$$\eta(\bar{u}) = \eta(u_\alpha)\eta(u_{\alpha'}) - \eta(u_{\alpha'})\eta(u_\alpha),$$

it follows from the above observations that if neither $\rho_\beta + \rho_\alpha$ nor $\rho_\beta - \rho_\alpha$ is a root of the characteristic equation of $\eta(u)$, then $\eta(\bar{u})$ operating on every vector of the root-space E_β is zero. Hence by theorem [44.1] $\rho_\beta(\bar{u}) = 0$, and $\bar{u}$ is a special vector of the sub-group U.

Suppose then we consider the possible sequence of roots

(44.2) $\rho_\beta - \mu\rho_\alpha, \cdots, \rho_\beta - 2\rho_\alpha, \rho_\beta - \rho_\alpha, \rho_\beta, \rho_\beta + \rho_\alpha,$

$$\rho_\beta + 2\rho_\alpha, \cdots, \rho_\beta + \lambda\rho_\alpha,$$

it being understood that $\rho_\beta + (\lambda + 1)\rho_\alpha$ and $\rho_\beta - (\mu + 1)\rho_\alpha$ are not roots, and denote by $\bar{E}_\beta$ the totality of the root-spaces of these roots. Then $\bar{E}_\beta$ is invariant under $\eta(u_\alpha)$ and $\eta(u_{\alpha'})$. Hence by theorem [40.1], the trace of the matrix $\eta(\bar{u})$ in $\bar{E}_\beta$ is zero. For the root-space of each root the trace is equal to the product of the root by the order of the root. Hence the trace is of the form $a\rho_\beta(\bar{u}) + b\rho_\alpha(\bar{u})$, where a and b are integers. Consequently we have:

[44.3] *If u_α and $u_{\alpha'}$ are vectors of two root-spaces whose roots ρ_α and $\rho_{\alpha'}$ differ only in sign, and $\bar{u} = (u_\alpha, u_{\alpha'})$ is a general vector of U, then $\rho_\beta(\bar{u})$ is a rational multiple of $\rho_\alpha(\bar{u})$.*

We are now in position to establish the sufficiency of Cartan's criterion for an integrable group (theorem [42.5]). We have observed that if a vector $\bar{u}$ of the derived group G'_{r_1} of G_r is in E_0, it is a linear combination of vectors (u, u'), where u and u' are in E_0, and of vectors $(u_\alpha, u_{\alpha'})$, where u_α and $u_{\alpha'}$ are vectors of root-spaces whose roots differ only in sign. The characteristic equation with reference to a vector of the first kind has only zero roots by theorem [43.8]. The non-zero roots of the characteristic equation with respect to one of the vectors of the second type are all rational multiples of one of them. Hence if $g_{ij}u^iu^j$, that is, the sum of the squares of the roots, is zero for a vector of the second kind, all of the roots are zero. From (43.5) it follows that any non-zero root for a vector u which is a linear combination of vectors of E_0 is the sum of the roots of the same root-space for the different vectors of the combination. Hence if $g_{ij}u^iu^j$ is zero for all vectors of the derived group which are in E_0, it follows that all the roots of the characteristic equation for any such vector are zero.

Since G_r is integrable when all the roots of its characteristic equation are zero by theorem [42.6], we have only to consider the case when some of the roots are not zero in which case we have a set of root-spaces for a general vector u_0. If the first derived group G'_{r_1} coincides with G_r, then each vector in E_0 is a vector of the derived group and if $g_{ij}u^iu^j$ vanishes for each of these vectors, then by the preceding paragraph all the roots are zero for the corresponding characteristic equation, and thus in particular for the vector u_0, which is contrary to the hypothesis that not all the roots are zero. Hence G'_{r_1} is a proper sub-group of G_r, and is an invariant sub-group.

Since G'_{r_1} is an invariant sub-group of G_r, if the generators of this sub-group are chosen as in §40, it follows that the sum of the squares of the roots of the characteristic equation of G'_{r_1} is equal to the sum of the squares of the roots of the characteristic equation of G_r for a general vector of G'_{r_1}. Hence we may apply to G'_{r_1} and the second derived group G''_{r_2} the same consideration which showed that G'_{r_1} was a proper sub-group of G_r. Accordingly by continuing the process we get a sequence of derived groups ending in the identity, and consequently G_r is integrable.

The results of the preceding paragraph were due to the fact that G'_{r_1} is a proper invariant sub-group of G_r and did not involve the fact that it is the derived group of G_r. Consequently we have as a corollary the criterion of Cartan:*

[44.4] *A necessary and sufficient condition that an invariant sub-group G_m of a G_r be integrable is that $g_{ij}u^iu^j$ vanish for all the vectors of the derived group of G_m.*

As an immediate consequence of this theorem we have that the invariant sub-groups defined by (41.4) and (41.6), if they exist, are integrable. For, in either case $g_{ij}u^iu^j$ is zero for all vectors of the sub-group, and consequently for its derived group. Consider now the invariant sub-group defined by (41.7), that is,

$$(44.3) \qquad c_{ijk}u^i = 0, \qquad c_{ijk} = c_{ij}^l g_{lk}.$$

If u_1 and u_2 are two vectors of this sub-group, we have

$$g_{ih}c^h_{jk}u_1^j u_2^k = g_{ih}\bar{u}^h = 0.$$

Since $\bar{u}$ is a vector of the derived group of the sub-group, the latter is integrable by the above theorem. Incidentally we have that either the sub-group (44.3) is Abelian, or its derived group is the sub-group (41.6).

45. Semi-simple groups. By definition a *semi-simple* group is one which does not contain an integrable invariant sub-group other than the identity. Since a simple group (§31) is one which has no

* 1894, 1, p. 47.

invariant sub-groups, other than itself and the identity, it follows that a simple group of order greater than one is semi-simple, so that theorems applying to the latter apply also to simple groups.

Since the sub-group (41.4), if it exists, is invariant and integrable, we have that a semi-simple group does not admit any exceptional sub-groups G_1. Since the sub-group (41.6), if it exists, is invariant and integrable (§44), the rank of the matrix $\|g_{ij}\|$ is r for a semi-simple group. Cartan has shown that this condition is also sufficient. In fact, if a G_r admits an integrable invariant sub-group G_m, either the latter is Abelian or one of its derived groups is Abelian. Since all of the derived groups of G_m are invariant sub-groups of G_r by theorem [35.5], in either case there is an invariant sub-group of G_r which is Abelian. Suppose its generators are $X_p f$ ($p = 1, \cdots, t$). Then, since $c_{pa}^{\ h} = 0$ for $a = 1, \cdots, r$ and $h > t$ by (31.3), we have

$$g_{ip} = c_{ik}^{\ l} c_{pl}^{\ k} = c_{iq}^{\ l} c_{pl}^{\ q} = c_{iq}^{\ s} c_{ps}^{\ q} \quad \begin{pmatrix} i = 1, \cdots, r; \\ p, q, s = 1, \cdots, t \end{pmatrix}.$$

But $c_{ps}^{\ q} = 0$, since the sub-group is Abelian, and consequently the rank of the matrix $\|g_{ij}\|$ is less than r. Hence we have the theorem of Cartan:*

[45.1] *A necessary and sufficient condition that a group be semi-simple is that the rank of the matrix $\|g_{ij}\|$ be r.*

When the rank of the matrix $\|g_{ij}\|$ is r, the quantities g_{ij} may be used to endow the vector space E_r with a metric, so that we may speak of lengths of vectors and angles between them. Thus if two vectors u and v are such that $g_{ij}u^i v^j = 0$, we say that they are *orthogonal*. Also if $g_{ij}u^i u^j = 0$, we say that u is a *null vector;* a null vector is self-orthogonal (cf. §47).

As a consequence of the preceding we prove that:

[45.2] *A semi-simple group is the direct product of a set of simple groups, the representations of which in E_r are mutually orthogonal.*

The theorem is trivial when the group is simple, there being only one group in the set, namely the group itself. If the group is semi-simple, by definition it admits at least one invariant non-integrable sub-group. If it admits more than one, there is at least one G_p

---

* 1894, 1, p. 52.

for $p < r$ which is not contained in an invariant sub-group other than G_r; also $p > 1$, otherwise the G_1 being integrable, the group G_r would not be semi-simple. Let u_a for $a = 1, \cdots, p$ be p independent vectors representing G_p, and denote by E_p its representative vector-space. The equations

$$g_{ij}u_a^i v^j = 0$$

define $r - p$ independent vectors orthogonal to E_p which determine an E_{r-p} orthogonal to E_p. Since g_{ij} is invariant under the adjoint group and E_p is invariant, it follows that E_{r-p} is invariant, and consequently represents an invariant sub-group G_{r-p}. E_p and E_{r-p} have one, or more vectors, u in common only in case $g_{ij}u^i u^j = 0$. In this case G_p and G_{r-p} would have in common an invariant sub-group of G_r by theorem [31.3] and it would be integrable in consequence of theorem [44.4] which is contrary to the hypothesis that G_r is semi-simple. Hence G_r is the direct product of G_p and G_{r-p}. Moreover, G_{r-p} is simple; otherwise it would contain an invariant sub-group. The product of the latter and G_p would give an invariant sub-group of G_r containing G_p, contrary to hypothesis.

Since G_p and G_{r-p} are invariant sub-groups of G_r and do not have a sub-group in common other than the identity, and if $X_h f$ for $h = 1, \cdots, p$ is the basis of G_p and $X_s f$ for $s = p + 1, \cdots, r$ the basis of G_{r-p}, then we have (§31)

$$c_{hk}^{\,s} = 0, \qquad c_{st}^{\,h} = 0, \qquad c_{hs}^{\,a} = 0$$

$$\begin{pmatrix} a = 1, \cdots, r; h, k = 1, \cdots, p; \\ s, t = p + 1, \cdots, r \end{pmatrix},$$

and consequently

$$g_{ks} = c_{ka}^{\,b}c_{sb}^{\,a} = c_{kh}^{\,b}c_{sb}^{\,h} = c_{kh}^{\,l}c_{sl}^{\,h} = 0,$$
$$g_{hk} = c_{hl}^{\,m}c_{km}^{\,l} \qquad\qquad (h, k, l, m = 1, \cdots, p).$$

From the first of these equations it follows that the determinant $|g_{ab}|$ for $a, b = 1, \cdots, r$ is equal to the product of the determinants $|g_{hk}|$ and $|g_{st}|$, and consequently both of these determinants are different from zero. From this result and the second equation it follows that G_p is semi-simple. If it is simple, the theorem is proved; if not, we proceed with G_p as with G_r and ultimately establish the theorem.

Consider any vector u of the sub-group U and a vector u_α of the root-space E_α. By theorem [44.2] it follows that there is a vector $\bar{u}_\alpha$ of this root-space such that $\eta(u)\bar{u}_\alpha = u_\alpha$. Hence we have

$$g_{ij}u^i u_\alpha^j = g_{ij}u^i c_{kl}^j u^k \bar{u}_\alpha^l = c_{kli}u^i u^k \bar{u}_\alpha^l = 0,$$

since c_{kli} is skew-symmetric in all of its indices. Accordingly we have:

[45.3] *For a semi-simple group the vectors of any root-space E_α are orthogonal to all the vectors of E_0.*

If u_β is a vector of a root-space other than E_0, we have

$$\begin{aligned}
g_{ij}u_\alpha^i u_\beta^j &= g_{ij}c_{\kappa l}{}^i u^\kappa \bar{u}_\alpha^l u_\beta^j = c_{\kappa lj}u^\kappa \bar{u}_\alpha^l u_\beta^j \\
(45.1) \qquad &= c_{\kappa jl}u^l \bar{u}_\alpha^\kappa u_\beta^j = g_{il}u^l c_{\kappa j}^i \bar{u}_\alpha^\kappa u_\beta^j = g_{il}u^l(\bar{u}_\alpha, u_\beta)^i.
\end{aligned}$$

If $\rho_\alpha + \rho_\beta$ is not a root, then $(\bar{u}_\alpha, u_\beta) = 0$ (§43) and u_α and u_β are orthogonal. If it is a root say $\rho_\gamma \neq 0$, then the last term is $g_{il}u^l u_\gamma^i$, which is zero by theorem [45.3]. Consider finally the case when $\rho_\beta = -\rho_\alpha$, in which case we call it $\rho_{\alpha'}$, as formerly. If this case did not arise, or if it does arise and $g_{ij}u_\alpha^i u_{\alpha'}^j = 0$ for every vector of $E_{\alpha'}$, we should have $g_{ij}u_\alpha^i v^j = 0$, where v is any vector of E_r, since in the above discussion u_α and u_β may be vectors of the same root-space. Hence we should have $g_{ij}u_\alpha^i = 0$, which is impossible since the rank of g_{ij} is r. Consequently we have:

[45.4] *For a semi-simple group if ρ_α is a root, so also is $-\rho_\alpha(\equiv \rho_{\alpha'})$, and for any u_α of the root-space E_α there is a $u_{\alpha'}$ of the root-space $E_{\alpha'}$ such that $g_{ij}u_\alpha^i u_{\alpha'}^j \neq 0$; the vectors of any two root-spaces are orthogonal, except when their roots differ only in sign.*

Since the fundamental vectors of each of the non-zero root-spaces are orthogonal to all of the fundamental vectors of E_0, if we put

$$(45.2) \qquad\qquad \bar{g}_{ij} = g_{\kappa l}e_i^\kappa e_j^l,$$

we have

$$(45.3) \qquad\qquad \bar{g}_{ap} = 0 \quad (a = 1, \cdots, \nu_0;\ p = \nu_0 + 1, \cdots, r).$$

If we define quantities $\bar{g}^{ij}$ by

$$(45.4) \qquad\qquad \bar{g}^{ij}\bar{g}_{jk} = \delta_k^i,$$

we have also

(45.5) $$\bar{g}^{ap} = 0.$$

Referring to (43.5), we see that if we put

(45.6) $$v_{\alpha a} = \frac{c_{ab}^{\;\;b}}{\nu_\alpha}, \qquad v_{\alpha p} = 0$$
$$(a = 1, \cdots, \nu_0; \; p = \nu_0 + 1, \cdots, r),$$

then, since $\rho_\alpha(u)$ is a scalar under change of basis and u is a contra-variant vector, $v_{\alpha i}$ is a covariant vector. If we put

(45.7) $$v_\alpha^{\;i} = \bar{g}^{ij} v_{\alpha j},$$

then $v_\alpha^{\;i}$ are the contravariant components of this vector. Since in consequence of (45.5),

(45.8) $$v_\alpha^{\;p} = \bar{g}^{pi} v_{\alpha i} = \bar{g}^{pa} v_{\alpha a} = 0,$$

we see that v_α is a vector of E_0, which we call the *root-vector* for the root ρ_α. Then (43.5) assumes the form

(45.9) $$\rho_\alpha(u) = v_{\alpha a} u^a = \bar{g}_{ab} u^a v_\alpha^{\;b}.$$

In the special coordinate system of §43 $\bar{g}_{ab} = g_{ab}$, so that the sum of the squares of the roots of the characteristic equation for a vector u of the sub-group U is $\bar{g}_{ab} u^a u^b$. Also from (45.9) we have that the sum of the squares of the roots is $\sum_\alpha^{1, \ldots, p} v_\alpha v_{\alpha a} v_{\alpha b} u^a u^b$. Since these two expressions must be equal for all vectors u of U, we have

$$\bar{g}_{ab} = \sum_\alpha' \nu_\alpha v_{\alpha a} v_{\alpha b}.$$

The rank of the matrix $\|\bar{g}_{ab}\|$ is ν_0 and this is at most equal to the rank of $\|v_{\alpha b}\|$, that is, the number of independent root-vectors, say l.[*] However, since these root-vectors are in E_0, l cannot be greater than ν_0. Hence we have:

[45.5] *The number of linearly independent root-vectors is equal to the order of the sub-space E_0, that is, the number of zero roots of the characteristic equation for a regular vector.*

---

[*] Cf, *Bôcher*, 1907, 1, p. 79.

From (45.9) it follows that a necessary and sufficient condition that there be a vector u of E_0 such that all the roots $\rho_\alpha(u)$ are zero is that $v_{\alpha a}u^a = 0$ for every α. This is impossible since the rank of $\|v_{\alpha a}\|$ is ν_0. Since all the roots of the characteristic equation for a vector of the derived group of U are zero by theorem [43.8] unless it is the identity, we have:

[45.6] *The sub-group U is Abelian.*

We have seen that if ρ_α is a root, so also is $-\rho_\alpha(\equiv \rho_{\alpha'})$. Consequently there are in E_0 vectors $\bar{u} = (u_\alpha, u_{\alpha'})$. None of these are null vectors. For, if one were, the sum of the squares of the roots of its characteristic equation would be zero by (40.5) and hence each root would be zero. Then $\bar{u}$ would be orthogonal to all the root-vectors by (45.9), which has just been seen to be impossible.

Let e_α be the first fundamental vector of the root-space E_α, in which case we have

$$(45.10) \qquad \eta_i^\kappa(u)e_\alpha^i = \rho_\alpha(u)e_\alpha^\kappa.$$

We know that there is at least one fundamental vector of the root-space $E_{\alpha'}$ not orthogonal to e_α, say $e_{\alpha'}$. If we multiply the above equation by $g_{\kappa l}e_{\alpha'}^l$ and sum for κ, we have

$$c_{hil}u^h e_\alpha^i e_{\alpha'}^l = \rho_\alpha g_{\kappa l}e_{\alpha'}^l e_\alpha^\kappa \equiv \rho_\alpha \sigma_\alpha \quad (\alpha \text{ not summed}),$$

where σ_α is a non-zero scalar. When this is written

$$g_{h\kappa}u^h \bar{u}^\kappa = \rho_\alpha(u)\sigma_\alpha, \qquad (\alpha \text{ not summed}),$$

where $\bar{u} = (e_\alpha, e_{\alpha'})$, we see that $\bar{u}$ is not a zero vector. Comparing this equation with (45.9) and observing that for the basis of §43 $\bar{g}_{hk} = g_{hk}$, we see that

$$(45.11) \qquad \bar{u} = (e_\alpha, e_{\alpha'}) = \sigma_\alpha v_\alpha \qquad (\alpha \text{ not summed}),$$

that is, $\bar{u}$ and v_α have the same direction. Furthermore, as shown in the preceding paragraph $g_{ij}\bar{u}^i\bar{u}^j \neq 0$. In consequence of (45.11) and (45.9) for the vector $\bar{u}$ we have

$$\sigma_\alpha \rho_\alpha(\bar{u}) = \bar{g}_{ab}\bar{u}^a\bar{u}^b \qquad (\alpha \text{ not summed})$$

and hence, since $\bar{u}$ is not a null vector,

$$\rho_\alpha(\bar{u}) \neq 0.$$

We are now in position to prove the following important theorem:

[45.7] *The non-zero roots of the characteristic equation of a semi-simple group are simple, and if ρ_α is a root, $m\rho_\alpha$, where m is an integer, is not a root unless $m = -1$.*[*]

For suppose that $\rho_\alpha, 2\rho_\alpha, \cdots, s\rho_\alpha$ and $(s + t)\rho_\alpha, (s + t + 1)\rho_\alpha, \cdots, (s + w)\rho_\alpha$ are roots, but not $(s + 1)\rho_\alpha, \cdots, (s + t - 1)\rho_\alpha$ and $(s + w + 1)\rho_\alpha$, and that the respective multiplicities of these roots are $\kappa_1, \kappa_2, \ldots, \kappa_s, \kappa_{s+t}, \ldots, \kappa_{s+w}$. We consider now the sum of the Jacobi relations

$$c_{aa'}^{i}c_{ip}^{p} + c_{a'p}^{i}c_{ja}^{p} + c_{pa}^{i}c_{ja'}^{p} = 0,$$

where a and a' are the indices of the fundamental vectors e_α and $e_{\alpha'}$ referred to above in the set of fundamental vectors of E_r, and p takes on the values of this set for the root-spaces of the above roots. Since $\bar{u}$ defined by (45.11) is in E_0, we have $c_{aa'}^{q} = 0$ for $q > \nu_0$. Similarly, if u_p is a vector of the root-space of a root $m\rho_\alpha$ then $(e_{\alpha'}, u_p)$ is a vector in the root-space of $(m - 1)\rho_\alpha$, if this is a root, and (e_α, u_p) is a vector of the root-space $(m + 1)\rho_\alpha$, if this is a root. Accordingly the above equation reduces to

$$c_{aa'}^{b}c_{bp}^{p} + c_{a'v}^{u}c_{ua}^{v} + c_{v'a}^{u'}c_{u'a'}^{v'} = 0 \quad (b = 1, \cdots, \nu_0),$$

where u, v, u', v' run through the indices belonging to the root-spaces of the following roots;

$$u:0, \rho_\alpha, \cdots, (s - 1)\rho_\alpha; \qquad (s + t)\rho_\alpha, \cdots, (s + w - 1)\rho_\alpha;$$
$$v:\rho_\alpha, \cdots, s\rho_\alpha \qquad ; \qquad (s + t + 1)\rho_\alpha, \cdots, (s + w)\rho_\alpha;$$
$$u':2\rho_\alpha, \cdots, s\rho_\alpha \qquad ; \qquad (s + t + 1)\rho_\alpha, \cdots, (s + w)\rho_\alpha;$$
$$v':\rho_\alpha, \cdots, (s - 1)\rho_\alpha \qquad ; \qquad (s + t)\rho_\alpha, \cdots, (s + w - 1)\rho_\alpha.$$

In consequence of these ranges and the skew-symmetry of the c's the above equation reduces to

$$c_{aa'}^{b}c_{bp}^{p} = c_{ab}^{v}c_{a'v}^{b} \quad (b = 1, \cdots, \nu_0).$$

Since $e_\alpha^{i} = \delta_\alpha^{i}$, it follows from (45.10) that $c_{ab}^{k} = 0$ unless $k = a$, and consequently the above equation reduces to

(45.12) $$c_{aa'}^{b}c_{bp}^{p} = c_{ba}^{a}c_{aa'}^{b} \quad (b = 1, \cdots, \nu_0).$$

The vector $\bar{u}$ defined by (45.11) is equal to $c_{aa'}^{b}$. Consequently the left-hand member of (45.12) is the trace of the matrix of $\eta(\bar{u})$ in the root-spaces of the given set of roots, which by (43.5) is equal to $[\kappa_1 + 2\kappa_2 + \cdots + s\kappa_s + (s + t)\kappa_{s+t} + \cdots + (s + w)\kappa_{s+w}]\rho_\alpha(\bar{u})$.

---

[*] *Cartan, 1894, 1, p. 55; Weyl 1926, 2, p. 364.*

From (45.10) for u replaced by $\bar{u}$ we have that the right-hand member of (45.12) is $\rho_\alpha(\bar{u})$. Equating these two expressions, we have that $\kappa_1 = 1$ and that the other κ's are zero. If there had been more than two groups of roots, the result would not have been altered and hence the theorem is proved.

46. Classification of the structure of simple and of semi-simple groups. Many of the results which have been presented in this chapter were derived in the process of the classification of semi-simple groups and in the determination of the normal forms of the members of the various classes. In 1885–1889 Lie* showed that there were four large classes of simple groups, those with the same structure as the general projective group in n variables; as the projective group of a linear complex in $2n - 1$ variables; and as the projective group of a hypersurface of the second degree in $2n$ and $2n - 1$ variables (cf. Exs. 14, 15, 16, p. 184). In 1889 Killing,† by making use of the characteristic equation of a group (§40), showed that in addition to these four classes of simple groups, there are only five other possible structures of simple groups, their orders being 14, 52, 78, 133 and 248. Cartan,‡ after deriving many of the results referred to in the preceding sections, established the existence of these five special types and showed that they are distinct. Weyl§ studied the same problem and gave a more geometrical setting to the problem. We shall give an outline of certain consequences of the results of the preceding sections which serve as a geometrical basis for classification.

We denote by $e_a^i (= \delta_a^i)$ for $a = 1, \cdots, l (= \nu_0)$ the independent vectors which determine the root-space E_0 of the zero root of the characteristic equation for a general vector. By theorem [45.7] the root-space of a non-zero root is one-dimensional, and if ρ_α is a root, so also is $-\rho_\alpha (\equiv \rho_{\alpha'})$, that is, the roots go in pairs. We indicate by $e_\alpha^i = \delta_\alpha^i$ and $e_{\alpha'}^i (= \delta_{\alpha'}^i)$ vectors of these root-spaces and understand that the indices α and α' for the various roots are chosen, so that $\alpha, \cdots$, take the values $l + 1$ to r. If we put (cf. §45)

$$(46.1) \qquad u_\alpha = (e_\alpha, e_{\alpha'}), \qquad g_{ij} e_\alpha^i e_{\alpha'}^j = \sigma_\alpha,$$

---

* 1885, 1, p. 130; 1886, 1, p. 413; 1889, 3, p. 325.

† 1889, 1, 48.

‡ 1894, 1, pp. 68–95.

§ 1925, 4; 1926, 2.

then by (45.11)

$$(46.2) \qquad u_\alpha = \sigma_\alpha v_\alpha \qquad (\alpha \text{ not summed}),$$

where the vector v_α defined by (45.6) and (45.7) with $\nu_\alpha = 1$ lies in E_0 and is the root-vector for the pairs of roots ρ_α and $-\rho_\alpha$. By a suitable choice of bases we make $\sigma_\alpha = -1$. From the results preceding theorem [45.5], from equations (46.1) and from theorem [45.4] we have

$$(46.3) \qquad g_{ab} = \sum_\alpha v_{\alpha a} v_{\alpha b}, \qquad g_{a\alpha} = 0, \qquad g_{\alpha\alpha'} = -1, \qquad g_{\alpha\beta} = 0$$

$$\begin{pmatrix} a = 1, \cdots, l; \\ \alpha, \beta = l + 1, \cdots, r; \beta \neq \alpha' \end{pmatrix},$$

from which we get

$$(46.4) \qquad g^{ab} = \sum_\alpha v_\alpha^a v_\alpha^b = \sum_\alpha u_\alpha^a u_\alpha^b, g^{a\alpha} = 0, g^{\alpha\beta} = 0 (\beta \neq \alpha'), g^{\alpha\alpha'} = -1.$$

Since the sub-group U is Abelian by theorem [45.6], we have $c_{ab}^e = 0$ for $a, b, e = 1, \cdots, l$, and by (41.8) and (46.3) $c_{abe} = 0$. From (46.1) and (46.2) we have

$$(46.5) \qquad u_\alpha^a = c_{\alpha\alpha'}^a = -v_\alpha^a, \qquad c_{\alpha\alpha'}^\beta = 0 \quad (\beta = l + 1, \cdots, r).$$

Hence we have

$$(46.6) \qquad c_{\alpha\alpha'a} = u_{\alpha a} = -v_{\alpha a},$$

and in consequence of (46.4)

$$(46.7) \qquad c_{\alpha\alpha}^\alpha = v_{\alpha a} = -u_{\alpha a} \qquad (\alpha \text{ not summed}).$$

By theorem [43.1] $c_{\alpha\beta}^\gamma = 0$ unless $\rho_\alpha + \rho_\beta = \rho_\gamma$. From this result and (46.4) we have $c_{\alpha\beta\gamma} = 0$ unless $\rho_\alpha + \rho_\beta + \rho_\gamma = 0$. It remains to be shown that these quantities are not zero under these conditions. In fact, Weyl,[*] by making use of a sequence of roots (44.2) has shown that

$$c_{\alpha\beta\gamma} c_{\alpha'\beta'\gamma'} = -\tfrac{1}{2}\lambda(\mu + 1)\rho_\alpha(u_\alpha).$$

Furthermore, Weyl[†] showed that by a suitable choice of basis we have $c_{\alpha'\beta'\gamma'} = c_{\alpha\beta\gamma}$, and consequently

$$(46.8) \qquad (c_{\alpha\beta\gamma})^2 = -\tfrac{1}{2}\lambda(\mu + 1)\rho_\alpha(u_\alpha).$$

---

[*] 1926, 2, p. 372.

[†] 1926, 2, p. 374.

It can be established that $\rho_\alpha(u_\alpha)$ is negative, so that $c_{\alpha\beta\gamma}$ is real, and Weyl* has shown that all the root-vectors are linear combinations with rational coefficients of any l independent ones. When the basis is chosen so that these are real, all of the coefficients $c_{ij}^{\ k}$ are real, and $g_{ab}u^a u^b$ is positive definite.

Referring again to a sequence (44.2) and remarking that each of these roots is simple, we have from the argument following (44.2) that

$$(\lambda + \mu + 1)\rho_\beta(u_\alpha) + \left(\frac{\lambda(\lambda + 1)}{2} - \frac{\mu(\mu + 1)}{2}\right)\rho_\alpha(u_\alpha) = 0,$$

and consequently

(46.9)
$$\rho_\beta(u_\alpha) = -\frac{\lambda - \mu}{2}\rho_\alpha(u_\alpha).$$

From (45.9) and (46.2) for $\sigma_\alpha = -1$ we get

(46.10)
$$\rho_\beta(u_\alpha) = -v_{\beta a}v_\alpha^a, \qquad \rho_\alpha(u_\alpha) = -v_{\alpha a}v_\alpha^a.$$

By a combination of these equations and (46.9) Schouten† was enabled to find the possible angles between root-vectors and their lengths. If the configuration of root-vectors of a group is called its *root-figure*, the problem of classification of the structures of semi-simple groups reduces to that of finding all possible and distinct root-figures. By theorem [45.2] a semi-simple group is the direct product of simple groups, the representations of which in E_r are mutually orthogonal. Hence the set of vector-spaces of these component simple groups are orthogonal, and consequently their root-figures. Therefore when root-figures are found which cannot be decomposed into mutually orthogonal figures, they are the root-figures of simple groups.

The problem of finding simple groups by the determination of all possible root-figures on the basis of Schouten's results was carried out by Graham,‡ who obtained from this point of view the results which Cartan had established by algebraic processes. In this investigation use was made of the fact that the reflection of a root-vector with respect to any other root-vector is also in the root-figure. These reflections for the root-figure of any group generate

---

* 1926, 2, p. 368.
† Unpublished lectures delivered at the University of Leyden.
‡ Not yet published.

a discrete group, which is a sub-group of the linear transformations leaving the root-figure invariant. This sub-group is the discrete group S of Weyl.*

In his study of semi-simple groups Weyl made use of the theory of *representation* of continuous groups. This theory for finite groups is due to Frobenius† and Burnside†† and was extended to continuous groups by Schur‡ and developed further by Weyl.‡‡ It consists in an association of a non-singular homogeneous transformation, say $A(a)$, with each set of values of the parameters a^α of a G_r, indicated by $y' = A(a)y$, which satisfies the law of composition (4.6) of G_r, denoted in this case by

$$A(a_2)A(a_1) = A(a_2a_1).$$

For example, the adjoint group is a representation of a G_r, when the parameters of the latter are canonical. When the representation is transitive, it is said to be *irreducible,* in the sense that there is no sub-space invariant under all the transformations of the representation. In certain cases a reducible representation may be considered as the sum of two or more irreducible representations. In such cases one of the principal problems concerns itself with the establishment of a complete set of irreducible representations of G_r such that any representation of G_r may be considered as the sum of certain members of the complete set. For a full treatment of this subject the reader is referred to the authors indicated. Also it should be remarked that recently the theory has been fruitful in the study of physical and chemical problems.§

Exercises

1. If a sub-group G_m of a G_r is integrable, it is contained in at least one sub-group of order $m + 1$ of G_r.

Lie-Scheffers, 1893, 1, p. 564.

2. If a G_r has an Abelian sub-group G_{r-1}, it contains an invariant Abelian sub-group G_{r-1}, and hence is integrable.

Lie-Scheffers, 1893, 1, p. 584.

3. If a G_r is of rank zero, it contains at least one exceptional sub-group G_1.

Killing, 1883, 3, p. 288; *Umlauf,* 1891, 4, p. 40.

---

* 1926, 2, p. 367.

† 1897, 2 and 1899, 1; ††1911, 2, pp. 231–242, 269–279.

‡ 1905, 5. ‡‡1925, 4 and 1926, 2.

§ Cf. *Weyl,* 1931, 1; also *Wigner,* 1931, 2.

4. If a G_r is of rank zero, its first derived group is at most of order $r - 2$.

Killing, 1888, 3, p. 288, *Umlauf*, 1891, 4, p. 40.

5. If a G_r is of rank zero, every sub-group G_2 is Abelian.

Umlauf, 1891, 4, p. 35.

6. A G_r contains at least ∞^{r-4} Abelian sub-groups G_2.

Lie-Engel, 1893, 2, p. 756.

7. In consequence of Ex. 11, p. 155, Ex. 6, p. 137 and the definition of an integrable group it follows that a G_4 which does not contain a simple sub-group G_3 is integrable.

Lie-Scheffers, 1893, 1, p. 574.

8. A group G_r is integrable, if and only if it does not contain a simple sub-group G_3.

Lie-Engel, 1893, 2, p. 757.

9. The direct product of two semi-simple groups is semi-simple.

10. If a group of order $r - m$ is isomorphic with a group G_r of rank zero, the former is of rank zero.

Umlauf, 1891, 4, p. 35.

11. A necessary and sufficient condition that a G_r be integrable is that

$$c_{i\lambda}^{\mu} c_{j\mu}^{\nu} c_{k\nu}^{\lambda} = c_{i\mu}^{\lambda} c_{j\nu}^{\mu} c_{k\lambda}^{\nu} \quad (i, j, k, \lambda, \mu, \nu = 1, \cdots, r).$$

Cartan, 1894, 1, p. 48.

12. The general linear group has the symbols p_i and $x^i p_j$, and has three invariant sub-groups with the symbols

$$p_i; \quad p_i, \quad x^i p_i; \quad p_i, \quad x^i p_j, \quad x^i p_i - x^j p_j \ (i, j \text{ not summed}; i \neq j).$$

Lie-Engel, 1888, 1, vol. 1, p. 562.

13. The general linear homogeneous group in n variables has the symbols $x^i p_j$, and has as invariant sub-groups the G_1 with the symbol $x^i p_i$ and the group with the symbols

$$x^i p_j, \qquad x^i p_i - x^j p_j \ (i \neq j; i, j \text{ not summed}).$$

The latter group is simple.

Lie-Engel, 1888, 1, vol. 1, pp. 560–1.

14. The general projective group in n variables has the symbols (cf. Ex. 9, p. 43).

$$p_i, \qquad x^i p_k, \qquad x^i x^j p_j$$

is of order $n(n + 2)$, and is simple.

Lie-Engel, 1888, 1, vol. 1, p. 560.

15. The group of order $n(n + 1)/2$ with the symbols

$$p_i - x^i x^j p_j, \qquad x^i p_j - x^j p_i \qquad (i, j = 1, \cdots, n)$$

leaves invariant the hyperquadric $\Sigma(x^i)^2 = 1$ and is simple except when $n = 3$ (cf. Ex. 10, p. 124).

Lie-Engel, 1893, 2, p. 354, 357.

16. The projective group of order $(n + 1)(2n + 3)$ in the $2n + 1$ variables z, x^i, y^i $(i = 1, \cdots, n)$ with the symbols

$$p_i - y^i r, \qquad q_i + x^i r, \qquad r, \qquad A + zr \quad \left(q_i = \frac{\partial f}{\partial y^i}; r = \frac{\partial f}{\partial z} \right),$$

$$x^i q_j + x^i q_i, \qquad x^i p_j - y^i q_i, \qquad y^i p_j + y^i p_i \qquad\qquad (i \neq j),$$
$$zp_i - y^i A, \qquad zq_i + x^i A, \qquad zA,$$

where $A = x^i p_i + y^i q_i + zr$, is simple and leaves invariant the linear complex defined by $dz + x^i dy^i - y^i dx^i = 0$.

Lie-Engel, 1888, 1, vol. 2, p. 522.

CHAPTER V

GEOMETRICAL PROPERTIES

47. Riemannian spaces. In the preceding chapters we have considered each set of values of the variables x^i as the coordinates of a point in an n-dimensional manifold V_n, and the transformations of a group G_r as transforming continuously a given point into other points of V_n. This geometry is a geometry of position and thus far, except in the case of semi-simple groups, there has been no basis for the determination of magnitude nor for a comparison of directions at two different points. In this section we define magnitude and parallelism, and in subsequent sections make use of these definitions in their relation to the theory of continuous groups.

Riemann* generalized the idea of element of length of euclidean 3-space, defining the element of length of a V_n by means of a quadratic differential form, thus $ds^2 = g_{ij}dx^i dx^j$, where in general the g's are functions of the x's, and the form is positive definite. Because the general theory of relativity and other physical theories make use of a quadratic differential form which is not positive definite, we take as the basis of the metric of V_n a real *fundamental quadratic form*

$$(47.1) \qquad \varphi = g_{ij}dx^i dx^j,$$

where the g's are functions of the x's such that $g_{ij} = g_{ji}$ and are subject only to the restriction that the determinant of the g's is not zero, that is,

$$(47.2) \qquad g \equiv |g_{ij}| \neq 0.$$

For a given set of differentials dx^i the *element of length ds* is defined by

$$(47.3) \qquad ds^2 = eg_{ij}dx^i dx^j,$$

where e is plus or minus one so that the right-hand member of (47.3) is positive. The letter e will be used in this sense hereafter;

---

* 1854, 1.

it is not to be confused with the e's as used in the equations of a group (§11). If ξ^i are the components of a contravariant vector, its magnitude ξ is defined by

$$(47.4) \qquad \xi^2 = eg_{ij}\xi^i\xi^j.$$

When $\xi^2 = 1$, the given vector is said to be a *unit* vector; when $\xi^2 = 0$ a *null* vector, except where all the components are zero in which case it is called a *zero* vector. The metric defined by (47.3) and (47.4) is called Riemannian, and a space with such a metric is called a *Riemannian space*.

If the coordinates are subjected to a non-singular transformation

$$(47.5) \qquad x'^i = \varphi^i(x), \qquad \left|\frac{\partial \varphi}{\partial x}\right| \neq 0,$$

and in these coordinates (47.1) is denoted by $g'_{ij}dx'^idx'^j$, then since

$$(47.6) \qquad dx'^i = \frac{\partial x'^i}{\partial x^j}dx^j,$$

we have that

$$(47.7) \qquad g'_{ij} = g_{lm}\frac{\partial x^l}{\partial x'^i}\frac{\partial x^m}{\partial x'^j}.$$

If $T^{i_1 \cdots i_r}_{j_1 \cdots j_s}$ are a set of functions of the x's and $T'^{\kappa_1 \cdots \kappa_r}_{l_1 \cdots l_s}$ of the x''s such that under a non-singular transformation (47.5) the relations

$$(47.8) \qquad T'^{\kappa_1 \cdots \kappa_r}_{l_1 \cdots l_s}\frac{\partial x^{i_1}}{\partial x'^{\kappa_1}} \cdots \frac{\partial x^{i_r}}{\partial x'^{\kappa_r}} = T^{i_1 \cdots i_r}_{j_1 \cdots j_s}\frac{\partial x^{j_1}}{\partial x'^{l_1}} \cdots \frac{\partial x^{j_s}}{\partial x'^{l_s}}$$

hold, these functions are said to be the components in the respective coordinate systems of a tensor, contravariant of order r and covariant of order s. Thus from (47.7) we see that g_{lm} and g'_{ij} are the components in the respective coordinate systems of a tensor covariant of order two, called the *fundamental tensor* of the space.*

If equations (47.7) are differentiated with respect to x'^k, the resulting equations can be solved for the second derivatives of the x's with respect to the x''s with the result

---

* For a full discussion of tensors, see 1926, 3, pp. 1–16.

$$(47.9) \qquad \frac{\partial^2 x^k}{\partial x'^i \partial x'^j} + \{{}^k_{lm}\}\frac{\partial x^l}{\partial x'^i}\frac{\partial x^m}{\partial x'^j} = \{{}^h_{ij}\}'\frac{\partial x^k}{\partial x'^h},$$

where, on defining quantities g^{ij} by

$$(47.10) \qquad g^{ij}g_{jk} = \delta^i_k,$$

$$(47.11) \quad \{{}^k_{lm}\} = g^{kh}[lm, h], \qquad [lm, h] = \frac{1}{2}\left(\frac{\partial g_{lh}}{\partial x^m} + \frac{\partial g_{mh}}{\partial x^l} - \frac{\partial g_{lm}}{\partial x^h}\right),$$

and $\{{}^h_{ij}\}'$ are similarly defined in terms of g'_{ij}.* A quantity g^{ij} as defined by (47.10) is the cofactor of g_{ij} in the determinant $|g_{ij}|$ divided by the determinant. When we express the condition of integrability of equations (47.9), we obtain the set of equations

$$(47.12) \qquad R^h_{ijk}\frac{\partial x^i}{\partial x'^a}\frac{\partial x^j}{\partial x'^b}\frac{\partial x^k}{\partial x'^c} = R'^d_{abc}\frac{\partial x^h}{\partial x'^d},$$

where

$$(47.13) \quad R^h_{ijk} = \frac{\partial\{{}^h_{ik}\}}{\partial x^j} - \frac{\partial\{{}^h_{ij}\}}{\partial x^k} + \{{}^l_{ik}\}\{{}^h_{lj}\} - \{{}^l_{ij}\}\{{}^h_{lk}\},$$

and R'^d_{abc} is of the same form in the g''s.† From (47.12) it is seen that R^h_{ijk} are the components of a tensor covariant of the third order and contravariant of the first order. It is called the *Riemannian curvature tensor*.‡

When all of the g's are constants, it follows from (47.13) that $R^h_{ijk} = 0$ and then from (47.12) it follows that the components of this tensor are zero in every coordinate system. It can be shown, conversely, that when all the components of this tensor are zero, there are coordinate systems for which all the g's are constants. This is clearly a generalization of the case of euclidean 3-space, referred to cartesian coordinates; we say that the space is *euclidean* or *flat*, when the curvature tensor is zero, and that the special coordinate systems just referred to are *cartesian*.

If equations (47.8) are differentiated with respect to x'^m, it is seen that the first derivatives of the T''s and T's do not satisfy equations of the form (47.8) and consequently are not components of a tensor. However, if use is made of equations (47.9), it can be shown that the quantities

---

* Cf. 1926, 3, pp. 17–19.

† Cf. 1926, 3, p. 19.

‡ For a geometrical interpretation of this tensor see 1926, 3, p. 81.

$$(47.14) \quad T^{i_1 \ldots i_r}_{j_1 \ldots j_s, k} \equiv \frac{\partial T^{i_1 \ldots i_r}_{j_1 \ldots j_s}}{\partial x^k} + \sum_{h}^{1, \ldots, r} T^{i_1 \ldots i_{h-1} l i_{h+1} \ldots i_r}_{j_1 \ldots j_s} \left\{ {}^{i_h}_{lk} \right\}$$

$$- \sum_{h}^{1, \ldots, s} T^{i_1, \ldots i_r}_{j_1 \ldots j_{h-1} l j_{h+1} \ldots j_s} \left\{ {}^{l}_{j_h k} \right\}$$

and similar expressions in the T''s are the components of a tensor contravariant of order r and covariant of order $s + 1$.[*] The tensor thus derived is said to be obtained from the given one by *covariant differentiation* with respect to g_{ij}. In consequence of (47.11) it follows that

$$(47.15) \quad g_{ij, k} = 0, \qquad g^{ij}_{, k} = 0,$$

that is, g_{ij} and g^{ij} behave like constants under covariant differentiation.

If covariant differentiation is applied to $T^{i_1 \ldots i_r}_{j_1 \ldots j_s, k}$, we obtain the *second covariant derivative* of components $T^{i_1 \ldots i_r}_{j_1 \ldots j_s, kl}$. These quantities are not symmetric in the indices k and l as in the case of ordinary second derivatives. But on eliminating the second derivatives from the two expressions, the resulting equations may be put in the form

$$(47.16) \quad T^{i_1 \ldots i_r}_{j_1 \ldots j_s, kl} - T^{i_1 \ldots i_r}_{j_1 \ldots j_s, lk} = \sum_{\alpha} T^{i_1 \ldots i_{\alpha-1} m i_{\alpha+1} \ldots i_r}_{j_1 \ldots j_s} R^m_{i_\alpha kl}.$$

$$- \sum_{\beta} T^{i_1 \ldots i_r}_{j_1 \ldots j_{\beta-1} m j_{\beta+1} \ldots j_r} R^{i_\beta}_{m k l}.$$

These are known as the *identities of Ricci* and, when covariant differentiation is used, these identities must be used in place of the usual condition of integrability for ordinary differentiation.[†]

In euclidean 3-space referred to cartesian coordinates the vectors of a vector-field of components ξ^i are parallel, if the ξ's be constants, in which case their first covariant derivatives are zero, and consequently they are zero in every coordinate system, that is,

$$(47.17) \quad \xi^i_{, j} \equiv \frac{\partial \xi^i}{\partial x^j} + \xi^h \left\{ {}^i_{hj} \right\} = 0.$$

---

[*] 1926, 3, p. 28.
[†] 1926, 3, p. 30.

We consider these equations in a general V_n.   When we express the condition of integrability of these equations, we obtain

$$(47.18) \qquad\qquad \xi^h R^i_{hjk} = 0.$$

Consequently the system (47.17) is completely integrable for a euclidean V_n and accordingly there exists a field of vectors parallel to any given vector; its components are the solutions of (47.17) determined by the components of the given vector as initial values for the values of the x's of the point where the given vector is.   If V_n is not euclidean, equations (47.18) are not satisfied identically, and consequently parallelism as understood in euclidean space does not hold in a general V_n.   However, if we consider any curve C in a given V_n defined by equations

$$x^i = \varphi^i(t),$$

where t is a parameter, the equations

$$(47.19) \qquad \xi^i_{,i} \frac{dx^j}{dt} \equiv \frac{d\xi^i}{dt} + \xi^h \left\{{}^{\ i}_{h j}\right\} \frac{dx^j}{dt} = 0$$

admit a solution determined by values of the ξ's for $t = 0$.   When such a solution is known, we have a set of vectors of components ξ^i at each point of the curve.   Following Levi-Civita* we say that these vectors are *parallel with respect to the curve* C.   In this sense parallelism is *relative*, whereas in euclidean space it is *absolute*.

From (47.15) it follows that for a solution of (47.19) $g_{ij}\xi^i\xi^j$ is a constant, and consequently the ξ's differ by a constant factor at most from those of a unit vector, if the initial values are chosen so that it is not a null vector.   Two vectors at a point are said to have the same *direction*, if corresponding components are proportional.   Accordingly, if a set of functions ξ^i satisfy (47.19), the vectors of components

$$\overline{\xi}^i = \xi^i \psi(t),$$

where ψ is any function of t, should be interpreted as parallel with respect to C.   From (47.19) we have

$$\frac{d\overline{\xi}^i}{dt} + \overline{\xi}^h \left\{{}^{\ i}_{h j}\right\} \frac{dx^j}{dt} = \overline{\xi}^i \frac{d \log \psi}{dt}.$$

---

* 1917, 1; another geometrical interpretation of parallelism as thus defined is given by *Levi-Civita;* cf. also, 1926, 3, pp. 62–65, 74.

Eliminating $\dfrac{d \log \psi}{dt}$ from these equations, we have as the general equations of vectors parallel with respect to a curve (dropping the bars)

$$(47.20) \quad \xi^j\left(\frac{d\xi^i}{dt} + \xi^h \left\{{i \atop hk}\right\} \frac{dx^k}{dt}\right) - \xi^i\left(\frac{d\xi^j}{dt} + \xi^h \left\{{j \atop hk}\right\} \frac{dx^k}{dt}\right) = 0.$$

Since $\dfrac{dx^i}{dt}$ are the components of the tangent to the curve C, a necessary and sufficient condition that the tangents be parallel with respect to C is that C be a curve satisfying the equations

$$(47.21) \quad \frac{dx^j}{dt}\left(\frac{d^2x^i}{dt^2} + \left\{{i \atop hk}\right\} \frac{dx^h}{dt} \frac{dx^k}{dt}\right) - \frac{dx^i}{dt}\left(\frac{d^2x^j}{dt} + \left\{{j \atop hk}\right\} \frac{dx^h}{dt} \frac{dx^k}{dt}\right) = 0.$$

If s is the arc of C, then $\dfrac{dx^i}{ds}$ is a unit vector and from (47.19) the equations which C must satisfy are

$$(47.22) \qquad\qquad \frac{d^2x^i}{ds^2} + \left\{{i \atop ik}\right\} \frac{dx^j}{ds} \frac{dx^k}{ds} = 0.$$

These curves, called the *geodesics* of the space, may be said to be the straightest curves in V_n and are the generalization of straight lines in euclidean space. In fact, if the coordinates in the latter are cartesian, equations (47.22) become $\dfrac{d^2x^i}{ds^2} = 0$, which are the equations of the straight lines. From the form of (47.22) it follows that an integral curve of (47.22) passes through each point in V_n in an arbitrary direction. They can be shown to be characterized by the property that for two points sufficiently near one another there is only one curve of the congruence through these points and the first variation of the arc between the points is zero.* When the fundamental form (47.1) is positive definite, this is the shortest distance between the points; when the form is indefinite, it may be the shortest or the greatest distance according to further distinguishing properties of the form.

---

* 1926, 3, pp. 48–50, 53.

48. Linearly connected manifolds. Linear connection determined by a simply transitive group. Let ξ_a^i be the components of the n vectors of a simply transitive group G_n in a V_n. Then functions ξ_i^a are defined uniquely by (21.9) for $q = n$, and as in (21.12) we have the functions

$$(48.1) \qquad \Lambda_{jk}^i = \xi_a^i \frac{\partial \xi_k^a}{\partial x^j} = -\xi_k^a \frac{\partial \xi_a^i}{\partial x^j} \quad (a, i, j, k = 1, \cdots, n).$$

If the x's undergo a non-singular transformation into coordinates x'^i and ξ'^i_a are the components of the vectors in the x''s, we have

$$(48.2) \qquad \xi_a^i = \xi'^i_a \frac{\partial x^i}{\partial x'^i}.$$

If we denote by Λ'^i_{jk} the functions in the ξ''s analogous to (48.1), we find that

$$(48.3) \qquad \frac{\partial^2 x^i}{\partial x'^i \partial x'^k} + \Lambda_{lm}^i \frac{\partial x^l}{\partial x'^i} \frac{\partial x^m}{\partial x'^k} = \Lambda'^h_{jk} \frac{\partial x^i}{\partial x'^h}.$$

We remark that these equations are similar to (47.9). From (21.15) we have

$$(48.4) \qquad \Lambda_{jk}^i - \Lambda_{kj}^i = c_{ab}^e \xi_k^a \xi_j^b \xi_e^i \quad (a, b, e, i, j, k = 1, \cdots, n),$$

and consequently Λ_{jk}^i are not symmetric in the indices j and k, unless the group is Abelian (§13), whereas $\{^k_m\}$ are symmetric in l and m as follows from (47.11).

Although (48.3) have been obtained from (48.1), we consider the general case when Λ_{lm}^i and Λ'^h_{jk} are two sets of functions in the relations (48.3) in their respective coordinate systems. We say that each set of these functions defines the same *linear connection;* they are called the *coefficients* of the connection. Comparing (47.13) and (21.17) we see that they are similar in form, so that the tensor of components Λ_{ijk}^h is called the *curvature tensor* for the given linear connection. Analogous to (47.20) are the equations

$$(48.5) \qquad \xi^i \left(\frac{d\xi^i}{dt} + \xi^h \Lambda_{hk}^i \frac{dx^k}{dt} \right) - \xi^i \left(\frac{d\xi^i}{dt} + \xi^h \Lambda_{hk}^i \frac{dx^k}{dt} \right) = 0.$$

A set of functions ξ^i which satisfy these equations for a curve C, defined by expressing the x's as functions of a parameter t, are said to be the components of vectors at points of C *parallel* with respect to C. And the curves defined by

$$(48.6) \quad \frac{dx^j}{dt}\left(\frac{d^2x^i}{dt^2} + \Lambda_{kl}^i\frac{dx^k}{dt}\frac{dx^l}{dt}\right) - \frac{dx^i}{dt}\left(\frac{d^2x^i}{dt^2} + \Lambda_{kl}^j\frac{dx^k}{dt}\frac{dx^l}{dt}\right) = 0$$

are called the *paths* of the manifold; they are a generalization of geodesics of a Riemannian space in that they are the straightest lines.* Thus by means of a linear connection we are enabled to compare directions at different points, but are in no position to handle metric properties.

From (21.19) it follows that for the linear connection defined by (48.1) in terms of a simply transitive group we have

$$(48.7) \qquad \Lambda_{ijk}^h = 0,$$

that is, the curvature tensor is zero. When any linear connection satisfies this condition, we say that the space is *flat* or that the *connection is of zero curvature*.

In consequence of (48.7) the equations (cf. equations (21.13))

$$(48.8) \qquad \frac{\partial \zeta^i}{\partial x^j} + \zeta^h\Lambda_{hj}^i = 0$$

are completely integrable, and consequently a solution is determined by arbitrary initial values of the ζ's. From (48.5) it follows that any two vectors of such a vector-field are parallel with respect to every curve joining their points of application and consequently we say that they are *absolutely parallel*. As in the case of euclidean space, there are n such independent vector fields, say ζ_a^i, and from §30 it follows that they are the vectors of the group reciprocal to the given group.

If we define quantities $\bar{\Lambda}_{jk}^i$ by the equations

$$(48.9) \qquad \bar{\Lambda}_{jk}^i = \Lambda_{kj}^i,$$

it follows from (48.3) that $\bar{\Lambda}_{jk}^i$ are the coefficients of a second linear connection. In consequence of (48.1) and (48.9) we have

$$(48.10) \qquad \frac{\partial \xi_a^i}{\partial x^k} + \xi_a^j\bar{\Lambda}_{jk}^i = 0.$$

---

* Cf. 1927, 1, pp. 4, 5, 13, 14.

These equations are necessarily completely integrable and consequently the curvature tensor of components $\bar{\Lambda}^h_{ijk}$, defined by expressions analogous to (21.17) is a zero tensor. Comparing (48.10) with (48.8), we see that the vectors ξ^i_a are parallel absolutely with respect to the connection $\bar{\Lambda}^i_{jk}$. Hence we have:

[48.1] *For the two linear connections determined by a simply transitive group the parallelism is absolute.* *

From (48.8) and (48.9) it follows that

$$(48.11) \qquad \bar{\Lambda}^i_{jk} = -\zeta^a_k \frac{\partial \zeta^i_a}{\partial x^j} = \zeta^i_a \frac{\partial \zeta^a_k}{\partial x^j},$$

where ζ^a_i are defined by equations of the form (21.9), that is,

$$(48.12) \qquad \zeta^a_i \zeta^j_a = \delta^j_i, \qquad \zeta^a_i \zeta^i_b = \delta^a_b.$$

Thus $\bar{\Lambda}^i_{jk}$ bear to the reciprocal group a relation analogous to that of Λ^i_{jk} to the given group.

If we denote by Γ^i_{jk} the symmetric part of the Λ^i_{jk}, that is,

$$(48.13) \qquad \Gamma^i_{jk} = \tfrac{1}{2}(\Lambda^i_{jk} + \Lambda^i_{kj}),$$

it follows from (48.3) that the Γ's satisfy equations of the form (48.3) and thus are the coefficients of a third linear connection. Since the Γ's are symmetric in the indices j and k, they are the coefficients of a *symmetric* linear connection.† In consequence of (48.9) Γ^i_{jk} is the symmetric part of $\bar{\Lambda}^i_{jk}$ also.

We define quantities g^{ij} and g_{ij} by

$$(48.14) \qquad g^{ij} = \zeta^i_a \zeta^j_a, \qquad g_{ij} = \zeta^a_i \zeta^a_j.$$

From these equations and (48.12) we have

$$(48.15) \qquad g^{ij} g_{ik} = \delta^j_k,$$

and

$$(48.16) \qquad g^{ij} \zeta^a_i = \zeta^j_a, \qquad g_{ij} \zeta^i_a = \zeta^a_j.$$

From the second set of these equations and (48.14) it follows that

$$g_{ij} \zeta^i_a \zeta^j_b = \delta_{ab}.$$

Consequently if g_{ij} are taken as the components of the fundamental

---

* Cf. 1925, 1; also *Mattioli*, 1930, 2.
† 1927, 1, p. 55.

tensor of V_n the vectors ζ_a^i are unit-vectors and any two are orthogonal to one another.*

From (48.14) and (48.8) we have

$$(48.17) \qquad \frac{\partial g^{ij}}{\partial x^k} + g^{hj}\Lambda_{hk}^i + g^{ih}\Lambda_{hk}^j = 0,$$

in consequence of which and (48.15) we have

$$(48.18) \qquad \frac{\partial g_{ij}}{\partial x^k} - g_{hj}\Lambda_{ik}^h - g_{ih}\Lambda_{jk}^h = 0.$$

Hence if we define covariant differentiation with respect to the linear connection by equations obtained from (47.14) on replacing $\{^i_{jk}\}$ by Λ_{jk}^i and the comma with a solidus, we have

$$(48.19) \qquad g^{ij}{}_{|k} = 0, \qquad g_{ij|k} = 0.$$

From (48.1), (48.18) and (21.9) we have

$$(48.20) \qquad \xi_a^k \frac{\partial g_{ij}}{\partial x^k} + g_{hj}\frac{\partial \xi_a^h}{\partial x^i} + g_{ih}\frac{\partial \xi_a^h}{\partial x^j} = 0,$$

the significance of which will be seen in §51.

49. Simply transitive groups determined by linear connections of zero curvature. We have seen that when the curvature of a linear connection is zero, there exists an ennuple of independent absolutely parallel vector-fields. We seek the conditions that these are the vectors of a simply transitive group.

We denote by Ω_{jk}^i the skew symmetric part of the coefficients Λ_{jk}^i of the linear connection, that is,

$$(49.1) \qquad \Omega_{jk}^i = \tfrac{1}{2}(\Lambda_{jk}^i - \Lambda_{kj}^i).$$

From these equations and (48.13) we have

$$(49.2) \qquad \Lambda_{jk}^i = \Gamma_{jk}^i + \Omega_{jk}^i.$$

From these expressions and (21.17) we have

$$(49.3) \qquad \Lambda_{ijk}^h = B_{ijk}^h + \Omega_{ijk}^h,$$

---

* 1926, 3, p. 38.

where

$$(49.4) \qquad B^h_{ijk} = \frac{\partial \Gamma^h_{ik}}{\partial x^j} - \frac{\partial \Gamma^h_{ij}}{\partial x^k} + \Gamma^l_{ik}\Gamma^h_{lj} - \Gamma^l_{ij}\Gamma^h_{lk}$$

and

$$(49.5) \qquad \Omega^h_{ijk} = \Omega^h_{ik|j} - \Omega^h_{ij|k} - \Omega^l_{ki}\Omega^h_{jl} - \Omega^l_{ij}\Omega^h_{kl} - 2\Omega^l_{jk}\Omega^h_{il};$$

as defined in §48, a solidus followed by an index indicates covariant differentiation with respect to the Λ's. B^h_{ijk} as defined by (49.4) are the components of the curvature tensor of the symmetric connection of coefficients Γ^i_{jk}; they satisfy the identities*

$$(49.6) \qquad\qquad\qquad B^h_{ijk} + B^h_{ikj} = 0,$$
$$B^h_{ijk} + B^h_{jki} + B^h_{kij} = 0.$$

When $\Lambda^h_{ijk} = 0$, we have from (49.3), (49.5) and (49.6)

$$(49.7) \qquad \Omega^h_{ij|k} + \Omega^h_{jk|i} + \Omega^h_{ki|j} + 2(\Omega^l_{ij}\Omega^h_{kl} + \Omega^l_{jk}\Omega^h_{il} + \Omega^l_{ki}\Omega^h_{jl}) = 0.†$$

If we denote by ξ^i_a the components of the n fields of absolutely parallel vectors and define ξ^i_i by (21.9), and put

$$(49.8) \qquad\qquad \Omega^i_{jk} = \tfrac{1}{2}c^e_{ab}\xi^a_j\xi^b_k\xi^i_e,$$

the quantities c^e_{ab} are scalars and

$$(49.9) \qquad\qquad c^e_{ab} + c^e_{ba} = 0.$$

From (49.8) it follows that

$$(49.10) \qquad\qquad \Omega^i_{jk|l} = 0,$$

when and only when the c's are constants. When (49.10) is satisfied, it follows from (49.7) that

$$(49.11) \qquad\qquad \Omega^l_{ij}\Omega^h_{kl} + \Omega^l_{jk}\Omega^h_{il} + \Omega^l_{ki}\Omega^h_{jl} = 0,$$

from which and (49.8) we have the Jacobi relations (7.4). Furthermore, from (49.1), (48.1) and (49.8) we have

$$\xi^i_a\frac{\partial \xi^j_b}{\partial x^i} - \xi^i_b\frac{\partial \xi^j_a}{\partial x^i} = c^e_{ab}\xi^j_e.$$

---

* 1927, 1, p. 55.
† Cf. *Einstein*, 1929, 2, p. 5.

In consequence of the second fundamental theorem we have that ξ_a^i are the vectors of a simply transitive group. Hence we have:

[49.1] *A necessary and sufficient condition that a linear connection of zero curvature determine a simply transitive group is that (49.10) be satisfied.*

We have seen in §48 that the coefficients Λ_{jk}^i and $\bar{\Lambda}_{jk}^i$ of the linear connections determined by a simply transitive group and its reciprocal are in the relation $\bar{\Lambda}_{jk}^i = \Lambda_{kj}^i$. From this and (49.2) it follows that

$$\bar{\Lambda}_{jk}^i = \Gamma_{jk}^i - \Omega_{jk}^i.$$

If, conversely, the coefficients of two linear connections of zero curvature are in this relation, we have equations (49.7) and the equations obtained from the latter on replacing each Ω_{jk}^i by its negative. Consequently we have (49.11) and

(49.12) $$\Omega_{ij|k}^h + \Omega_{jk|i}^h + \Omega_{ki|j}^h = 0.$$

Since B_{ijk}^h are the same for the two connections, we have $\overline{\Omega}_{ijk}^h = \Omega_{ijk}^h$, where $\overline{\Omega}_{ijk}^h$ is obtained from Ω_{ijk}^h on replacing each Ω_{jk}^i by its negative, from which, in consequence of (49.5), we have

$$\Omega_{ij|k}^h - \Omega_{ik|j}^h = 0.$$

From these equations and (49.12) we obtain (49.10) and consequently we have:

[49.2] *A necessary and sufficient condition that an asymmetric linear connection of zero curvature determine a simply transitive group is that the connection of coefficients $\bar{\Lambda}_{jk}^i = \Lambda_{kj}^i$ be of zero curvature; in this case the second connection determines the simply transitive group reciprocal to the given one.*

When $\Lambda_{ijk}^h = 0$ and (49.10) hold, it follows from (49.3), (49.5) and (49.11) that

(49.13) $$B_{ijk}^h = \Omega_{jk}^l \Omega_{il}^h.$$

Conversely, if this condition is satisfied, we have (49.11) satisfied in consequence of (49.6). Hence if (49.10) is satisfied, it follows from (49.5) and (49.3) that $\Lambda_{ijk}^h = 0$. Accordingly we have:

[49.3] *A necessary and sufficient condition that an asymmetric linear connection determine a simply transitive group is that (49.10) and (49.13) be satisfied.*

If we indicate by a semi-colon covariant differentiation with respect to the Γ's, that is, if we replace $\{{}^i_{jk}\}$ by Γ^i_{jk} and the comma by a semi-colon in (47.14), we have

$$(49.14) \qquad \Omega^h_{ij;k} = \frac{\partial \Omega^h_{ij}}{\partial x^k} + \Omega^l_{ij}\Gamma^h_{lk} - \Omega^h_{lj}\Gamma^l_{ik} - \Omega^h_{il}\Gamma^l_{jk}$$
$$= \Omega^h_{ij|k} + (\Omega^l_{ij}\Omega^h_{kl} + \Omega^l_{jk}\Omega^h_{il} + \Omega^l_{ki}\Omega^h_{jl}).$$

Consequently for a simply transitive group we have

$$(49.15) \qquad\qquad \Omega^h_{ij;k} = 0.$$

Conversely, if this condition and (49.13) are satisfied, we obtain (49.10) from (49.14) and (49.11). In consequence of (49.15) we have from (49.13)

$$(49.16) \qquad\qquad B^h_{ijk;l} = 0.$$

50. Geometry of the group-space. Since the parameter groups of a group are simply transitive and reciprocal (§30), the results of the preceding sections may be applied to define two asymmetric linear connections and a symmetric one for the group-space. In fact, we have made use of the coefficients of the first two connections in §8, defining them by

$$(50.1) \qquad \begin{aligned} L^\alpha_{\beta\gamma} &= -A^b_\beta\frac{\partial A^\alpha_b}{\partial a^\gamma} = A^\alpha_b\frac{\partial A^b_\beta}{\partial a^\gamma}, \\ \bar L^\alpha_{\beta\gamma} &= -\bar A^b_\beta\frac{\partial \bar A^\alpha_b}{\partial a^\gamma} = \bar A^\alpha_b\frac{\partial \bar A^b_\beta}{\partial a^\gamma}. \end{aligned}$$

As thus defined, the vectors A^α_b are absolutely parallel with respect to the first connection and $\bar A^\alpha_b$ with respect to the second. Moreover the respective curvature tensors of components $L^\alpha_{\beta\gamma\delta}$ and $\bar L^\alpha_{\beta\gamma\delta}$, defined in the L's and $\bar L$'s by (21.17), are zero tensors, that is,

$$(50.2) \qquad\qquad L^\alpha_{\beta\gamma\delta} = 0, \qquad \bar L^\alpha_{\beta\gamma\delta} = 0.$$

From the results of §48 it follows that $\bar L^\alpha_{\beta\gamma} = L^\alpha_{\gamma\beta}$. Hence the paths for the two connections, as defined by (48.6), are the same for both and are the integral curves of the equations

$$(50.3) \qquad \frac{da^\beta}{dt}\left(\frac{d^2a^\alpha}{dt^2} + \Gamma^\alpha_{\gamma\delta}\frac{da^\gamma}{dt}\frac{da^\delta}{dt}\right) - \frac{da^\alpha}{dt}\left(\frac{d^2a^\beta}{dt^2} + \Gamma^\beta_{\gamma\delta}\frac{da^\gamma}{dt}\frac{da^\delta}{dt}\right) = 0,$$

where

(50.4) $$\Gamma^\alpha_{\beta\gamma} = \tfrac{1}{2}(L^\alpha_{\beta\gamma} + L^\alpha_{\gamma\beta})$$

are the coefficients of the symmetric connection determined by the parameter groups. For each path we have

$$\frac{d^2a^\alpha}{dt^2} + \Gamma^\alpha_{\gamma\delta}\frac{da^\gamma}{dt}\frac{da^\delta}{dt} = \varphi(t)\frac{da^\alpha}{dt},$$

where φ is a determinate function of t. If we define a parameter s by

$$\frac{ds}{dt} = e^{\int\varphi dt},$$

the preceding equations become

(50.5) $$\frac{d^2a^\alpha}{ds^2} + \Gamma^\alpha_{\gamma\delta}\frac{da^\gamma}{ds}\frac{da^\delta}{ds} = 0.$$

Thus along each path a parameter s, called the *affine parameter*, can be chosen so that the equations of the paths are (50.5). Comparing these equations with (47.22), we note that s is the analogue of the arc in a Riemannian space.

In this section we consider the geometrical properties of the group-space S determined by the linear connections thus defined. Many of the results are due to Cartan and Schouten,[*] who developed this theory following the announcement by the author[†] of the results of §48. They introduced the terms $(+)$-parallelism, $(-)$-parallelism and (0)-parallelism for the connections with the coefficients $L^\alpha_{\beta\gamma}$, $\bar{L}^\alpha_{\beta\gamma}$ and $\Gamma^\alpha_{\beta\gamma}$ respectively; we use this terminology. We recall that the first two of these parallelisms are absolute, but not the third. We refer to these connections as respectively the $(+)$-*connection*, the $(-)$-*connection* and the (0)-*connection*.

Because of the conditions (8.4) we have that at the point of coordinates a^α_0 of S, representing the identity in G_r, the vectors A^α_a and $\bar{A}^\alpha_a$ coincide. Consequently the vector fields A^α_a and $\bar{A}^\alpha_a$ throughout S may be thought of as obtained from the r vectors $A^\alpha_a(a_0)(=\bar{A}^\alpha_a(a_0))$ by $(+)$-parallel and $(-)$-parallel displacement respectively. We have related the vectors at a general point by the equation (8.14). Comparing these equations and (39.2) we see that the vectors of the first set at a point are transformed into those of the second set by the adjoint group.

---

[*] Cf. 1926, 1; *Cartan*, 1927, 2; *Schouten*, 1929, 1.

[†] 1925, 1.

If a point of coordinates a^α is transformed into a point of coordinates a'^α by a transformation of the second parameter group, it follows from (6.7) and (9.2) that

(50.6) $$\frac{\partial a'^\alpha}{\partial a^\beta} = A_b^\alpha(a')A_\beta^b(a).$$

The transform of the vector A_a^β by a transformation of the second parameter group is $A_a^\beta \dfrac{\partial a'^\alpha}{\partial a^\beta}$ which reduces by means of (50.6) to $A_a^\alpha(a')$; that is, the vectors of each vector-field A_a^α are transformed into one another by the second parameter group. Similarly those of the second set are transformed into one another by the first parameter group. Hence we have:

[50.1] *The vector-fields A_a^α and $\bar{A}_a^\alpha$ are transformed into themselves by the transformations of the second and first parameter groups respectively.*

In §12 it was shown that the trajectories of the first parameter group are integral curves of equations (50.5), and that the equations of the path through the point of coordinates a^α are (12.1). Since the paths for the two connections are the same, it follows that the equations of the above path as a trajectory of the second parameter group are expressible in the form

(50.7) $$a'^\alpha = a^\alpha + t\bar{e}^a\bar{A}_a a^\alpha + \cdots,$$

where necessarily

$$\bar{e}^a\bar{A}_a^\alpha(a) = e^a A_a^\alpha(a), \qquad \bar{A}_a f = \bar{A}_a^\alpha \frac{\partial f}{\partial a^\alpha}.$$

The paths which are $(+)$-parallel to the given one form a congruence which ordinarily is not the same as the congruence of paths which are $(-)$-parallel to the given one. From theorem [50.1] it follows that the curves of the first congruence are transforms of one another by the second parameter group and of the second congruence by the first parameter group.

A one-parameter group of G_r is defined, as we have seen in §11, by taking for the a's the values given by (12.2) for assigned values of the e's; these are the equations of a path through the *origin*, the point a_0^i which defines the identity in G_r. If b^i is any point of S not on this path, the equation

(50.8) $$T_{b'} = T_a T_b$$

defines a trajectory of the first parameter group through b. It is an integral curve of

$$\frac{db'^\alpha}{dt} = \frac{\partial b'^\alpha}{\partial a^\beta}\frac{da^\beta}{dt} = A_a^\alpha(b')A_\beta^a(a)\frac{da^\beta}{dt}$$
$$= A_a^\alpha(b')A_\beta^a(a)e^b A_b^\beta(a) = e^a A_a^\alpha(b').$$

Since similar results hold for the second parameter group we have:

[50.2] *The trajectories of a one-parameter sub-group of the first (second) parameter group are $(+)$-parallel $((-)$-parallel) to the trajectory through the origin (a_0^i) which determines the sub-group.*

Consider two ordered point-pairs a^i, b^i and a'^i, b'^i of the group-space S such that the transformations of the group satisfy the condition

(50.9) $$T_b T_a^{-1} = T_{b'} T_{a'}^{-1}.$$

When a^i is the origin a_0^i, it follows from (50.9) that $T_{b'} = T_b T_{a'}$, from which and (50.8) we have that the point b'^α is the transform of the point a'^α by a transformation of the first parameter group. If the point b^α describes a trajectory of the first parameter group, in which case we have a one-parameter sub-group of this group, the point b'^α describes a trajectory of this group, and the two trajectories are $(+)$-parallel, as follows from theorem [50.2]. The same result holds for every point-pair a', b' in the relation (50.9) to a_0, b, and any two such point-pairs are in the relation (50.9). Since the paths are the generalization of straight lines in euclidean space, we may speak of the segments $\overrightarrow{ab}$ and $\overrightarrow{a'b'}$ of the paths as vectors. In the terminology of Cartan* two vectors in the relation (50.9) are *equipollent of the first kind.*

When the two ordered point-pairs a^i, b^i and a'^i, b'^i are such that

(50.10) $$T_a^{-1}T_b = T_{a'}^{-1}T_{b'},$$

___
* 1927, 2; the reader is referred to this paper for a complete treatment of the subject.

it follows by reasoning similar to the above, that ab and $a'b'$ are segments of paths of the second parameter group. In this case the vectors $\overrightarrow{ab}$ and $\overrightarrow{a'b'}$ are said to be *equipollent of the second kind.* If each term of equation (50.9) is multiplied on the left by T_b^{-1} and on the right by $T_{a'}$ we obtain $T_a^{-1}T_{a'} = T_b^{-1}T_{b'}$. Consequently we have:*

[50.3]　*If $\overrightarrow{ab}$ and $\overrightarrow{a'b'}$ are equipollent of the first kind, $\overrightarrow{aa'}$ and $\overrightarrow{bb'}$ are equipollent of the second kind, and vice-versa.*

As a consequence of theorem [50.2] we have:

[50.4]　*If two vectors are equipollent of the first or second kind, they are segments of $(+)$-parallel or $(-)$-parallel paths.*

Suppose that G_p is a sub-group of G_r and that any one parameter sub-group of the former is a linear combination (constant coefficients) of the symbols $e_s^a X_a f$ for $s = 1, \cdots, p$. In terms of general parameters a^α a given path of the first parameter group for G_p through a point a^α is defined by equations (12.1) when the e's are replaced by a suitable combination of e_s^a. The locus of all these paths through the point a^α is a sub-space of p dimensions, say S_p. In terms of canonical parameters the equations of the S_p through the origin, say S_{0p}, are (cf. (12.6))

(50.11) $$u^\alpha = c^s e_s^\alpha t,$$

where the c's are arbitrary parameters. A point in S_{0p} other than the origin has coordinates $\bar{c}^s e_s^\alpha \bar{t}$, where the $\bar{c}$'s and $\bar{t}$ have definite values. From (12.6) it follows that the paths of the first parameter sub-group through this point have the equations

$$u'^\alpha = \bar{c}^s e_s^\alpha \bar{t} + c^s e_s^\alpha t.$$

Let A be a point on this path corresponding to the values c_1^s and t_1. If we take for c'^s and t' values satisfying the equations

$$c'^s t' = \bar{c}^s \bar{t} + c_1^s t_1,$$

the path through the origin with the equations $u'^\alpha = c'^s e_s^\alpha t'$ passes through A. Consequently for points sufficiently near the origin all the paths of the sub-group of the first parameter group lie in

---

* *Cartan*, 1927, 2, p. 5.

S_{0p}. Thus S_{0p} is a generalization of a plane in euclidean 3-space in the sense that all the paths through a point in directions of the sub-space lie entirely in the sub-space; accordingly we say that S_{0p} is *flat*. In consequence of theorem [50.1] the S_p through any point P is obtained by the transformation of the second parameter group which sends the origin into P. Since each path in S_{0p} is transformed into a path $(+)$-parallel to the given one, the S_p through P is flat, and it is parallel to S_{0p}, that is, the paths of S through any point of S_p which are parallel to a path of S_{0p} lie in S_p. These flat sub-spaces are a generalization also of totally geodesic sub-spaces of a Riemannian space.*

The preceding results may be seen in another way. If T_u is the general transformation of the sub-group G_p, then $T_u T_a$ defines the S_p through the point a. For two particular values of u, say u_1 and u_2, we have

$$T_{a_1} = T_{u_1} T_a, \qquad T_{a_2} = T_{u_2} T_a.$$

Let b be a point not in S_p, and in the S_p through b we have the point pair b_1, b_2 defined by

$$T_{b_1} = T_{u_1} T_b, \qquad T_{b_2} = T_{u_2} T_b.$$

From these we have

$$T_{a_2} T_{a_1}^{-1} = T_{u_2} T_a T_a^{-1} T_{u_1}^{-1} = T_{u_2} T_{u_1}^{-1} = T_{b_2} T_{b_1}^{-1}.$$

Consequently $\overrightarrow{a_1 a_2}$ and $\overrightarrow{b_1 b_2}$ are equipollent of the first kind and are segments of $(+)$-parallel paths.

In a similar manner we get a set of flat S_p's $(-)$-parallel to S_{0p}. In general these two sets are different. If they are to be the same, we must have for every a and every u relations of the form $T_u T_a = T_a T_{u'}$, that is, G_p is an invariant sub-group. Hence we have:†

[50.5] *If G_r admits a sub-group G_p, there are in S two sets of flat varieties of the pth order, those of the first set being $(+)$-parallel to one another and of the second set $(-)$-parallel; the varieties of the two systems are the same, when and only when G_p is an invariant sub-group.*

---

* 1926, 3, pp. 183–186.

† *Cartan* and *Schouten*, 1926, 1, p. 813.

For the connection of coefficients $\Gamma^\alpha_{\beta\gamma}$ we have a set of equations analogous to (48.5), namely

$$\xi^\beta\left(\frac{d\xi^\alpha}{dt} + \xi^\delta\Gamma^\alpha_{\delta\gamma}\frac{da^\gamma}{dt}\right) - \xi^\alpha\left(\frac{d\xi^\beta}{dt} + \xi^\delta\Gamma^\beta_{\delta\gamma}\frac{da^\gamma}{dt}\right) = 0,$$

a solution of which ξ^α, when we put

$$(50.12) \qquad\qquad a^\alpha = \varphi^\alpha(t),$$

gives at each point of the curve C defined by (50.12) a vector, and all of these vectors are (0)-parallel with respect to C. If we have such a solution and we replace ξ^α by $\xi^\alpha\psi(t)$, where $\psi(t)$ is suitably chosen, we have that the new functions ξ^α satisfy the equations

$$(50.13) \qquad\qquad \frac{d\xi^\alpha}{dt} + \xi^\beta\Gamma^\alpha_{\beta\gamma}\frac{da^\gamma}{dt} = 0.$$

We say that a solution ξ^α of these equations determined by initial values ξ^α_0 defines at each point of C the vector obtained from ξ^α_0 by (0)-parallel displacement of ξ^α_0.

If S_p is a sub-space of p dimensions, the coordinates a^α can be chosen so that S_p is defined by

$$(50.14) \qquad\qquad a^\sigma = 0 \qquad\qquad (\sigma = p + 1, \cdots, r).$$

If this is to be a flat sub-space, equations (50.5) must admit solutions which satisfy (50.14), and consequently we must have

$$(50.15) \qquad \Gamma^\sigma_{\lambda\mu} = 0 \qquad \text{for } a^\sigma = 0 \qquad \left(\begin{matrix}\lambda, \mu = 1, \cdots, p; \\ \sigma = p + 1, \cdots, r\end{matrix}\right).$$

For a vector tangential to S_p we must have $\xi^\sigma = 0$. It follows from (50.15) that, if the initial values of a solution of (50.13) satisfy these conditions, they are satisfied by the solution and we have:*

[50.6] *If a vector tangential to a flat sub-space is transported by (0)-parallelism along any curve of the sub-space, it remains tangential to the latter.*

In the S_{0p} of a sub-group, as defined, a vector at any point P of S_{0p} (+)-parallel to a vector tangential at any other point Q is

---

* Cf. *Cartan*, 1925, 3, p. 40.

tangential to S_{0p} at P. Conversely, suppose that there is a flat-space S_{0p} through the origin possessing this property; then for determinate values of the e's the parallel vector-fields $e_s^a A_a^\alpha$ at points of S_{0p} are tangential to it. In accordance with theorem [50.6] a vector (0)-parallel to any of these vectors is tangential to S_{0p}; that is, $e_s^a A_a^\alpha (e_t^b A_b^\beta)_{;\alpha}$ is a vector of the field and so also is

$$e_s^a A_a^\alpha (e_t^b A_b^\beta)_{;\alpha} - e_t^b A_b^\alpha (e_s^a A_a^\beta)_{;\alpha}.$$

Hence we must have

$$e_s^a A_a^\alpha \frac{\partial}{\partial a^\alpha}(e_t^b A_b^\beta) - e_t^b A_b^\alpha \frac{\partial}{\partial a^\alpha}(e_s^a A_a^\beta) = \gamma_{st}^u e_u^a A_a^\beta,$$

from which it follows that $e_s^a A_a f$ are the generators of a sub-group of the first parameter group, and consequently $e_s^a X_a f$ of a sub-group of G_r. Since similar results follow for $(-)$-parallelism, we have:[*]

[50.7] *A necessary and sufficient condition that the manifold of the paths through the origin in a bundle of directions of order p represent a sub-group of order p is that the manifold be flat, and that the directions at any point $(+)$-parallel or $(-)$-parallel to tangential directions at any other point be tangential.*

The results of §§48, 49 can be applied immediately to the group-space S. If we denote by $\Omega_{\beta\gamma}^\alpha$ the skew-symmetric part of $L_{\beta\gamma}^\alpha$, we have from (8.8)

$$(50.16) \qquad \Omega_{\beta\gamma}^\alpha = \tfrac{1}{2} c_{ab}^e A_\beta^a A_\gamma^b A_e^\alpha.$$

From equations analogous to (49.13) we have

$$B_{\beta\gamma} \equiv B_{\beta\gamma\alpha}^\alpha = \Omega_{\beta\epsilon}^\alpha \Omega_{\gamma\alpha}^\epsilon,$$

from which and from (50.16), (5.7) and (40.5) we have

$$(50.17) \qquad B_{\beta\gamma} = \tfrac{1}{4} g_{ab} A_\beta^a A_\gamma^b = \tfrac{1}{4} g_{\beta\gamma},$$

where $g_{\beta\gamma}$ is thus defined.

If the group G_r is semi-simple, the determinant $|g_{ab}|$ is of rank r by Cartan's criterion (§44) and likewise the rank of $|g_{\alpha\beta}|$. Consequently $g_{\alpha\beta}$ serves as the fundamental tensor of a Riemannian

---

[*] Cf. *Cartan* and *Schouten*, 1926, 1, p. 813; *Schouten*, 1929, 1, p. 266; *Cartan*, 1927, 2, p. 18.

metric for the group space.   From (49.16) and (50.17) it follows that $g_{\alpha\beta;\epsilon} = 0$, in which case it can be shown that the coefficients $\Gamma_{\beta\gamma}^{\alpha}$ are the Christoffel symbols as defined by (47.11), and consequently the curvature tensor is the Riemannian curvature tensor.   Furthermore $B_{\beta\gamma}$ is the Ricci tensor $R_{\beta\gamma}$ and in this case equations (50.17) are $R_{\beta\gamma} = \frac{1}{4}g_{\beta\gamma}$ so that the space is an Einstein space.*   Hence we have the theorem of Cartan and Schouten:†

[50.8]   *For a semi-simple group the (0)-connection of the group-space is Riemannian, the fundamental tensor being*

$$(50.18) \qquad g_{\alpha\beta} = g_{ab}A_{\alpha}^{a}A_{\beta}^{b};$$

and it is an Einstein space.

If we put $A_{a|\alpha} = g_{\alpha\beta}A_{a}^{\beta}$, we have

$$\begin{aligned} A_{a|\beta;\gamma} &= g_{\alpha\beta}A_{a;\gamma}^{\alpha} = g_{\alpha\beta}(-A_{a}^{\delta}L_{\delta\gamma}^{\alpha} + A_{a}^{\delta}\{{}_{\delta\gamma}^{\alpha}\}) = g_{\alpha\beta}A_{a}^{\delta}\Omega_{\gamma\delta}^{\alpha} \\ &= g_{bc}A_{\alpha}^{b}A_{\beta}^{c}A_{a}^{\delta}c_{ef}^{h}A_{\gamma}^{e}A_{\delta}^{f}A_{h}^{\alpha} = g_{hc}c_{ea}^{h}A_{\beta}^{c}A_{\gamma}^{e} \\ &= c_{eac}A_{\beta}^{c}A_{\gamma}^{e}, \end{aligned}$$

where c_{eac} is defined by (41.8).   Since these constants are skew-symmetric in all the indices, we have

$$(50.19) \qquad A_{a|\beta;\gamma} + A_{a|\gamma;\beta} = 0,$$

and from (50.18)

$$(50.20) \qquad g_{\alpha\beta}A_{a}^{\alpha}A_{a}^{\beta} = g_{aa}.$$

These results are interpreted by theorem [52.6].

For any group equations (50.18), in which g_{ab} are any constants such that the determinant $|g_{ab}|$ is not zero, may be used to assign a Riemannian metric to the group-space.   If we calculate the Christoffel symbols for $g_{\alpha\beta}$, we obtain

$$\{{}_{\beta\gamma}^{\alpha}\} = \Gamma_{\beta\gamma}^{\alpha} + g^{\alpha\delta}(g_{\beta\epsilon}\Omega_{\delta\gamma}^{\epsilon} + g_{\gamma\epsilon}\Omega_{\delta\beta}^{\epsilon}).$$

From the above results it follows that for a semi-simple group the Γ's are equal to the Christoffel symbols, and consequently the Riemannian geodesics coincide with the paths.   Furthermore, we have

$$g_{\beta\epsilon}\Omega_{\delta\gamma}^{\epsilon} + g_{\gamma\epsilon}\Omega_{\delta\beta}^{\epsilon} = 0.$$

---

* 1926, 3, pp. 113–114.
† 1926, 1, p. 810; cf. also, *Schouten*, 1929, 1, p. 271.

If we multiply these equations by $g^{\beta\delta}$ and sum for β and δ we obtain

$$\Omega^{\delta}_{\delta\gamma} = 0,$$

and from (50.16) it follows that $c^a_{ab} = 0$. In order that the Riemannian geodesics for any group coincide with the paths the same results must hold.*

<h3 style="text-align:center">Exercises</h3>

1. Show that the components of the curvature tensor satisfy the identities

$$R^h_{ijk} = -R^h_{ikj}, \qquad R^h_{ijk} + R^h_{jki} + R^h_{kij} = 0.$$

1926, 3, p. 21.

2. Show that $R^h_{hjk} = 0$ and that $R_{ij} \equiv R^h_{ijh}$ is a symmetric tensor of the second order; it is called the Ricci tensor.

1926, 3, p. 21.

3. If we put $R_{hijk} = g_{hl}R^l_{ijk}$, then

$$R_{hijk} = -R_{ihjk} = -R_{hikj} = R_{jkhi},$$

and

$$g^{hl}R_{lijk} = R^h_{ijk}, \qquad g^{ii}R_{hijk} = R_{hk}, \qquad g^{hk}R_{hijk} = R_{ij}.$$

1926, 3, p. 21.

4. When $R_{hijk} = \rho(g_{hk}g_{ij} - g_{hj}g_{ik})$, ρ must be a constant, and V_n is called a space of constant curvature.

1926, 3, pp. 25, 83.

5. When the fundamental quadratic form of a Riemannian space is positive definite, none of the principal minors of any order of the matrices $\|g_{ij}\|$ and $\|g^{ij}\|$ are zero.

1932, 2, p. 202.

6. Equipollence of the first or second kind (§50) possesses the following properties:

1° A vector equipollent to a zero vector, that is, one whose extremity coincides with its origin, is a zero vector;

2° Any vector is equipollent to itself;

3° If a vector is equipollent to a second vector, the latter is equipollent to the first;

4° If two vectors are equipollent, so also are their opposites;

5° Every point in space is the origin of one, and only one vector, equipollent to a given vector;

6° Two vectors equipollent to a third are equipollent to one another;

7° If $\overrightarrow{ab}$ is equipollent to $\overrightarrow{a'b'}$ and $\overrightarrow{bc}$ to $\overrightarrow{b'c'}$, the vector $\overrightarrow{ac}$ is equipollent to $\overrightarrow{a'c'}$.

Cartan, 1927, 2, p. 4.

---

* *Cartan* and *Schouten*, 1926, 1.

7. If in a space S there is defined an equipollence of vectors possessing the seven properties of Ex. 6, then S may be regarded as the group-space of a group for which the equipollence is of the first kind for the group.

Cartan, 1927, 2, p. 7.

8. A necessary and sufficient condition that two groups of the same order be isomorphic is that a point correspondence can be established between the group-spaces of the two groups transforming an equipollence of either kind of one space into an equipollence of either kind of the second space.

Cartan, 1927, 2, p. 11.

51. Groups of motions. Equations of Killing.

Each transformation of the group of motions in euclidean 3-space preserves lengths and angles, and thus leaves invariant the metric properties of the space. We generalize this conception and say that a Riemannian space admits a group of motions into itself, when each transformation leaves invariant the metric properties of the space. In order to obtain the conditions for a group of motions, we consider infinitesimal transformations.

If a V_n with the fundamental form

$$(51.1) \qquad \varphi = g_{ij}dx^i dx^j$$

is subjected to an infinitesimal transformation

$$(51.2) \qquad x'^i = x^i + \xi^i \delta t,$$

we have

$$(51.3) \qquad \delta dx^i = d\delta x^i = \frac{\partial \xi^i}{\partial x^j}dx^j \delta t, \qquad \delta g_{ij} = \frac{\partial g_{ij}}{\partial x^k}\xi^k \delta t.$$

Consequently in order that $\delta\varphi = 0$, it is necessary and sufficient that

$$(51.4) \qquad \xi^k\frac{\partial g_{ij}}{\partial x^k} + g_{ik}\frac{\partial \xi^k}{\partial x^i} + g_{jk}\frac{\partial \xi^k}{\partial x^i} = 0,$$

which in consequence of (47.15) may be written

$$(51.5) \qquad \xi_{i,j} + \xi_{j,i} = 0, \qquad where \qquad \xi_i = g_{ik}\xi^k.$$

These equations of condition were first obtained by Killing* and are known as the *equations of Killing*.

---

* 1892, 1, p. 167.

The angle between two directions defined by d_1x^i and d_2x^i is given by*

$$\cos\,\alpha = \frac{g_{ij}d_1x^id_2x^j}{\sqrt{(e_1g_{ij}d_1x^id_1x^j)(e_2g_{kl}d_2x^kd_2x^l)}}.$$

In consequence of (51.3) we have

$$\delta(g_{ij}d_1x^id_2x^j) = \left(\xi^k\frac{\partial g_{ij}}{\partial x^k} + g_{ik}\frac{\partial \xi^k}{\partial x^j} + g_{jk}\frac{\partial \xi^k}{\partial x^i}\right)d_1x^id_2x^j\delta t,$$

and hence $\delta \cos \alpha = 0$ when (51.4) hold. Accordingly we have that equations (51.4) are necessary and sufficient conditions that (51.2) define an *infinitesimal motion*.

If (51.4) are satisfied and the coordinate system is chosen so that

$$(51.6) \qquad\qquad \xi^i = \delta^i_1,$$

then equations (51.4) reduce to

$$(51.7) \qquad\qquad \frac{\partial g_{ij}}{\partial x^1} = 0.$$

Hence the g's are independent of x^1 and consequently the fundamental form (51.1) is unaltered by the finite equations

$$(51.8) \qquad x'^1 = x^1 + t, \qquad x'^i = x^i \quad (j = 2, \cdots, n)$$

of the G_1 generated by (51.2). Hence we have:

[51.1] *When a space admits an infinitesimal motion, it admits the group G_1 of motions generated by the infinitesimal motion.*

Conversely when (51.7) are satisfied, equations (51.4) are satisfied by the quantities (51.6), and we have:

[51.2] *A necessary and sufficient condition that a V_n admit a group G_1 of motions is that there exists a coordinate system for which all of the coefficients g_{ij} do not involve one coordinate, say x^1; then the curves of parameter x^1 are the trajectories of the motion.*

From the definition of a motion and the property of geodesics stated at the close of §47 it follows that:

---

* 1926, 3, p. 38.

[51.3]   *In a motion of a space into itself geodesics go into geodesics.*

A necessary and sufficient condition that the trajectories of two motions be the same is that (51.5) be satisfied by $\bar{\xi}^i$ and ξ^i, where $\bar{\xi}^i = \rho \xi^i$. For this to be true we must have $\dfrac{\partial \rho}{\partial x^i} \xi_j + \dfrac{\partial \rho}{\partial x^j} \xi_i = 0$. Replacing j by k and eliminating ξ_i from these two equations. we obtain

$$(51.9) \qquad \frac{\partial \rho}{\partial x^i}\left(\frac{\partial \rho}{\partial x^k} \xi_j - \frac{\partial \rho}{\partial x^j} \xi_k \right) = 0.$$

From these two sets of equations it follows that ρ is a constant and we have:

[51.4]   *Two groups G_1 of motions of a V_n cannot have the same trajectories.*

It follows from (51.4) that:

[51.5]   *If ξ_a^i for $a = 1, \cdots, r$ are the components of infinitesimal motions of a V_n, so also is $c^a \xi_a^i$ where the c's are arbitrary constants.*

As a consequence we have:

[51.6]   *If each of the r generators of a group G_r generate a G_1 of motions, every transformation of G_r is a motion.*

If geodesics are drawn normal to a hypersurface V_{n-1} in a V_n and equal lengths are laid off on these geodesics from V_{n-1}, the locus of the end points is a hypersurface normal to the geodesics.   In this way we obtain a family of hypersurfaces which are said to be *geodesically parallel.** If the coordinates are chosen so that this family of hypersurfaces are the spaces $x^1 = $ const. and we choose for the coordinate x^1 the distance measured from the given V_{n-1} along the normal geodesics, we have

$$(51.10) \qquad\qquad g_{11} = e_1 \qquad g_{1j} = 0 \qquad (j = 2, \cdots, n).$$

We are in position to prove the theorem:

[51.7]   *If a space V_n admits an intransitive group G_r of motions and a hypersurface V_{n-1} is an invariant variety, the hypersurfaces geodesically parallel to V_{n-1} are invariant varieties.*

---

* Cf. 1926, 3, p. 57.

For, if V_n is referred to the coordinate system just referred to, then $x^1 = 0$ is an invariant variety and from (19.6) it follows that $\xi_a^1 = 0$ for $a = 1, \cdots, r$, when $x^1 = 0$. From (51.4) for $i = j = 1$ and (51.10) we have $\dfrac{\partial \xi_a^1}{\partial x^1} = 0$ and consequently $\xi_a^1 = 0$ for all values of x^1, and the theorem is proved.

If the rank of the matrix $\|\xi_a^i\|$ of a group G_r of motions is $n - 1$, the minimum invariant varieties for ordinary points are hypersurfaces, that is, varieties of order $n - 1$. If we choose the coordinate system so that these hypersurfaces are the coordinate spaces $x^1 = $ const., and the other coordinate spaces are orthogonal to them, then $g_{11} \neq 0$ and $g_{1j} = 0$ for $j = 2, \cdots, n$. In this case $\xi_a^1 = 0$ and equations (51.4) for $i = j = 1$ reduce to $\xi_a^k \dfrac{\partial g_{11}}{\partial x^k} = 0$ for $k = 2, \cdots, n$. Since the rank of the matrix $\|\xi_a^i\|$ is $n - 1$, we have that g_{11} is a function of x^1 alone. By choosing x^1 suitably we have (51.10) holding and consequently we have:

[51.8] *When the minimum invariant varieties of a group G_r of motions are hypersurfaces, they are geodesically parallel.*

52. Translations. A particular sub-class of motions are those for which the trajectory of each point is a geodesic. They are called *translations* and are an evident generalization of translations in euclidean space.

If the coordinates are chosen so that the components of the translation have the values (51.6), then the trajectories (51.8) are to be geodesics. In this case we have from equations (47.21) and (47.11) that

$$(52.1) \qquad \{{}_{11}^{i}\} = g^{ik}\left(\frac{\partial g_{1k}}{\partial x^1} - \frac{1}{2}\frac{\partial g_{11}}{\partial x^k}\right) = 0 \qquad \left(\begin{matrix} j = 2, \cdots, n; \\ k = 1, \cdots, n \end{matrix}\right).$$

In consequence of (51.7) this condition reduces to

$$(52.2) \qquad g^{ik}\frac{\partial g_{11}}{\partial x^k} = 0 \qquad (j, k = 2, \cdots, n).$$

From (47.10) it follows that the determinant of these equations is equal to g_{11}/g, where g is the determinant $|g_{ij}|$. Hence from (52.2) and (51.7) it follows that g_{11} is a constant, different from or equal to zero. Conversely, if g_{11} is a constant and (51.7) holds, we have that $\{^i_{11}\} = 0$ and consequently the trajectories (51.8) are geodesics. Since $g_{ij}\xi^i\xi^j$ is an invariant, we have that it is a constant in any coordinate system. From (51.2) it follows that the square of the displacement in any motion is $eg_{ij}\xi^i\xi^j(\delta t)^2$. Hence we have:

[52.1] *A necessary and sufficient condition that a motion be a translation is that*

$$g_{ij}\xi^i\xi^j = \text{const.,}$$

in which case every point is moved the same distance.

This property is an evident generalization of a characteristic property of translations in euclidean space. Since $a\xi^i$, where a is any constant not zero, are vectors of the same G_1 as ξ^i, we have:

[52.2] *A necessary and sufficient condition that the vector ξ^i of a G_1 of motions to within a constant multiplier be a unit or a null vector is that the motions be translations.*

A characteristic property of translations is given by the theorem:

[52.3] *A necessary and sufficient condition that a field of unit or null vectors determine a translation is that the vectors along any geodesic make a constant angle with the geodesic.*

In order to prove the theorem we take the arc s for the parameter along a geodesic C, so that the geodesic satisfies equations (47.22).

The cosine of the angle between the vector ξ^i and C is equal to $\xi_i\dfrac{dx^i}{ds}$.[*]

For this to be a constant we must have

$$\frac{d}{ds}\left(\xi_i\frac{dx^i}{ds}\right) = \frac{\partial\xi_j}{\partial x^k}\frac{dx^k}{ds}\frac{dx^i}{ds} + \xi_i\frac{d^2x^i}{ds^2}$$
$$= \xi_{j,k}\frac{dx^i}{ds}\frac{dx^k}{ds} = 0.$$

Since this condition must be satisfied for every geodesic, we must have (51.5) and, since ξ^i is a unit or null vector, the motion is a translation. As a consequence of theorems [52.2] and [52.3] we have:

[52.4] *The trajectories of two G_1's of translations meet under constant angle.*[†]

---

[*] In the case of a null vector this is taken as the definition of the cosine of the angle. Cf. 1926, 3, p. 38.

[†] *Bianchi*, 1918, 1, p. 501.

From theorems [51.5], [52.1] and [52.4] we obtain:

[52.5] *If ξ_a^i for $a = 1, \cdots, p$ are the vectors of translations, so also is $c^a \xi_a^i$, where the c's are arbitrary constants.*

From (50.19) and (50.20) it follows that the infinitesimal transformations of the first parameter group in the group-space S of a semi-simple group are translations. Since similar results hold for the second parameter group, we have the theorem of Cartan and Schouten:*

[52.6] *The infinitesimal transformations of the first and second parameter groups of a semi-simple group are translations of the group-space.*

53. The determination of groups of motions. Theorem [51.2] supplies a criterion for determining whether or not a given Riemannian space admits a one-parameter group of motions. Instead of proceeding in this manner, we consider under what conditions equations (51.4), or their equivalent (51.5), admit one or more solutions.

Differentiating equation (51.5) covariantly, we obtain

(53.1) $$\xi_{i,jk} + \xi_{j,ik} = 0.$$

If these equations are satisfied, so also are

$$\xi_{i,jk} + \xi_{j,ik} + \xi_{i,kj} + \xi_{k,ij} - (\xi_{j,ki} + \xi_{k,ji}) = 0.$$

From (47.16) it follows that

(53.2) $$\xi_{i,jk} - \xi_{i,kj} = \xi_h R^h_{ijk},$$

so that the preceding equations are equivalent to

$$2\xi_{i,jk} + \xi_h(R^h_{ikj} + R^h_{jik} + R^h_{kij}) = 0.$$

In consequence of the identities of Ex. 1, p. 207 these equations reduce to

(53.3) $$\xi_{i,jk} = -\xi_h R^h_{kij},$$

and when these equations are satisfied, so also are (53.1) and (53.2).

---

* 1926, 1, p. 811.

The conditions of integrability of these equations obtained from the Ricci identities (cf. (47.16))

$$\xi_{i,jkl} - \xi_{i,jlk} = \xi_{h,j}R^h_{ikl} + \xi_{i,h}R^h_{jkl}$$

are

(53.4) $\quad \xi_h(R^h_{kij,l} - R^h_{lij,k}) + \xi_{h,l}R^h_{kij} - \xi_{h,k}R^h_{lij} + \xi_{h,j}R^h_{ikl}$
$$+ \xi_{i,h}R^h_{jkl} = 0.$$

If we write (53.3) in the form

(53.5) $\qquad \dfrac{\partial \xi_{i,i}}{\partial x^k} = \xi_{h,j}\{^h_{ik}\} + \xi_{i,h}\{^h_{jk}\} - \xi_h R^h_{kij},$

and observe that by definition

(53.6) $\qquad \dfrac{\partial \xi_i}{\partial x^j} = \xi_h\{^h_{ij}\} + \xi_{i,j},$

we see that the solution of (53.3) reduces to the solution of the system of equations (53.5), (53.6) in the $n(n + 1)$ quantities ξ_i and $\xi_{i,j}$, the latter being subject to the conditions (51.5); consequently the system is mixed, equations (51.5) being the set F_0 in the terminology of §1, and (53.4) are the set F_1. If (53.4) are a consequence of (51.5), then since there are $n(n + 1)/2$ of the latter, the general solution of the above system involves $n(n + 1)/2$ parameters. If (53.4) is not a consequence of (51.5), we differentiate (53.4) and proceed as in §1. The question of whether there is any solution, and if so how many parameters are involved reduces to the question of the consistency of a sequence of equations as in §1 (cf. Ex. 2, p. 12). Moreover, if there are p independent solutions, any solution is a linear combination (constant coefficients) of these.

We consider further the case where (53.4) is a consequence of (51.5). As a first consequence we have

(53.7) $\qquad\qquad R^h_{kij,l} - R^h_{lij,k} = 0,$

and from the other terms of (53.4) we have

$$\xi_{h,p}(\delta^p_l R^h_{kij} - \delta^p_k R^h_{lij} + \delta^p_j R^h_{ikl} - \delta^p_i R^h_{jkl}) = 0,$$

from which because of (51.5) it follows that

(53.8) $\quad \delta^p_l R^h_{kij} - \delta^h_l R^p_{kij} - \delta^p_k R^h_{lij} + \delta^h_k R^p_{lij} + \delta^p_j R^h_{ikl} - \delta^h_j R^p_{ikl}$
$$- \delta^p_i R^h_{jkl} + \delta^h_i R^p_{jkl} = 0.$$

Contracting for l and p, we have in consequence of Exs. 1, 2, p. 207

$$(53.9) \qquad R^h_{kij} = \frac{1}{n-1}(\delta^h_j R_{ik} - \delta^h_i R_{jk}).$$

Multiplying by g_{hl} and summing for h, we have (cf. Ex. 3, p. 207)

$$(53.10) \qquad R_{lkij} = \frac{1}{n-1}(g_{jl}R_{ik} - g_{il}R_{jk}).$$

Multiply (53.10) by g^{ki} and sum for k and i; the result reduces to

$$nR_{jl} = Rg_{jl},$$

so that from (53.10) we have

$$(53.11) \qquad R_{lkij} = \frac{R}{n(n-1)}(g_{jl}g_{ik} - g_{il}g_{jk}),$$

where $R = g^{ij}R_{ij}$. Hence V_n is a space of constant curvature (cf. Ex. 4, p. 207). From (53.9) we have

$$R^h_{kij} = \frac{R}{n(n-1)}(\delta^h_j g_{ik} - \delta^h_i g_{jk}),$$

which values satisfy (53.8) identically, and also (53.7) in consequence of (47.15). Hence we have:

[53.1] *When and only when V_n is a space of constant curvature, the equations of Killing admit solutions involving $n(n+1)/2$ parameters; in all other cases there are fewer parameters.*

We shall show that, if X_1f and X_2f are the symbols of two one-parameter groups of motions, so also is $(X_1, X_2)f$. If we denote by ξ^i the vector of the latter, we have

$$(53.12) \qquad \xi^i = \xi^k_1 \frac{\partial \xi^i_2}{\partial x^k} - \xi^k_2 \frac{\partial \xi^i_1}{\partial x^k} = \xi^k_1 \xi^i_{2,k} - \xi^k_2 \xi^i_{1,k}.$$

If we put $\xi_i = g_{ij}\xi^j$, we obtain in consequence of (51.5)

$$\xi_i = -\xi^k_1 \xi_{2k,i} + \xi^k_2 \xi_{1k,i}.$$

By means of (53.3) and the properties of R_{hijk} (Ex. 3, p. 207) we have

$$\xi_{i,j} = \xi_1^k \xi_2^h (R_{kjih} - R_{kijh}) - \xi_{1,j}^k \xi_{2k,i} + \xi_{2,j}^k \xi_{1k,i},$$

from which follows (51.5).   Hence we have:

[53.2]   *If $X_a f$ for $a = 1, \cdots , p$ are generators of p one-parameter groups of motions, so also are each of the commutators $(X_a, X_b)f$, for $a, b = 1, \cdots , p$.*

From this result and the remarks following equations (53.5) and (53.6) we have:

[53.3]   *If $X_a f$ for $a = 1, \cdots , p$ are generators of the complete set of one-parameter groups of motions, they are the generators of a group G_p of motions.*

Combining this result with theorem [51.1], we have:

[53.4]   *A space of constant curvature V_n admits a group of motions involving $n(n + 1)/2$ parameters, and only in this case does the group of motions of a space admit as large a number of parameters.*

Exercises

1. Prove analytically that geodesics go into geodesics when a V_n undergoes a motion into itself.

2. Show that, if (47.5) are the equations of a motion of V_n, the functions g'_{ij} in (47.7) are the same functions of the x''s as g_{ij} are of the x's, and conversely that, if these equations are satisfied by an infinitesimal transformation, the equations of Killing are satisfied.

3. Show that equations (53.3) may be written in the form

$$\xi^i_{,jk} = \xi^h R^i_{jkh}.$$

4. The transformations of a group of motions are of order zero or one at any point.

5. A V_3 with the fundamental form

$$\varphi = e_1 (dx^1)^2 + X_1 [e_2 (dx^2)^2 + e_3 (dx^3)^2],$$

where the e's are plus or minus one and X_1 is a function of x^1 alone, admits the intransitive group G_3 of motions whose generators are

$$p_i, \qquad e_j x^j p_i - e_i x^i p_j \qquad (i = 2, 3; i \neq j; i, j \text{ not summed}).$$
$$\text{Bianchi, 1918, 1, p. 545.}$$

6. A V_4 with the fundamental form

$$\varphi = e_1 (dx^1)^2 + X_1 [e_2 (dx^2)^2 + e_3 (dx^3)^2 + e_4 (dx^4)^2],$$

where the e's are plus or minus one and X_1 is a function of x^1 alone, admits the intransitive group G_6 of motions whose generators are

$$p_i, \qquad e_j x^i p_i - e_i x^i p_j \quad (i, j = 2, 3, 4; i \neq j; i, j \text{ not summed}).$$

Fubini, 1904, 2, p. 64.

7. A V_4 with the fundamental form

$$\varphi = (dx^1)^2 + X_1\{(dx^2)^2 + e^{2x^2}[(dx^3)^2 + (dx^4)^2]\},$$

where X_1 is a function of x^1 alone, admits the intransitive group G_6 of motions whose generators are

$$p_i, \qquad x^j p_i - x^i p_j, \qquad -p_2 + x^3 p_3 + x^4 p_4,$$
$$-x^i p_2 + \tfrac{1}{2}[(x^i)^2 - (x^j)^2 - e^{-2x^2}]p_i + x^i x^j p_j$$
$$(i, j = 3, 4; i \neq j; i, j \text{ not summed}).$$

Fubini, 1904, 2, p. 64.

8. A V_2 admits a translation, when and only when its curvature is zero.

9. If a space admits a field of absolutely parallel vectors, they are the vectors of a translation.

1926, 3, p. 239.

10. The vectors of a G_1 of motions make constant angle with any non-minimal geodesic.

54. Equations of Killing in another form.

By means of the formulas of §21 we give the equations of Killing a form which is adapted to the solution of certain problems.

If the generic rank q of the matrix

$$(54.1) \qquad\qquad M = \left\| \xi_a^i \right\|$$

is less than r, we arrange the ξ's so that the matrix (54.1) for $a = 1, \cdots, q$ is of rank q. When $q < n$ we use the special coordinate system for which (21.8) hold, and the equations (51.4) for $a = 1, \cdots, q$ become

$$(54.2) \qquad \xi_h^l \frac{\partial g_{ij}}{\partial x^l} + g_{il}\frac{\partial \xi_h^l}{\partial x^j} + g_{lj}\frac{\partial \xi_h^l}{\partial x^i} = 0 \quad \begin{pmatrix} h, l = 1, \cdots, q; \\ i, j = 1, \cdots, n \end{pmatrix}.$$

If these equations are multiplied by ξ_m^h, defined by (21.9), and h is summed, we have in consequence of (21.9) and (21.12)

$$(54.3) \qquad \frac{\partial g_{ij}}{\partial x^m} = g_{il}\Lambda_{jm}^l + g_{lj}\Lambda_{im}^l \quad \begin{pmatrix} l, m = 1, \cdots, q; \\ i, j = 1, \cdots, n \end{pmatrix}.$$

If $q = n$, we have equations (54.2) for $h, l, m = 1, \cdots, n$.

When $q = r$, equations (54.2) or their equivalent (54.3) are the only conditions. However, when $q < r$, we have in addition to (54.2) the equations (51.4) where a takes the values $q + 1, \cdots, r$. When in these equations we substitute from (21.2) and (21.8), we obtain with the aid of (54.3) the following equivalent system:

$$(54.4) \qquad \xi_h^l\left(g_{il}\frac{\partial \varphi_s^h}{\partial x^i} + g_{lj}\frac{\partial \varphi_s^h}{\partial x^i}\right) = 0 \qquad \begin{pmatrix} l, h = 1, \cdots, q; \\ s = q + 1, \cdots, r; \\ i, j = 1, \cdots, n \end{pmatrix}.$$

When $q < r$, equations (54.3) and (54.4) are the conditions of the problem.

The conditions of integrability of equations (54.3) are

$$(54.5) \qquad g_{il}\Lambda_{jmp}^l + g_{lj}\Lambda_{imp}^l = 0 \qquad \begin{pmatrix} l, m, p = 1, \cdots, q; \\ i, j = 1, \cdots, n \end{pmatrix},$$

where Λ_{jmp}^l are defined by (21.17). When $q = r$ and $r \leqslant n$, we have from (21.19) that (54.5) are satisfied identically, and consequently (54.3), which are the only conditions in this case, are completely integrable. Hence a solution is determined by arbitrary initial values of the g's, which when $n > r$ may be arbitrary functions of the variables $x^{r+1}, \ldots, x^n$, and we have:[*]

[54.1] *Any r-parameter continuous group in n variables such that the generic rank of the matrix $\|\xi_a^i\|$ is $r(\leqslant n)$ is the group of motions of a Riemannian space whose fundamental tensor g_{ij} involves $n(n + 1)/2$ arbitrary functions.*

When $n = r$, that is, when the group is simply transitive, the solution involves $n(n + 1)/2$ arbitrary constants, as was shown by Bianchi.[†] In particular, if the group is simply transitive and Abelian, the coordinates may be chosen so that $\xi_a^i = \delta_a^i$ (§13), and then from (51.4) we have that the g's are constants and therefore V_n is euclidean.[‡]

We met with the above result in another form in §48; in fact (48.20) are the equations of Killing, the g's being defined by (48.14) in terms of the vectors of the reciprocal group. The determination

---

[*] Cf. *Fubini*, 1903, 1, p. 54.

[†] 1897, 1, p. 291.

[‡] Cf. *Bianchi*, 1918, 1, p. 521.

of the latter involve n^2 arbitrary constants, but from the definition of the g's it follows that the latter involve only $n(n + 1)/2$ arbitrary constants.

Before proceeding with the general case when $q < r$, we consider intransitive Abelian groups. As shown in §13 the coordinates can be chosen so that

$$(54.6) \qquad \xi_h^l = \delta_h^l, \qquad \xi_h^s = 0 \qquad \binom{h, l = 1, \cdots, q;}{s = q + 1, \cdots, n},$$

q being the generic rank of the matrix (54.1). From (21.5) we have that the functions Φ_{as}^h are zero, so that from (21.11) it follows that the functions φ_s^h do not involve $x^1, \ldots, x^q$. In this case as follows from (54.4) each set of φ's, that is, φ_s^h for s fixed and $h = 1, \cdots, q$, satisfies the equations

$$(54.7) \qquad g_{hi}\frac{\partial \varphi^h}{\partial x^j} + g_{hj}\frac{\partial \varphi^h}{\partial x^i} = 0.$$

When i and j take the values $1, \ldots, q$, these equations are satisfied in view of the above results. When i takes the values 1 to q and j from $q + 1, \cdots, n$, we get the first set of the following equations, and when $i = j = q + 1, \cdots, n$, we get the second set in which t is not summed:

$$(54.8) \qquad g_{hi}\frac{\partial \varphi^h}{\partial x^t} = 0, \qquad g_{ht}\frac{\partial \varphi^h}{\partial x^t} = 0 \qquad \binom{h, l = 1, \cdots, q;}{t = q + 1, \cdots, n}.$$

From the first set it follows that the determinant $|g_{hl}|$ is zero, otherwise the φ's are constants which is contrary to hypothesis. In order that the second set also may hold, we must have relations of the form

$$g_{hi} = g_{hl}A_i^l \qquad (h, l = 1, \cdots, q; i = 1, \cdots, n).$$

From these equations we have

$$|g_{hs}| = |g_{hl}| \cdot |A_s^l|,$$

when s takes any q values of the set $1, \ldots, n$. Hence the matrix $\|g_{hi}\|$ is of rank $r - 1$ at most and consequently $|g_{ij}| = 0$ for $i, j = 1, \cdots, n$, contrary to hypothesis. Accordingly $q = r$ and we have with the aid of theorem [54.1]:

[54.2] *When and only when the rank of the matrix $\|\xi_a^i\|$ of an intransitive Abelian group G_r is r, it is a group of motions, and every Abelian group satisfying this condition is a group of motions.*[*]

We consider now the case of non-Abelian groups for which $q < r$, and observe that we have the mixed system consisting of equations (54.3) and the set (54.4), the latter being the set F_0 in the sense of §1. In these equations the independent variables are $x^1, \ldots, x^q$ and $x^{q+1}, \ldots, x^n$ enter as parameters when the group is intransitive.

From (21.18) it follows that (54.5) are satisfied in consequence of (54.4). We introduce the notation

$$A_{sij} = \xi_h^l \left(g_{il} \frac{\partial \varphi_s^h}{\partial x^j} + g_{jl} \frac{\partial \varphi_s^h}{\partial x^i} \right).$$

Because of (54.3), (21.13) and (21.15) we have

$$\frac{\partial}{\partial x^m}(\xi_h^l g_{il}) = \xi_p^l \xi_m^{l_1}(c_{l_1 h}^p + c_{l_1 h}^t \varphi_t^p) g_{il} + \xi_h^p g_{pl} \Lambda_{im}^p$$

for $h, l, l_1, m, p = 1, \cdots, q; t = q + 1, \cdots, r; i = 1, \cdots, n$. Also from (21.11), (21.5) and (21.14) we have

$$\frac{\partial^2 \varphi_s^h}{\partial x^i \partial x^m} = \xi_m^{l_1} \left[(c_{l_1 s}^t - c_{l_1 p}^t \varphi_s^p) \frac{\partial \varphi_t^h}{\partial x^i} - (c_{l_1 p}^{\cdot h} + c_{l_1 p}^t \varphi_t^h) \frac{\partial \varphi_s^p}{\partial x^i} \right] + \xi_p^{l_1} \Lambda_{jm}^p \Phi_{l_1 s}^h.$$

Making use of these results we have

$$\frac{\partial A_{sij}}{\partial x^m} = \Lambda_{im}^p A_{spj} + \Lambda_{jm}^p A_{spi} + \xi_m^{l_1}(c_{l_1 s}^t - c_{l_1 p}^t \varphi_t^h) A_{tij}.$$

From the form of these equations it follows that, when G_r is intransitive, if for particular values of $x^1, \ldots, x^n$, say x_0^i, the equations (54.4) are consistent and we take a set of values of g_{il} for $i = 1, \cdots, n; l = 1, \cdots, q$ from these equations as the initial values of g_{il} such that $\|g_{il}\|$ is of rank q and any arbitrary initial values for g_{ab}, where $a, b = q + 1, \cdots, n$, such $|g_{ij}| \neq 0$, then the solutions g_{ij} of (54.3) so determined will satisfy (54.4) for the values of the variables in the domain about the initial values for which these functions are defined. Hence we have the following theorem:

---

[*] Cf. *Bianchi*, 1918, 1, p. 521.

[54.3] *When for an intransitive group G_r the generic rank q of the matrix $\|\xi_a^i\|$ is less than r, if in the coordinate system for which $\xi_a^t = 0(t = q + 1, \cdots, n)$ the equations (54.4) are consistent in the g's when the coordinates are given particular values, such that the matrix $\|g_{il}\|$ is of rank q, there exist Riemannian spaces for which G_r is a group of motions and the components of the tensor g_{ij} involve $(n - q)(n - q + 1)/2$ arbitrary constants at least.*

When the group is multiply transitive $q = n < r$, in the equations (54.4) l takes the values $1, \ldots, n$, and all of the initial values must satisfy these equations. In this case we have:

[54.4] *When for a multiply transitive group there are particular values of the x's for which equations (54.4) are consistent and admit solutions g_{ij} such that the determinant $|g_{ij}|$ is not zero, each such solution determines a Riemannian space for which the given group is a group of motions.*

55. Intransitive groups of motions.

The problem of intransitive groups of motions is reduced to that of transitive groups by means of the following theorem due to Fubini:*

[55.1] *If a space V_n admits an intransitive group of motions G_r, the group induced on a minimum invariant variety V_q, where q is the generic rank of the matrix $\|\xi_a^i\|$, has r parameters, and the finite equations of G_r are reducible by a suitable choice of coordinates to those of a transitive group on q variables.*

In proving the first part of this theorem we recall from §20 that each ordinary point of V_n is in a minimum invariant variety V_q and that if the induced group on any V_q is not of order r, there is a sub-group G_σ of G_r leaving this V_q point-wise invariant. Assume that there is an invariant variety V_q^0 which is left point-wise invariant by a sub-group G_σ of G_r and let P be an ordinary point not in V_q^0. Consider now the V_{q+1} consisting of infinity of V_q's including V_q^0 and the V_q through P; this evidently is an invariant variety of G_r and in particular of the subgroup G_σ leaving V_q^0 point-wise invariant. Any motion in G_r induces a motion in V_{q+1} which sends geodesics into geodesics and preserves angles and distances (§51). The

---

* 1903, 1, p. 40; the proof of the second part of the theorem as here given is due to the author, cf. 1932, 2, pp. 199–201.

point P is on a geodesic in V_{q+1} normal to V_q^0. This geodesic remains fixed under the group induced in V_{q+1} by G_σ, because angles are preserved and geodesics go into geodesics. Hence P remains fixed, because distances are preserved. Since P is any point, G_σ must be the identity and the first part of the theorem is proved.

In order to prove the second part of the theorem we make use of the results of §54 and the formulas of §21, noting in particular that the coordinates x^i are those for which (21.8) hold. We seek a non-singular transformation of coordinates

$$(55.1) \qquad x^l = \varphi^l(x'^1, \cdots, x'^n), \qquad x^s = x'^s$$

$$\begin{pmatrix} l = 1, \cdots, q; \\ s = q+1, \cdots, n \end{pmatrix}$$

such that the components $\xi_h'^l$ for $l, h = 1, \cdots, q$ are independent of $x'^{q+1}, \ldots, x'^n$. We indicate by

$$(55.2) \qquad x'^l = \varphi'^l(x^1, \cdots, x^n), \qquad x'^s = x^s$$

the inverse of (55.1). From the equations (48.2) and (21.8) we have $\xi_h'^s = 0$, and

$$(55.3) \qquad \xi_h^l = \xi_h'^\lambda \frac{\partial x^l}{\partial x'^\lambda}, \qquad \xi_h'^\lambda = \xi_h^l \frac{\partial x'^\lambda}{\partial x^l}. \, *$$

Assuming that the new coordinate system has the desired property and differentiating the first set of (55.3) with respect to x'^σ, on making use of (21.13), the second set of (55.3) and the fact that the determinant $\left| \xi_h^l \right|$ is not zero, we obtain

$$(55.4) \qquad \frac{\partial x'^\lambda}{\partial x^m} \frac{\partial^2 x^l}{\partial x'^\lambda \partial x'^\sigma} + \Lambda_{im}^l \frac{\partial x^i}{\partial x'^\sigma} = 0.$$

For the transformation (55.1) we have

$$\frac{\partial x'^\lambda}{\partial x^m} \frac{\partial x^m}{\partial x'^\mu} = \delta_\mu^\lambda.$$

Consequently if we multiply (55.4) by $\dfrac{\partial x^m}{\partial x'^\mu}$ and sum for m we obtain

---

* In these and subsequent equations h, l, m, p, λ, μ take the values 1 to q; $\sigma, \tau = q+1, \cdots, n; i, \alpha = 1, \cdots, n$.

$$(55.5) \qquad \frac{\partial^2 x^l}{\partial x'^\mu \partial x'^\sigma} + \Lambda^l_{im} \frac{\partial x^i}{\partial x'^\sigma} \frac{\partial x^m}{\partial x'^\mu} = 0.$$

When we differentiate this equation with respect to x'^λ, reduce the resulting equation by means of (55.5) and subtract from the result the equation obtained from it on interchanging λ and μ, we get

$$(55.6) \qquad \Lambda^l_{imp} \frac{\partial x^i}{\partial x'^\sigma} \frac{\partial x^m}{\partial x'^\mu} \frac{\partial x^p}{\partial x'^\lambda} = 0,$$

where Λ^l_{imp} is defined by (21.17). When we differentiate equation (55.5) with respect to x'^τ for $\tau = q + 1, \cdots, n$, reduce the resulting equation by means of (55.5) and subtract from the result the equation obtained from it on interchanging σ and τ, we obtain an equation which vanishes identically because of (21.16). Hence (55.6) are the only conditions of integrability of the system (55.5). We replace (55.5) by the system

$$(55.7) \qquad \begin{aligned} \frac{\partial x^l}{\partial x'^\alpha} &= \psi^l_\alpha, \\ \frac{\partial \psi^l_\mu}{\partial x'^\sigma} = \frac{\partial \psi^l_\sigma}{\partial x'^\mu} &= -\Lambda^l_{pm}\psi^p_\sigma\psi^m_\mu - \Lambda^l_{\sigma m}\psi^m_\mu, \end{aligned}$$

of which the conditions of integrability are

$$(55.8) \qquad (\Lambda^l_{hmp}\psi^h_\sigma + \Lambda^l_{\sigma mp})\psi^m_\mu\psi^p_\lambda = 0.$$

These equations are of the kind discussed in §1 and (55.8) is the set F_1 in the terminology of §1.

When $q = r$, equations (55.8) are satisfied identically because of (21.19), and consequently the system (55.7) is completely integrable. Since the system does not contain derivatives of ψ^l_μ with respect to $x^1, \ldots, x^q$, these and higher derivatives may be chosen arbitrarily. In choosing the initial values of the ψ's, we must insure the requirement that the transformation be non-singular. Hence we have the theorem:

[55.2] *When the rank of the matrix $\|\xi^i_a\|$ of an intransitive group G_r is r, there exist coordinate systems for V_n in each of which the ξ's are functions of $x^1, \ldots, x^r$ at most, the minimum invariant varieties being defined by $x^{r+1} = \text{const.}, \cdots, x^n = \text{const.}$*

When $q < r$, we have equations (54.4) to be taken into account. If we multiply these by g^{it} for $t = q + 1, \cdots, n$ and sum for i, we obtain, because $\xi_h^t = 0$,

$$\xi_h^l g^{it} g_{lj} \frac{\partial \varphi_s^h}{\partial x^i} = 0.$$

If we multiply this equation by g^{im}, sum for j and note that the determinant $|\xi_h^l|$ is not zero, we find that the $q(r - q)$ functions φ_s^h are solutions of the $n - q$ equations

(55.9) $$g^{it} \frac{\partial \varphi}{\partial x^i} = 0 \qquad \begin{pmatrix} i = 1, \cdots, n; \\ t = q + 1, \cdots, n \end{pmatrix}.$$

Consequently at most q of the φ's are independent.

Suppose that q of the φ's are independent and denote them by $\varphi^1, \cdots, \varphi^q$, then the matrix $\left\| \dfrac{\partial \varphi^h}{\partial x^i} \right\|$ for $h = 1, \cdots, q$; $i = 1, \cdots, n$ is of rank q. Assume that the jacobian $\left| \dfrac{\partial \varphi^h}{\partial x^l} \right|$ for $l = 1, \cdots, q$ is zero. If so, there exist functions A_h such that

(55.10) $$A_h \frac{\partial \varphi^h}{\partial x^l} = 0.$$

Then from (55.9) we have

(55.11) $$g^{tu} A_h \frac{\partial \varphi^h}{\partial x^u} = 0 \qquad (t, u = q + 1, \cdots, n).$$

Hence if the determinant $|g^{tu}|$ is not zero, we have $A_h \dfrac{\partial \varphi^h}{\partial x^u} = 0$ and the φ's are not independent. By Ex. 5 p. 207 this determinant cannot be zero, if the fundamental form is positive definite. Consequently the jacobian $\left| \dfrac{\partial \varphi^h}{\partial x^l} \right|$ is not zero in this case, and accordingly a non-singular transformation is defined by

$$x'^l = \varphi^l, \qquad x'^s = x^s \qquad (l = 1, \cdots, q; \, s = q + 1, \cdots, n).$$

In this new coordinate system (dropping primes) we have from the equations (21.11) for φ^l equations of the form

$$\Phi_m^h \xi_l^m = \delta_l^h,$$

where Φ_m^h are certain of the functions Φ_{ls}^h, which being functions of

φ^l are independent of $x^{q+1}, \ldots, x^n$. From these equations and (21.9) it follows that

$$(55.12) \qquad \xi_h^l = \Phi_h^l,$$

from which and (21.2) we have that all of the ξ's are functions of $x^1, \ldots, x^q$ at most. This result holds accordingly when q of the functions φ_s^h are independent and the fundamental form is positive definite; if the form is indefinite, it holds when the jacobian $\left|\dfrac{\partial \varphi^h}{\partial x^l}\right|$ is not zero.

We consider next the case when $p(<q)$ of the functions φ_s^h are independent, and denote them by $\varphi^1, \ldots, \varphi^p$. If the jacobian matrix of these functions with respect to $x^1, \ldots, x^q$ is of rank less than p, we have equations of the form (55.10) and (55.11) in which h takes the values 1 to p. Hence unless the determinant $|g^{tu}|$ is zero, we have that the rank of the matrix $\left\|\dfrac{\partial \varphi^h}{\partial x^i}\right\|$ is less than p, contrary to hypothesis. If the fundamental form is definite, in which case $|g^{tu}| \neq 0$, there is no loss in generality in assuming that the jacobian $\left|\dfrac{\partial \varphi^h}{\partial x^l}\right|$ for $h, l = 1, \cdots, p$ is of rank p. When we effect the transformation

$$x'^{l_1} = \varphi^{l_1}, \qquad x'^{l_2} = x^{l_2}, \qquad x'^t = x^t \quad \begin{pmatrix} l_1 = 1, \cdots, p; \\ l_2 = p+1, \cdots, q; \\ t = q+1, \cdots, n \end{pmatrix},$$

in the new coordinate system (dropping primes) all the functions φ_s^h are functions of $x^1, \ldots, x^p$. Proceeding as above, we have in place of (55.12) $\xi_h^{l_1} = \Phi_h^{l_1}$, where the Φ's are functions of $x^1, \cdots, x^p$. We apply a new transformation

$$x^{l_1} = x'^{l_1}, \qquad x^{l_2} = \psi^{l_2}(x'^1, \cdots, x'^n), \qquad x^s = x'^s,$$

which is a particular case of (55.1). In place of (55.5) we have

$$(55.13) \qquad \frac{\partial^2 x^{l_2}}{\partial x'^\mu \partial x'^\sigma} + \Lambda_{im}^{l_2} \frac{\partial x^i}{\partial x'^\sigma} \frac{\partial x^m}{\partial x'^\mu} = 0.$$

In this case we have a system of equations (55.7) where l takes the values $p+1, \cdots, q$, and the equations corresponding to (55.8) are

$$(\Lambda_{m_2 mp}^{l_2} \psi_\sigma^{m_2} + \Lambda_{\sigma mp}^{l_2}) \psi_\mu^m \psi_\lambda^p = 0,$$

where l_2, $m_2 = p + 1$, $\cdots$, q. Since all the quantities φ_s^h are functions of x^1, $\ldots$, x^p, it follows from (21.18) that the above equations are satisfied identically, and consequently equations (55.13) are completely integrable. Accordingly there exists a set of coordinates, such that $\xi_h^{l_2}$ are functions of x^1, $\ldots$, x^p at most and then from (21.2) it follows that all of the vectors ξ_a^i are functions of these coordinates at most. Gathering together the preceding results we have the theorems:

[55.3] *When a space V_n with positive definite fundamental quadratic form admits an intransitive group of motions G_r and the generic rank of the matrix $\|\xi_a^i\|$ is q less than r, there exist coordinate systems in each of which the ξ's are functions of x^1, $\ldots$, x^q at most.*

[55.4] *When a space V_n with an indefinite fundamental quadratic form admits an intransitive group of motions G_r and the generic rank of the matrix $\|\xi_a^i\|$ is q less than r, $p(\leqq q)$ of the functions φ_s^h are independent; if in a coordinate system for which $\xi_a^t = 0$ for $t = q + 1$, $\ldots$, n, the jacobian matrix of the φ's with respect to x^1, $\ldots$, x^q is of rank p, there exist coordinate systems in each of which the components ξ_a^i are functions of x^1, $\ldots$, x^q at most.*

56. Spaces V_2 which admit a group of motions. We consider first the case of a group of motions G_1 of a V_2, take the components in the form (51.6) and choose the curves of parameter x^2 orthogonal to the paths. Then $g_{12} = 0$, and from (51.4) we find that g_{11} and g_{22} are independent of x^1, so that by a suitable choice of x^2, we have

$$(56.1) \qquad \varphi = g_{11}(dx^1)^2 + e_2(dx^2)^2,$$

that is, V_2 is applicable to a surface of revolution, if φ is definite.

In order to determine whether a V_2 can admit more than one motion, we consider the equations of Killing for the form (56.1). They reduce to

$$\xi^2 \frac{\partial g_{11}}{\partial x^2} + 2g_{11} \frac{\partial \xi^1}{\partial x^1} = 0, \qquad g_{11} \frac{\partial \xi^1}{\partial x^2} + e_2 \frac{\partial \xi^2}{\partial x^1} = 0, \qquad \frac{\partial \xi^2}{\partial x^2} = 0.$$

From the third of these equations, we have $\xi^2 = X_1$, where X_1 is a function of x^1 alone. Indicating by primes derivatives with respect to the argument, from the first two we have

$$(56.2) \qquad \frac{\partial \xi^1}{\partial x^1} = -X_1 \frac{\partial \log \sqrt{g_{11}}}{\partial x^2}, \qquad \frac{\partial \xi^1}{\partial x^2} = -\frac{e_2}{g_{11}} X_1',$$

of which the condition of consistency is

(56.3) $$g_{11}\frac{\partial^2 \log \sqrt{g_{11}}}{\partial x^{2^2}} = e_2\frac{X_1''}{X_1} = c,$$

where c is a constant, since the first and second terms of this equation are independent of x^1 and x^2 respectively. Equating to zero the derivatives of the first term with respect to x^2, we find from the resulting equation that $\dfrac{1}{\sqrt{g_{11}}}\dfrac{\partial^2 \sqrt{g_{11}}}{\partial x^{2^2}} = k$, where k is a constant. Then from (Ex. 3, p. 207) and (47.13) we have $R_{2112} = g_{11}k$, that is, V_2 is of constant curvature. For a given V_2 the constant c in (56.3) is determined, and the general solution of $X_1'' = ce_2X_1$ involves two arbitrary constants. Another is introduced in the determination of ξ^1 from (56.2). Hence the general group is a G_3, and since the rank of $\|\xi_a^i\|$ is two, the group is transitive. Thus we have the theorem, well-known for the case when φ is definite:*

[56.1] *The fundamental form of any surface admitting a continuous deformation is reducible to (56.1) where g_{11} is independent of x^1, and the group involves one parameter, unless the surface is of constant curvature; in the latter case the complete group is a G_3.*

We inquire whether a V_2 admits a sub-group G_2 of motions. In §54 it is shown that when and only when V_2 is euclidean, it admits an Abelian G_2. Its generators are

(56.4) $$X_1f = p_1, \qquad X_2f = p_2.$$

From theorem [16.1] it follows that in the non-Abelian case the basis can be chosen so that

(56.5) $$(X_1, X_2)f = X_1f.$$

Because of theorem [51.4] we may choose the paths of X_1f and X_2f as coordinate lines, and as a result $\xi_1^2 = \xi_2^1 = 0$. Then from (56.5) we have

$$\frac{\partial \xi_2^2}{\partial x^1} = 0, \qquad \frac{\partial \log \xi_1^1}{\partial x^2} = -\xi_2^2.$$

Hence the coordinates can be chosen so that $\xi_2^2 = 1$, $\xi_1^1 = e^{-x^2}$. From (51.4) we have

---

* 1909, 1, pp. 323, 326.

$$\frac{\partial g_{ij}}{\partial x^2} = 0, \qquad \frac{\partial g_{11}}{\partial x^1} = 0, \qquad \frac{\partial g_{12}}{\partial x^1} = g_{11}, \qquad \frac{\partial g_{22}}{\partial x^1} = 2g_{12},$$

and consequently

(56.6) $g_{11} = a, \qquad g_{12} = ax^1 + b, \qquad g_{22} = a(x^1)^2 + 2bx^1 + c,$

where a, b and c are constants. The generators are

(56.7) $X_1 f = e^{-x^2} p_1, \qquad X_2 f = p_2.$*

In this case the curvature of V_2 is $a/(b^2 - ac)$†.

57. Spaces V_3 admitting a G_2 of motions. A group G_2 of a V_3 is intransitive and from theorem [51.4] it follows that the minimum invariant varieties are V_2's. From §55 we have that the induced group on each of these varieties is a G_2 and from §56 that their curvature is constant. From theorem [51.8] it follows that they are geodesically parallel, and that if they are taken for the surfaces $x^3 = $ const., then $\xi_\sigma^3 = 0$ for $\sigma = 1, 2$ and ξ_σ^i for $i = 1, 2$ are independent of x^3 by theorem [55.2]. We may write the fundamental form as follows (cf. §51):

(57.1) $\varphi = g_{ij} dx^i dx^j + e_3 (dx^3)^2$ $(i, j = 1, 2).$

For a particular surface $x^3 = $ const. the infinitesimal transformations are given by (56.4) or (56.7), and from the preceding observations these are the symbols of a G_2 in V_3.

In order that equations (51.4) be satisfied by the transformations (56.4) it is necessary and sufficient that g_{ij} be functions of x^3 alone, subject only to the condition $g_{11}g_{22} - g_{12}^2 \neq 0$. In order that equations (51.4) be satisfied by the transformations (56.7), it is necessary and sufficient that

(57.2) $g_{11} = \alpha, \qquad g_{12} = \alpha x^1 + \beta, \qquad g_{22} = \alpha(x^1)^2 + 2\beta x^1 + \gamma,$

where α, β, γ are arbitrary functions of x^3 such that $\alpha\gamma - \beta^2 \neq 0$. In the former case the curvature of the surfaces $x^3 = $ const. is zero, and in the latter $\alpha/(\beta^2 - \alpha\gamma)$ (cf. §56).‡

By means of these results we shall show that a V_3 cannot admit a complete group G_5 of motions. The group cannot be intransitive, otherwise a family of surfaces (the minimum varieties) would admit a G_5, which is impossible since $5 > 2 \cdot 3/2$ (§53). Hence the group must be transitive, and the sub-group of stability (§§18, 20) of any

---

    * Cf. *Bianchi*, 1918, 1, p. 510.

    † Cf. 1909, 1, p. 155.

    ‡ Cf. *Bianchi*, 1918, 1, p. 542.

point P_0 is of order $5 - 3 = 2$. If there were such a G_2, the points at a constant geodesic distance from P_0 would constitute a minimum invariant variety, and thus we should have a family of geodesically parallel invariant varieties. This is the case just considered, and from the form of the transformations (56.4) and (56.7) it follows that all the transformations of such a G_2 are of order zero (§18), and consequently there cannot be an invariant point.*

We are now in position to prove the following theorem due to Fubini.†

[57.1] *A V_n for $n > 2$ cannot admit a complete group of motions of order $\frac{1}{2}n(n + 1) - 1$.*

We prove this theorem by induction, assuming it to hold for a V_{n-1}. If a V_n admits a G_r with $r = \frac{1}{2}n(n + 1) - 1$, it must be transitive; otherwise by theorem [55.1] a variety of order $n - 1$, or less, would admit a group of this order, which is impossible since a V_{n-1} can admit at most a group of order $n(n - 1)/2$. If G_r is transitive, there is a sub-group of order

$$r_1 = r - n = \tfrac{1}{2}n(n - 1) - 1$$

leaving a point P_0 fixed, which is a group of motions of $\infty^1 V_{n-1}$'s, the loci of points at constant geodesic distance from P_0. But this is contrary to the assumption that the theorem holds for a V_{n-1}. Since we have shown that the theorem holds for a V_3, the proof is complete.

58. Motions in a linearly connected manifold. We have remarked in §50 that, if a point of coordinates a^α in the group-space S of a group G_r is transformed into a point of coordinates a'^α by a transformation of the second parameter group of G_r, we have equations (50.6), that is,

$$(58.1) \qquad \frac{\partial a'^\alpha}{\partial a^\beta} = A_b^\alpha(a')A_\beta^b(a).$$

Differentiating these equations with respect to a^γ, we can reduce the resulting equations by means of (50.1) and (58.1) to

$$(58.2) \qquad \frac{\partial^2 a'^\alpha}{\partial a^\beta \partial a^\gamma} + L_{\delta\epsilon}^\alpha(a')\frac{\partial a'^\delta}{\partial a^\beta}\frac{\partial a'^\epsilon}{\partial a^\gamma} = L_{\beta\gamma}^\delta(a)\frac{\partial a'^\alpha}{\partial a^\delta}.$$

---

* Cf. *Bianchi*, 1918, 1, p. 540.
  † 1903, 1, p. 54.

These equations are similar to (48.3) but in this case

(58.3) $$L'^{\alpha}_{\delta\epsilon}(a') = L^{\alpha}_{\delta\epsilon}(a'),$$

that is, the L''s are the same functions of the a''s as the L's are of the a's. Consequently the geometry of S in the neighborhood of each of the two points as determined by the $(+)$-connection is the same. Accordingly we say that each transformation of the second parameter group determines an *automorphism* of the space S.

If the coefficients Λ^i_{jk} of any linear connection are such that there exists a solution of equations

(58.4) $$\frac{\partial^2 x'^i}{\partial x^j \partial x^k} + \Lambda^i_{lm}(x')\frac{\partial x'^l}{\partial x^j}\frac{\partial x'^m}{\partial x^k} = \Lambda^l_{jk}(x)\frac{\partial x'^i}{\partial x^l},$$

we say that the solution determines an automorphism in V_n. A solution of these equations transforms paths into paths as follows from (48.6). When such a solution involves one or more arbitrary continuous parameters, we say that the space admits a *motion* in the sense that paths are carried continuously into paths as in the case of a motion in a Riemannian space (cf. [51.3]). Thus the group-space S admits the second parameter group as a group of motions. Furthermore the trajectories of this group are the paths of S, and thus the motions are a generalization of translations in a Riemannian space (cf. §52). In particular, when G_r is simple or semi-simple, by theorem [52.6] we have that these motions are translations in the Riemannian sense.

In like manner it can be shown that the first parameter group establishes an automorphism of S for the $(-)$-connection, whose coefficients are $\bar{L}^{\alpha}_{\beta\gamma} = L^{\alpha}_{\gamma\beta}$ (cf. §§48, 50). Hence we have:

[58.1] *The first and second parameter groups of a G_r are groups of translations of the group-space S in the sense that they determine automorphisms of the geometry of S as defined by the $(-)$-connection and $(+)$-connection respectively of S.*

We remark that in accordance with Ex. 2, p. 216 a group of motions in a Riemannian space is a group of automorphisms of the space.

From (50.4) and (50.16) we have

(58.5) $$L^{\alpha}_{\beta\gamma} = \Gamma^{\alpha}_{\beta\gamma} + \tfrac{1}{2}c^{e}_{ab}A^{a}_{\beta}A^{b}_{\gamma}A^{\alpha}_{e}.$$

When these expressions are substituted in (58.2), the resulting equations are reducible in consequence of (58.1) to

$$(58.6) \qquad \frac{\partial^2 a'^{\alpha}}{\partial a^{\beta} \partial a^{\gamma}} + \Gamma^{\alpha}_{\delta \epsilon}(a') \frac{\partial a'^{\delta}}{\partial a^{\beta}} \frac{\partial a'^{\epsilon}}{\partial a^{\gamma}} = \Gamma^{\delta}_{\beta \gamma}(a) \frac{\partial a'^{\alpha}}{\partial a^{\delta}}.$$

If we proceed in like manner with the equations for $(-)$-connection analogous to (58.2), we obtain (58.6). Hence we have:

[58.2] *The first and second parameter groups of a G_r are groups of automorphisms of the geometry of the (0)-connection of the group-space and they are groups of translations of the space.*

We turn to the general case of motions, that is, equations (58.4). Denoting by Γ^i_{jk} and Ω^i_{jk} the symmetric and skew-symmetric parts of Λ^i_{jk}, we may replace (58.4) by

$$(58.7) \qquad \frac{\partial^2 x'^i}{\partial x^j \partial x^k} + \Gamma^i_{lm}(x') \frac{\partial x'^l}{\partial x^j} \frac{\partial x'^m}{\partial x^k} = \Gamma^l_{jk}(x) \frac{\partial x'^i}{\partial x^l}$$

and

$$(58.8) \qquad \Omega^i_{lm}(x') \frac{\partial x'^l}{\partial x^j} \frac{\partial x'^m}{\partial x^k} = \Omega^l_{jk}(x) \frac{\partial x'^i}{\partial x^l}.$$

In order that the infinitesimal transformation

$$(58.9) \qquad x'^i = x^i + \xi^i \delta t$$

be a motion, we have on substituting in (58.7) and (58.8) and neglecting terms in the second and higher powers of δt

$$(58.10) \qquad \frac{\partial^2 \xi^i}{\partial x^j \partial x^k} + \frac{\partial \Gamma^i_{jk}}{\partial x^l} \xi^l + \Gamma^i_{lk} \frac{\partial \xi^l}{\partial x^j} + \Gamma^i_{jl} \frac{\partial \xi^l}{\partial x^k} - \Gamma^l_{jk} \frac{\partial \xi^i}{\partial x^l} = 0$$

and

$$(58.11) \qquad \frac{\partial \Omega^i_{jk}}{\partial x^l} \xi^l + \Omega^i_{lk} \frac{\partial \xi^l}{\partial x^j} + \Omega^i_{jl} \frac{\partial \xi^l}{\partial x^k} - \Omega^l_{jk} \frac{\partial \xi^i}{\partial x^l} = 0.$$

If the coordinate system is chosen so that $\xi^i = \delta^i_1$, we have that

$$\frac{\partial \Gamma^i_{jk}}{\partial x^1} = \frac{\partial \Omega^i_{jk}}{\partial x^1} = 0.$$

When these conditions are satisfied, equations (58.4) are satisfied by the finite equations (51.8) of the group G_1 generated by the infinitesimal transformation (58.9). Hence we have:

[58.3] *When a space with a linear connection admits an infinitesimal motion, it admits as a group G_1 of motions the group generated by the infinitesimal one;*

and

[58.4]   *A necessary and sufficient condition that a space with a linear connection admit a group G_1 of motions is that there exist a coordinate system in which Λ^i_{jk} are independent of one of the coordinates, say x^1; then the curves of parameter x^1 are the trajectories of the motion.*

Indicating by a semi-colon covariant differentiation with respect to the Γ's, equations (58.10) may be written in the form

(58.12) $$\xi^i_{;jk} = \xi^h B^i_{jkh},$$

where B^i_{jkl} are defined by (49.4); these equations are a generalization of (53.3) (cf. Ex. 3, p. 216).   Because of (49.6) the Ricci identities (cf. (47.16))

$$\xi^i_{;jk} - \xi^i_{;kj} = -\xi^l B^i_{ljk}$$

are satisfied identically.   Differentiating (58.12) covariantly with respect to x^l and substituting in the Ricci identity

$$\xi^i_{;jkl} - \xi^i_{;jlk} = \xi^i_{;h}B^h_{jkl} - \xi^h_{;j}B^i_{hkl},$$

we obtain

(58.13) $$\xi^h B^i_{jkl;h} - \xi^i_{;h}B^h_{jkl} + \xi^h_{;j}B^i_{hkl} + \xi^h_{;k}B^i_{jhl} + \xi^h_{;l}B^i_{jkh} = 0,$$

in consequence of (49.6) and of the generalized identities of Bianchi

(58.14) $$B^i_{jkl;h} + B^i_{jlh;k} + B^i_{jhk;l} = 0.*$$

Equations (58.11) may be written in the form

(58.15) $$\Omega^i_{jk;h}\xi^h + \Omega^i_{hk}\xi^h_{;j} + \Omega^i_{jh}\xi^h_{;k} - \Omega^h_{jk}\xi^i_{;h} = 0.$$

The problem of finding one-parameter groups of motions reduces accordingly to the solution of the system of equations

(58.16)
$$\frac{\partial \xi^i}{\partial x^j} = \xi^i_{;j} - \xi^h\Gamma^i_{hj},$$

$$\frac{\partial \xi^i_{;j}}{\partial x^k} = \xi^h B^i_{jkh} - \xi^h_{;j}\Gamma^i_{hk} + \xi^i_{;h}\Gamma^h_{jk},$$

in the $n(n + 1)$ quantities ξ^i and $\xi^i_{;j}$, subject to the conditions (58.15), which is the set F_0 of the mixed system consisting of the latter and (58.16).

Let ξ^i_1 and ξ^i_2 be two solutions of these equations, then

(58.17) $$\xi^i = \xi^h_1\frac{\partial \xi^i_2}{\partial x^h} - \xi^h_2\frac{\partial \xi^i_1}{\partial x^h} = \xi^h_1\xi^i_{2;h} - \xi^h_2\xi^i_{1;h}$$

are the components of the vector of the commutator $(X_1, X_2)f$.

---

* Cf. 1927 1, p. 56.

We shall show that this also is a solution. From (58.17) and (58.12) we have, in consequence of (49.6),

$$(58.18) \qquad \xi^i_{;j} = \xi^h_{1;j}\xi^i_{2;h} - \xi^h_{2;j}\xi^i_{1;h} + \xi^h_1\xi^l_2 B^i_{jhl}.$$

Differentiating these equations covariantly with respect to x^k and in the reduction making use of (58.13) and (58.14), we find that (58.12) is satisfied. From (58.15) we have by covariant differentiation and reduction by means of (58.12)

$$\Omega^i_{jk;hl}\xi^h_1 + \Omega^i_{jk;h}\xi^h_{1;l} + \Omega^i_{hk;l}\xi^h_{1;j} + \Omega^i_{jh;l}\xi^h_{1;k} - \Omega^h_{jk;l}\xi^i_{1;h}$$
$$+ \xi^m_1(\Omega^i_{hk}B^h_{jlm} + \Omega^i_{jh}B^h_{klm} - \Omega^h_{jk}B^i_{hlm}) = 0.$$

If we multiply by ξ^l_2 and sum for l, then subtract the equation obtained by interchanging the sub-scripts 1 and 2, and in the result make use of the Ricci identity

$$\Omega^i_{jk;hl} - \Omega^i_{jk;lh} = \Omega^i_{mk}B^m_{jhl} + \Omega^i_{jm}B^m_{khl} - \Omega^m_{jk}B^i_{mhl},$$

the resulting equations are equivalent to the conditions that ξ^i and $\xi^i_{;j}$ given by (58.17) and (58.18) satisfy (58.15). Accordingly we have for motions in a space with a linear connection theorems [53.2] and [53.3].

If we substitute the expressions (58.9) in (58.4), we obtain the equations resulting from (58.10), when the Γ's are replaced by the Λ's. In consequence of (49.1) the resulting equations may be written in the form

$$(58.19) \qquad (\xi^i_{|j} + 2\Omega^i_{jh}\xi^h)_{|k} - \xi^h\Lambda^i_{jkh} = 0,$$

where, as in §49, a solidus followed by a subscript indicates covariant differentiation with respect to the Λ's. Equations (58.11) may be written

$$(58.20) \quad \xi^h[\Omega^i_{jk|h} + 2(\Omega^l_{jk}\Omega^i_{hl} + \Omega^l_{kh}\Omega^i_{jl} + \Omega^l_{hj}\Omega^i_{kl})] + \Omega^i_{hk}\xi^h_{|j}$$
$$+ \Omega^i_{jh}\xi^h_{|k} - \Omega^h_{jk}\xi^i_{|h} = 0.$$

In order that the mixed system (58.15) and (58.16) in the $n(n + 1)$ quantities ξ^i and $\xi^i_{;j}$ admit a solution involving $n(n + 1)$ parameters, it is necessary that

$$\Omega^i_{jk} = 0, \qquad B^i_{jkl} = 0,$$

that is, the connection is symmetric and of zero curvature. In this case there exists a coordinate system for which $\Gamma^i_{jk} = 0$,[*]

---

[*] 1927, 1, p. 81.

and consequently in this coordinate system equations (58.12) reduce to $\dfrac{\partial^2 \xi^i}{\partial x^j \partial x^k} = 0$, so that

$$\xi^i = a^i + a^i x^j,$$

and the group $G_{n(n+1)}$ of motions is the general linear group. Hence we have:

[58.5] *A necessary and sufficient condition that a linearly connected manifold admit a group of motions of maximum order is that the connection be symmetric and of zero curvature; the group is the general linear group.*

If the linear connection is asymmetric and of zero curvature, there is an ennuple of absolutely parallel vector-fields ζ^i_a in terms of which and ζ^a_i, defined by (48.12), we have

$$\Lambda^i_{jk} = \zeta^i_a \frac{\partial \zeta^a_j}{\partial x^k} = -\zeta^a_j \frac{\partial \zeta^i_a}{\partial x^k}.$$

By means of these expressions equations (58.4) may be put in the form

$$\frac{\partial}{\partial x^k}\left(\frac{\partial x'^i}{\partial x^j} \zeta^j_b(x) \zeta^a_i(x') \right) = 0,$$

from which we have

(58.21) $$\zeta^i_a(x') = a^b_a \zeta^j_b(x) \frac{\partial x'^i}{\partial x^j},$$

where the a's are constants. These equations may be taken in place of (58.4) as the equations defining a motion for a linearly connected space of zero curvature.* In this case equations (58.19) are equivalent to

(58.22) $$\xi^i_{|j} + 2\Omega^i_{jh} \xi^h = \alpha_{ab} \zeta^a_j \zeta^i_b,$$

where the α's are constants. If we substitute the expressions (58.9) in (58.21) and put

(58.23) $$a^b_a = \delta^b_a - \alpha_{ab} \delta t,$$

we obtain (58.22).

If equations (58.22) admit a solution when $\alpha_{ab} = 0$, for these values of the ξ's equations (58.20) reduce to

(58.24) $$\Omega^i_{jk|h} \xi^h = 0.$$

---

* Cf. *Robertson*, 1932, 3, p. 501.

Since the curvature is zero, we have*

$$\xi^i_{|jk} - \xi^i_{|kj} = 2\xi^i_{|h}\Omega^h_{kj}.$$

From (58.22) for the α's zero these equations reduce to (58.24) in consequence of (49.7). Since the trajectories of a G_1 are the integral curves of

$$\frac{dx^i}{dt} = \xi^i,$$

it follows from (58.22) for the α's zero that the trajectories are paths and consequently the motions are translations.

If X_1f and X_2f are symbols of one-parameter groups of translations, we find from (58.17) that $(X_1, X_2)f$ is the symbol of such a group. Consequently all the translations, if any, form a group, which is either the complete group of motions or a subgroup of it. Since any motion carries paths into paths (cf. Ex. 9, p. 236), we have:

[58.6] *The group of translations of a linearly connected space of zero curvature, if any, is an invariant subgroup of the complete group of motions of the space, or it is the complete group.*

The group of translations is at most of order n. In this case the linear connection is determined by this simply transitive group in the sense of §48 and the ennuple of absolutely parallel vector-fields defines the reciprocal group.

Exercises

1. If a Riemannian space admits a G_1 of translations, a surface formed by an infinity of the trajectories is of zero curvature.

Bianchi, 1918, 1, p. 501.

2. When a Riemannian space admits a system of coordinates for which $g_{ii} = $ const. for $i = 1, \cdots, r$ and the other g's are independent of $x^1, \cdots, x^r$, the space admits a group G_r of translations, the curves of parameter x^i being the trajectories.

3. When a Riemannian space admits an intransitive group G_r of motions, where $r = n(n-1)/2$, the minimum invariant varieties are geodesically parallel hypersurfaces of constant curvature.

Bianchi, 1918, 1, p. 544.

4. A necessary and sufficient condition that an infinitesimal transformation carry geodesics into geodesics is that

---

* Cf. 1927, 1, p. 7.

$$\xi_{i,jk} = \xi^h R_{ijkh} + g_{ij}\frac{\partial \psi}{\partial x^k} + g_{ik}\frac{\partial \psi}{\partial x^j},$$

where ψ is a function of the x's.

1926, 3, p. 229.

5. A necessary and sufficient condition that a Riemannian space V_4 admit the group G_3 of rotations (Ex. 5, p. 43) is that the fundamental form be reducible to

$$A_1 dt^2 - A_2 dr^2 - r^2 A_3 (d\theta^2 + \sin^2\theta d\varphi^2),$$

where the A's are at most functions of t and r.

Eiesland, 1925, 5, p. 222.

6. Any Riemannian space in geodesic correspondence with one which admits a G_r of motions also admits a G_r of motions; the vectors ξ_a^i of the latter are linear combinations (constant coefficients) of the vectors of the given group multiplied by a scalar.

Knebelman, 1930, 3, p. 281.

7. If a Riemannian space V_n admits an intransitive group of motions G_r, there exist $n - r$ functionally independent spaces conformal to V_n admitting the same group of motions.

Knebelman, 1930, 3, p. 282.

8. A necessary and sufficient condition that a space with a symmetric connection be the space of a group is that its group of automorphisms admit a simply transitive sub-group of translations.

Cartan, 1927, 2, p. 95.

9. Show that, if ξ^i is a set of solutions of (58.12), the transformations of the group G_1 with the symbol $\xi^i p_i$ carry paths into paths.

Eisenhart and Knebelman, 1927, 3, p. 42.

10. The fundamental form whose coefficients are defined by (48.14) is positive definite. If we put

$$g_{ij} = \sum_a e_a \zeta_i^a \zeta_j^a \qquad g^{ij} = \sum_a e_a \zeta_a^i \zeta_a^j,$$

where each e_a has an assigned value which is plus or minus one, the fundamental form may be made indefinite. The above expressions satisfy (47.10) and from them follow

$$g_{ij}\zeta_a^i = e_a \zeta_j^a, \qquad g^{ij}\zeta_i^a = e_a \zeta_a^j, \qquad g_{ij}\zeta_a^i \zeta_b^j = e_a \delta_{ab} \quad (a \text{ not summed}),$$

and consequently the vectors ζ form an orthogonal ennuple (cf. 1926, 3, p. 40).

11. The Christoffel symbols of the second kind formed with respect to g_{ij} of Ex. 10 are of the form

$$\left\{{}_{jk}^{i}\right\} = \Gamma_{jk}^i + g^{ih}(g_{jl}\Omega_{hk}^l + g_{kl}\Omega_{hj}^l).$$

12. In order that equations (58.21) define a motion in the space with the metric of Ex. 10 (cf. Ex. 2, p. 216), it is necessary and sufficient that the a's satisfy the conditions

$$\sum_a e_a a_a^b a_a^c = e_b \delta^{bc} \qquad\qquad (b \text{ not summed}),$$

and for an infinitesimal motion satisfying (58.22) the condition is

$$(i) \qquad e_c \alpha_{bc} + e_b \alpha_{cb} = 0.$$

When these conditions are satisfied, on the elimination of the α's from (58.22) the equations of Killing (51.4) are obtained with the aid of (48.18).

Robertson, 1932, 3, p. 503.

13. From (58.20) and (58.22) we have

$$\xi^h \Omega^i_{jkl h} + \alpha_{ab}(\Omega^i_{hk} \zeta^a_j \zeta^h_b + \Omega^i_{jh} \zeta^a_k \zeta^h_b - \Omega^h_{jk} \zeta^a_h \zeta^i_b) = 0.$$

Hence for the group of maximum order, it is necessary that $\Omega_{jkl h} = 0$, and from the results of §49

$$\Omega^i_{jk} = \tfrac{1}{2} c^e_{ab} \zeta^a_j \zeta^b_k \zeta^i_e,$$

where the c's are constants satisfying (7.3) and (7.4). The further conditions are reducible by means of Ex. 10 to

$$(i) \qquad \sum_{a,b} e_a \alpha_{ab} [(\Omega^i_{hk} \zeta^a_j + \Omega^i_{jh} \zeta^a_k) g^{hl} \zeta^b_l - \Omega^h_{jk} \zeta^a_h g^{li} \zeta^b_l] = 0.$$

14. Show that the maximum order of a group of motions in the space with the metric of Ex. 10 is $n(n + 1)/2$ in consequence of Ex. 12; also that in consequence of Ex. 13 in this case

$$e_e c^a_{de} + (n - 2) e_a c^e_{da} = e_a c^b_{db} \delta_{ae} \qquad (a, e \text{ not summed}),$$

from which it follows that, unless $n = 1$ or 3, the group is the group of motions of a euclidean space.

Robertson, 1932, 3, p. 511.

15. Show for the hypersurface of Ex. 14, p. 138 $ds^2 = A^2 \Sigma (dy^\alpha)^2$, where $A = 1/(1 + r^2 k/4)$, and that for the vectors ζ^α_a and $\bar{\zeta}^\alpha_a$ of Ex. 15, p. 139

$$\zeta^a_\alpha = \zeta^\alpha_a A^2, \qquad \bar{\zeta}^a_\alpha = \bar{\zeta}^\alpha_a A^2,$$

where $\zeta^a_\alpha \zeta^\beta_a = \bar{\zeta}^a_\alpha \bar{\zeta}^\beta_a = \delta^\beta_\alpha$. Show also that

$$\Lambda^\alpha_{\beta\gamma} = -\zeta^a_\beta \frac{\partial \zeta^\alpha_a}{\partial y^\gamma} = -\frac{k}{2} A \left(\delta^\alpha_\beta y^\gamma + \delta^\alpha_\gamma y^\beta - \delta_{\beta\gamma} y^\alpha + \frac{2}{\sqrt{k}} \epsilon_{\alpha\beta\gamma} \right)$$

and

$$\bar{\Lambda}^\alpha_{\beta\gamma} = -\bar{\zeta}^a_\beta \frac{\partial \bar{\zeta}^\alpha_a}{\partial y^\gamma} = \Lambda^\alpha_{\gamma\beta}.$$

Consequently for the connections of coefficients $\Lambda^\alpha_{\beta\gamma}$ and $\bar{\Lambda}^\alpha_{\beta\gamma}$ the respective sets of vectors $\bar{\zeta}^\alpha_a$ and ζ^α_a determine translations; the parallelisms determined by the two sets of coefficients are the parallelisms of Clifford.

Cartan, 1924, 2, pp. 307–308; *Bortolotti*, 1925, 7, pp. 828–831.

CHAPTER VI

CONTACT TRANSFORMATIONS

59. Definition of homogeneous contact transformations. The set of equations

$$(59.1) \qquad \bar{x}^i = \varphi^i(x^1, \cdots, x^n),$$

for which the rank of the jacobian $\left| \dfrac{\partial \varphi}{\partial x} \right|$ is n may be looked upon either as a transformation of coordinates or as a transformation of the points of the V_n of coordinates x^i among themselves. The same applies to the inverse of (59.1) which we denote by

$$(59.2) \qquad x^i = \bar{\varphi}^i(\bar{x}^1, \cdots, \bar{x}^n).$$

Let p_i, functions of the x's, denote the components of a covariant vector-field so that the components of the same field in $\bar{x}^i$ are given by

$$(59.3) \qquad \bar{p}_i = p_j \frac{\partial x^j}{\partial \bar{x}^i}; \qquad p_i = \bar{p}_i \frac{\partial \bar{x}^i}{\partial x^j} \cdot {}^{*}$$

From these follow the equation

$$(59.4) \qquad \bar{p}_i d\bar{x}^i = p_j dx^j.$$

By means of (59.1) and (59.2) equations (59.3) may be written

$$(59.5) \qquad \begin{aligned} \bar{p}_i &= \psi_i(x^1, \cdots, x^n; p_1, \cdots, p_n); \\ p_i &= \bar{\psi}_i(\bar{x}^1, \cdots, \bar{x}^n; \bar{p}_1, \cdots, \bar{p}_n). \end{aligned}$$

At a point P the equation

$$(59.6) \qquad p_i dx^i = 0$$

defines $n - 1$ directions, and so we may say that p_i define an elemental V_{n-1}, or hyperplane, at P. Accordingly we may think

---

* The quantities p_i used extensively in this chapter must not be confused with the notation for $\dfrac{\partial f}{\partial x^i}$ used in the preceding chapters; when the latter is used, it will be so stated.

of (59.1) and (59.5) as defining a transformation of points and elements at them into points and elements. When, in particular, $p_i = \dfrac{\partial f}{\partial x^i}$, the hypersurfaces $f(x) = $ const. and their tangential hyperplanes are transformed into the hypersurfaces $\bar{f}(\bar{x}) = $ const. and their tangentiallhyperplanes, where

$$(59.7) \qquad \bar{f}(\bar{x}) = f(\bar{\varphi}), \qquad \bar{f}(\varphi) = f(x).$$

In this case p_i may be called the components of the covariant normal to these hypersurfaces. It is evident that, if any two hypersurfaces are tangent at a point P, their transforms are tangent at $\bar{P}$, the transform of P.

We consider a generalization of the transformations (59.1) and (59.5) of the form

$$(59.8) \qquad \begin{aligned} \bar{x}^i &= \varphi^i(x^1, \cdots, x^n; p_1, \cdots, p_n), \\ \bar{p}_i &= \psi_i(x^1, \cdots, x^n; p_1, \cdots, p_n) \end{aligned}$$

in the $2n$ variables x^i and p_i, such that equations (59.4) hold, that is,

$$(59.9) \qquad \psi_i\left(\frac{\partial \varphi^i}{\partial x^i}dx^i + \frac{\partial \varphi^i}{\partial p_i}dp_i\right) = p_k dx^k,$$

for arbitrary values of the differentials dx^i and dp_i. A transformation (59.8) satisfying this condition is called a *homogeneous contact transformation;* the significance of this term will appear later.[*] From (59.9) it follows that a necessary and sufficient condition for such a transformation is that φ^i and ψ_i satisfy the equations

$$(59.10) \qquad \psi_i\frac{\partial \varphi^i}{\partial x^i} = p_i, \qquad \psi_i\frac{\partial \varphi^i}{\partial p_i} = 0.$$

From the second set of these equations it follows that the rank of the jacobian $\left|\dfrac{\partial \varphi^i}{\partial p_i}\right|$ must be less than n. If the rank is zero, we have the case (59.1) and (59.5). If the rank is $n - r$, on the elimination of the p's from the first set of equations (59.8) we obtain r independent equations

$$(59.11) \qquad F_\alpha(\bar{x}^1, \cdots, \bar{x}^n; x^1, \cdots, x^n) = 0 \quad (\alpha = 1, \cdots, r).$$

---

[*] The following presentation of this subject is substantially the same as made by the author several years ago, 1929, 4.

From (59.11) it follows that the differentials at a transformed point corresponding to differentials at a point $P(x^i)$ are related by the system of equations

$$(59.12) \qquad \frac{\partial F_\alpha}{\partial \bar{x}^i} d\bar{x}^i + \frac{\partial F_\alpha}{\partial x^j} dx^j = 0.$$

Since the conditions (59.4) must be satisfied as a consequence of (59.12), we must have

$$(59.13) \qquad \bar{p}_i = \rho^\alpha \frac{\partial F_\alpha}{\partial \bar{x}^i}, \qquad p_i = -\rho^\alpha \frac{\partial F_\alpha}{\partial x^i},$$

where the ρ's are parameters such that, when these expressions for p_i are substituted in the first set of equations (59.8), the latter are satisfied identically because of (59.11). From the second set of equations (59.13) it follows that the rank of the jacobian matrix $\left\| \dfrac{\partial F_\alpha}{\partial x^i} \right\|$ must be r. Consequently r of these equations can be solved for the ρ's, which as thus obtained are linear functions of r of the p's, and when these are substituted in the remaining $n - r$ equations, we obtain $n - r$ equations involving x^i, $\bar{x}^i$ and p_i, the latter entering linearly and homogeneously. Since on solving these new equations and (59.11) for the $\bar{x}$'s, we obtain the first set of (59.8), it follows that φ^i are homogeneous of degree zero in the p's, and consequently

$$(59.14) \qquad p_i \frac{\partial \varphi^i}{\partial p_i} = 0.$$

Also from the first set of equations (59.13) we have that the functions $\psi_i (= \bar{p}_i)$ are homogeneous of degree one in the p's.

In order that the first set of equations (59.10) admit a unique solution in the ψ's, it is necessary and sufficient that the jacobian of the φ's with respect to the x's be of rank n, that is,

$$(59.15) \qquad \left| \frac{\partial \varphi^j}{\partial x^i} \right| \neq 0.$$

When this condition is satisfied and the solutions ψ_i of these equations are substituted in the second set, we have the n equations of condition

$$\begin{vmatrix} \dfrac{\partial \varphi^1}{\partial p_i} & \cdots & \dfrac{\partial \varphi^n}{\partial p_i} & 0 \\[2mm] \dfrac{\partial \varphi^1}{\partial x^1} & \cdots & \dfrac{\partial \varphi^n}{\partial x^1} & p_1 \\[2mm] \cdot & \cdot & \cdot & \cdot \\ \cdot & \cdot & \cdot & \cdot \\[1mm] \dfrac{\partial \varphi^1}{\partial x^n} & \cdots & \dfrac{\partial \varphi^n}{\partial x^n} & p_n \end{vmatrix} = 0 \qquad (i = 1, \cdots, n).$$

From these equations it follows that

$$(59.16) \qquad \frac{\partial \varphi^k}{\partial p_i} = \lambda^{ij} \frac{\partial \varphi^k}{\partial x^j}, \qquad \lambda^{ij} p_j = 0,$$

where the λ's are so defined. From the first of these sets of equations we have

$$\frac{\partial^2 \varphi^k}{\partial p_h \partial p_i} = \left(\frac{\partial \lambda^{ij}}{\partial p_h} + \lambda^{il} \frac{\partial \lambda^{hj}}{\partial x^l} \right) \frac{\partial \varphi^k}{\partial x^j} + \lambda^{ij} \lambda^{hl} \frac{\partial^2 \varphi^k}{\partial x^j \partial x^l}.$$

Hence because of (59.15) the conditions of integrability of the first set of equations (59.16) are

$$\frac{\partial \lambda^{ij}}{\partial p_h} - \frac{\partial \lambda^{hj}}{\partial p_i} + \lambda^{il} \frac{\partial \lambda^{hj}}{\partial x^l} - \lambda^{hl} \frac{\partial \lambda^{ij}}{\partial x^l} = 0.$$

Multiplying by p_j, summing for j and noting that from the second set of (59.16) we have

$$\frac{\partial \lambda^{ij}}{\partial x^l} p_j = 0, \qquad \frac{\partial \lambda^{ij}}{\partial p_h} p_j + \lambda^{ih} = 0,$$

we find that $\lambda^{ih} = \lambda^{hi}$, that is, λ^{ij} is symmetric in the indices. From (59.16) we have

$$\frac{\partial \varphi^k}{\partial p_i} \frac{\partial \varphi^l}{\partial x^i} = \lambda^{ij} \frac{\partial \varphi^k}{\partial x^j} \frac{\partial \varphi^l}{\partial x^i}.$$

Since λ^{ij} is symmetric in i and j, we have the identities

$$(59.17) \qquad (\varphi^k, \varphi^l) \equiv \frac{\partial \varphi^k}{\partial p_i} \frac{\partial \varphi^l}{\partial x^i} - \frac{\partial \varphi^k}{\partial x^i} \frac{\partial \varphi^l}{\partial p_i} = 0.$$

Conversely, if φ^i are any functions of x^i and p_i satisfying (59.14), (59.15) and (59.17) and ψ_i are functions defined by the first set of (59.10), then these functions are such that

$$\psi_j \frac{\partial \varphi^j}{\partial p_i} \frac{\partial \varphi^k}{\partial x^i} = \psi_j \frac{\partial \varphi^j}{\partial x^i} \frac{\partial \varphi^k}{\partial p_i} = p_i \frac{\partial \varphi^k}{\partial p_i} = 0.$$

Because of (59.15) it follows that these functions ψ_i satisfy the second set of (59.10). Hence we have:

[59.1] *A necessary and sufficient condition that a set of functions φ^i determine a homogeneous contact transformation* (59.8) *for which the ψ's are uniquely determined is that the functions φ^i be homogeneous of degree zero in the p's, that the jacobian of the φ's with respect to the x's be of rank n and that* (59.17) *be satisfied.* *

60. Geometrical properties of homogeneous contact transformations. We consider now the geometrical significance of equations (59.11) and begin with the case when there is a single equation

$$(60.1) \qquad F(\bar{x}^1, \cdots, \bar{x}^n; x^1, \cdots, x^n) = 0$$

arising from the elimination of the p's from the first set of equations (59.8). In this case a point $P(x^i)$ is transformed into the points of the hypersurface $\bar{\Sigma}$, whose equation is given by (60.1) when the x's are given the values at P. Now equations (59.13) become

$$(60.2) \qquad \bar{p}_i = \rho \frac{\partial F}{\partial \bar{x}^i}, \qquad p_i = -\rho \frac{\partial F}{\partial x^i}.$$

Each choice of p_i in (59.8) determines an elemental V_{n-1} at P, and for this choice there is a determined point $\bar{P}$ on the hypersurface $\bar{\Sigma}$. Moreover, from the first of (60.2) it is seen that the transform of the vector p_i is the covariant normal to the hypersurface $\bar{\Sigma}$ at $\bar{P}$. Also if the $\bar{x}$'s in (60.1) are given the values at $\bar{P}$, this equation in the x's defines a hypersurface Σ, all of whose points transform into $\bar{P}$. In particular, it follows from the second of (60.2) that the vector p_i chosen at P is the covariant normal to Σ at P.

Consider a hypersurface S, whose equation is

$$(60.3) \qquad f(x^1, \cdots, x^n) = 0.$$

To each of its points there corresponds a hypersurface $\bar{\Sigma}$ by means of (60.1). When, in particular, we put $p_i = \dfrac{\partial f}{\partial x^i}$ in the first of (59.8) these equations define a hypersurface $\bar{S}$, the $\bar{x}$'s being defined in

---

* This theorem was established in a different manner by *Lie-Engel*, 1888, 1, vol. 2, p. 137.

terms of the x's as parameters, which are subject to the condition (60.3). Thus each point $P(x^i)$ of S is transformed into a definite point $\bar{P}(\bar{x}^i)$ of $\bar{S}$. A displacement on $\bar{S}$ is given by

$$d\bar{x}^i = \left(\frac{\partial \varphi^i}{\partial x^j} + \frac{\partial \varphi^i}{\partial p_k} \frac{\partial p_k}{\partial x^j} \right) dx^j$$

for each variation of the x's subject to the condition $\dfrac{\partial f}{\partial x^i} dx^i = 0$. Because of this consideration and (59.10) it follows that $\bar{p}_i$ given by (59.8) for this case satisfy the condition $\bar{p}_i d\bar{x}^i = 0$. Hence if $\bar{f}(\bar{x}^1, \cdots, \bar{x}^n) = 0$ is the equation of $\bar{S}$, resulting from the elimination of the x's from the first of (59.8) $\left(\text{in which } p_i = \dfrac{\partial f}{\partial x^i} \right)$ and (60.3), it follows that $\bar{p}_i = \lambda \dfrac{\partial \bar{f}}{\partial \bar{x}^i}$. If we take the hypersurface $\overline{\Sigma}$ corresponding to the point $P(x^i)$, the point on it determined by $p_i = \dfrac{\partial f}{\partial x^i}$ is the point $\bar{P}$ referred to above and $\bar{p}_i$ given by (59.8) for these p's are the components of the covariant normal to $\overline{\Sigma}$. Hence $\overline{\Sigma}$ and $\bar{S}$ are tangent at $\bar{P}$.

We consider now the general case when $r > 1$ in (59.11). From (59.13) it follows that the jacobian matrix of the F's with respect to the x's is necessarily of rank r, and likewise the matrix of the F's with respect to the $\bar{x}$'s. Geometrically this means that a point $P\left(x^i\right)$ is transformed into the points of a space of $n - r$ dimensions, say $\overline{\Sigma}_{n-r}$, and that a point $\bar{P}(\bar{x}^i)$ is the transform of the points of a space of $n - r$ dimensions, say Σ_{n-r}.* Each choice of p_i at a point P determines an elemental V_{n-1} at P, and for this choice there is a unique point $\bar{P}$ of $\overline{\Sigma}_{n-r}$ defined by the first set of equations (59.8). Any displacement in $\overline{\Sigma}_{n-r}$ at $\bar{P}$ satisfies the conditions $\dfrac{\partial F_\alpha}{\partial \bar{x}^i} d\bar{x}^i = 0$ and consequently $\bar{p}_i d\bar{x}^i = 0$, as follows from (59.13). Hence $\bar{p}_i$ are the components of a covariant normal to $\overline{\Sigma}_{n-r}$, in the sense that a covariant vector λ_i in a V_n which satisfies the condition $\lambda_i dx^i = 0$ for all displacements at a point in a sub-space of V_n is called a

---

* These observations are consequences of the fundamental theorem on the existence of solutions of a set of equations involving implicit functions, cf. *Fine*, 1927, 4, p. 253.

covariant normal to the sub-space at the point. When p_i are chosen, $\bar{x}^i$ are determined as we have seen and then the corresponding Σ_{n-r}, and from the second set of (59.13) it is seen that p_i are the components of a covariant normal to this Σ_{n-r} at P.

As in the case when $r = 1$, each point of a hypersurface (60.3) is transformed into a $\bar{\Sigma}_{n-r}$ and each of these varieties is tangent to the hypersurface whose equation is obtained, when we replace p_i by $\dfrac{\partial f}{\partial x^i}$ in the first of (59.8) and eliminate the x's from these equations and (60.3).

We call the space $\bar{\Sigma}_{n-r}$ defined by (59.11) for a point the *fundamental variety* of the transformation for the point. Also we say that two sub-spaces of orders p and q respectively $(p \geqq q)$ of a space V_n are tangent at a point P, if every covariant normal to the former at P is a covariant normal to the latter at P. Accordingly we have:

[60.1] *When the points of a hypersurface and all the elements at these points undergo a homogeneous contact transformation, the fundamental varieties of the transformation for the points are tangent to the hypersurface which is the transform of the points for elements tangential to the given hypersurface; and the transformed elements are tangential to the hypersurface.*

As a corollary of this theorem we have that, if two hypersurfaces are tangent at a point, that is, have the same covariant normal at the point, the envelopes of the two sets of varieties $\bar{\Sigma}_{n-r}$ for the points of the two given hypersurfaces will be tangent at the point which is the transform of the given point. The foregoing results justify the term contact transformations.

Consider a sub-space S_{n-p} of $n - p$ dimensions, defined by the equations

$$(60.4) \qquad\qquad f_\sigma(x^1, \cdots, x^n) = 0 \qquad (\sigma = 1, \cdots, p).$$

With each point $P(x^i)$ of S_{n-p} there is an associated $\bar{\Sigma}_{n-r}$. If in the first set of equations (59.8) we put

$$(60.5) \qquad\qquad p_i = u^\sigma \frac{\partial f_\sigma}{\partial x^i},$$

where the u's are parameters, we obtain the coördinates of a point $\bar{P}(\bar{x}^i)$ of $\bar{\Sigma}_{n-r}$. Since the functions φ^i are homogeneous of degree zero in the p's, when we eliminate the x's and u's from (59.8), (60.4) and (60.5), we get a single equation in the $\bar{x}$'s, that is, the locus of the points is a hypersurface $\bar{S}$. Its equations in parametric form are

$$(60.6) \qquad \bar{x}^i = \varphi^i\left(x^1, \cdots, x^n; u^\sigma \frac{\partial f_\sigma}{\partial x^1}, \cdots, u^\sigma \frac{\partial f^\sigma}{\partial x^n}\right),$$

the parameters being the x's, subject to the conditions (60.4), and the ratio of the u's. Any displacement on $\bar{S}$ at $\bar{P}$ is given by

$$(60.7) \qquad d\bar{x}^i = \frac{\partial \varphi^i}{\partial x^j} dx^j + \frac{\overline{\partial \varphi^i}}{\partial p_k} d\left(u^\sigma \frac{\partial f_\sigma}{\partial x^k}\right),$$

for any variations of the u's and for variations of the x's subject to the conditions $\dfrac{\partial f_\sigma}{\partial x^i} dx^i = 0$; in these equations $\dfrac{\overline{\partial \varphi^i}}{\partial p_k}$ indicates the expression which results when p_j in $\dfrac{\partial \varphi^i}{\partial p_k}$ are replaced by the expressions (60.5). Since (59.10) hold whatever be p_i, we have

$$\bar{p}_i d\bar{x}^i = p_j dx^j = u^\sigma \frac{\partial f_\sigma}{\partial x^j} dx^j = 0.$$

Since this equation holds for all displacements in $\bar{S}$, it follows that $\bar{p}_i$ are the components of the covariant normal to $\bar{S}$. When the coordinates of a point $P(x^i)$ of S_{n-p} are substituted in the right-hand member of (60.5), the p's so defined for all values of the u's are the covariant normals to S_{n-p} at P. Hence we have:

[60.2] *When the points of a sub-space of dimensions $n - p$ and all the elements at these points undergo a homogeneous contact transformation, the fundamental varieties of the transformation for the points are tangent to the hypersurface which is the transform of the points and the vectors which are the covariant normals to the sub-space; and the transformed elements are tangent to the hypersurface.*

61. Definition of homogeneous contact transformations by means of fundamental varieties. Suppose that in a V_n of coordinates x^i we have r equations of the form

$$(61.1) \qquad F_\alpha(\bar{x}^1, \cdots, \bar{x}^n; x^1, \cdots, x^n) = 0 \quad (\alpha = 1, \cdots, r),$$

where the $\bar{x}$'s are parameters. If and only if the jacobian matrix

$\left\|\dfrac{\partial F_\alpha}{\partial x^i}\right\|$ is of rank r because of the equations (61.1), then for each set of values of the $\bar{x}$'s equations (61.1) define a variety of $n - r$ dimensions, say Σ_{n-r}. Any covariant normal to Σ_{n-r} has components of the form

$$(61.2) \qquad\qquad p_i = u^\alpha \frac{\partial F_\alpha}{\partial x^i}$$

for suitable values of the parameters u^α; and any choice of the u's determines a covariant normal. A necessary and sufficient condition that the $n + r$ equations (61.1), (61.2) can be solved for the $\bar{x}$'s and u's uniquely is that the jacobian of these equations with respect to these quantities does not vanish because of (61.1) and (61.2); this follows from the general theorem concerning implicit functions.* This jacobian is

$$(61.3) \qquad
\begin{vmatrix}
\dfrac{\partial F_1}{\partial \bar{x}^1} & \cdots & \dfrac{\partial F_1}{\partial \bar{x}^n} & 0 & \cdots & 0 \\[2ex]
\cdots & \cdots & \cdots & \cdots & \cdots & \cdots \\[1ex]
\dfrac{\partial F_r}{\partial \bar{x}^1} & \cdots & \dfrac{\partial F_r}{\partial \bar{x}^n} & 0 & \cdots & 0 \\[2ex]
u^\alpha \dfrac{\partial^2 F_\alpha}{\partial x^1 \partial \bar{x}^1} & \cdots & u^\alpha \dfrac{\partial^2 F_\alpha}{\partial x^1 \partial \bar{x}^n} & \dfrac{\partial F_1}{\partial x^1} & \cdots & \dfrac{\partial F_r}{\partial x^1} \\[2ex]
\cdots & \cdots & \cdots & \cdots & \cdots & \cdots \\[1ex]
u^\alpha \dfrac{\partial^2 F_\alpha}{\partial x^n \partial \bar{x}^1} & \cdots & u^\alpha \dfrac{\partial^2 F_\alpha}{\partial x^n \partial \bar{x}^n} & \dfrac{\partial F_1}{\partial x^n} & \cdots & \dfrac{\partial F_r}{\partial x^n}
\end{vmatrix}.$$

Since this determinant does not involve the p's, it must not vanish because of (61.1) alone. Since the latter do not involve the u's, it must not vanish for arbitrary values of the u's and as a result of (61.1). If we apply the rule of Laplace to this determinant, we see that the rank of the matrix $\left\|\dfrac{\partial F_\alpha}{\partial \bar{x}^i}\right\|$ must be r.

Since we are concerned only with the determination of the $\bar{x}$'s, we may arrive at our result by solving (61.1) and the equations resulting from (61.2) on the elimination of the u's. By a suitable numbering

---

* It is assumed that the functions F_α and their first and second derivatives are continuous for the domain under consideration. Cf. *Fine*, 1927, 4, p. 253.

of the x's, we have that the determinant $\left|\dfrac{\partial F_\alpha}{\partial x^\beta}\right|$ for $\alpha, \beta = 1, \cdots, r$ is of rank r, in consequence of the assumption concerning the matrix $\left\|\dfrac{\partial F_\alpha}{\partial x^i}\right\|$. Then by the elimination of the u's from (61.2) we have $n - r$ independent equations

$$(61.4) \qquad \begin{vmatrix} \dfrac{\partial F_1}{\partial x^1} & \cdots & \dfrac{\partial F_1}{\partial x^r} & \dfrac{\partial F_1}{\partial x^\sigma} \\ \cdot & \cdots & \cdot & \cdot \\ \dfrac{\partial F_r}{\partial x^1} & \cdots & \dfrac{\partial F_r}{\partial x^r} & \dfrac{\partial F_r}{\partial x^\sigma} \\ p_1 & \cdots & p_r & p_\sigma \end{vmatrix} = 0 \quad (\sigma = r + 1, \cdots, n).$$

In order that these equations and (61.1) be solvable for the $\bar{x}$'s, it is seen that the matrix $\left\|\dfrac{\partial F_\alpha}{\partial \bar{x}^i}\right\|$ must be of rank r for the equations (61.1), as previously seen from the considerations concerning the determinant (61.3). It is evident from (61.4) that these solutions, say

$$(61.5) \qquad \bar{x}^i = \varphi^i(x; p),$$

are such that the φ's are homogeneous of degree zero in the p's. When these expressions are substituted in (61.1), we have identities in the x's and p's. Differentiating these identities, we have

$$(61.6) \qquad \frac{\partial F_\alpha}{\partial x^i} + \left(\frac{\partial F_\alpha}{\partial \bar{x}^j}\right)\frac{\partial \varphi^j}{\partial x^i} = 0, \qquad \left(\frac{\partial F_\alpha}{\partial \bar{x}^j}\right)\frac{\partial \varphi^j}{\partial p_i} = 0,$$

where $\left(\dfrac{\partial F_\alpha}{\partial \bar{x}^j}\right)$ indicates the result of replacing the $\bar{x}$'s by the φ's in $\dfrac{\partial F_\alpha}{\partial \bar{x}^j}$. If we denote by $\bar{u}^\alpha$ the functions obtained by solving (61.2) for the u's after the $\bar{x}$'s have been replaced by the φ's, and put

$$(61.7) \qquad \psi_i = -\bar{u}^\alpha\left(\frac{\partial F_\alpha}{\partial \bar{x}^i}\right),$$

then from (61.2) and (61.6) it follows that ψ_i satisfy the conditions

(59.10), and consequently the functions φ^i and ψ_i define a contact transformation. Hence we have:

[61.1] *If F_α are r functions of $2n$ variables, $\bar{x}^i$ and x^i, such that the determinant (61.3) is of rank $n + r$ for the equations $F_\alpha = 0$ for all values of the parameters u^α, equations (61.5) and (61.7) define a homogeneous contact transformation.* *

We consider a hypersurface S defined by (60.3), and in place of (61.2) take

$$(61.8) \qquad \frac{\partial f}{\partial x^i} + \lambda^\alpha \frac{\partial F_\alpha}{\partial x^i} = 0.$$

When equations (61.1) and the equations resulting from (61.8) on the elimination of the λ's are solved for the $\bar{x}$'s, we get the equations in parametric form of the hypersurface $\bar{S}$ defined in §60, the x's being the parameters subject to the condition (60.3). Eliminating the x's from these equations, we get the equation $\bar{f}(\bar{x}^1, \cdots, \bar{x}^n) = 0$ of $\bar{S}$, which would likewise be obtained by eliminating directly the x's and λ's from (61.1), (61.8) and (60.3), in accordance with the usual theory of envelopes.

When we have a contact transformation defined by (59.8) and obtain the corresponding set of equations (59.11), the above mentioned conditions upon the corresponding determinant (61.3) are necessarily satisfied. We observe, furthermore, that in the form of (61.3) the variables x^i and $\bar{x}^i$ enter symmetrically. Consequently, if we interchange the roles of the x's and $\bar{x}$'s in the above process, we obtain a unique contact transformation

$$(61.9) \qquad x^i = \bar{\varphi}^i(\bar{x}; \bar{p}), \qquad p_i = \bar{\psi}_i(\bar{x}; \bar{p}).$$

This transformation is the inverse of the former. For, if we take any point $P(x^i)$ and a point $\bar{P}(\bar{x}^i)$, such that (61.1) are satisfied, and in (61.5) replace p_i by the quantities (61.2), the resulting equations are satisfied identically for all values of u^α, and $\bar{p}_i$ are given by

$$(61.10) \qquad \bar{p}_i = -u^\alpha \frac{\partial F_\alpha}{\partial \bar{x}_i},$$

as follows from (59.13). A similar result is true, if we proceed with (61.9) by means of (61.10). Hence we have:

---

* Cf. *Lie-Engel*, 1888, 1, vol. 2, p. 150.

[61.2]   *Every homogeneous contact transformation admits a unique inverse contact transformation.*

If we have two contact transformations, namely (59.8) and

$$x_1^i = \bar{\varphi}^i(\bar{x};\ \bar{p}), \qquad p_{1i} = \bar{\psi}_i(\bar{x};\ \bar{p}),$$

we have

$$\bar{\psi}_i d\bar{\varphi}^i = \bar{p}_i d\bar{x}^i, \qquad \psi_i d\varphi^i = p_i dx^i.$$

Replacing $\bar{p}_i$ and $\bar{x}_i$ in the first by ψ_i and φ^i, we have $\bar{\psi}_i d\bar{\varphi}^i = p_i dx^i$. Consequently we have the theorem of Lie:*

[61.3]   *The totality of homogeneous contact transformations in* $2n$ *quantities* x^i *and* p_i *form an infinite group.*

For the inverse of the transformation (59.8) we have analogously to (59.10)

$$p_k \frac{\partial x^k}{\partial \bar{x}^i} = \bar{p}_i, \qquad p_k \frac{\partial x^k}{\partial \bar{p}_j} = 0.$$

Differentiating the first with respect to $\bar{x}^j$, we have from the resulting expression that

$$\{\bar{x}^i,\ \bar{x}^j\} \equiv \frac{\partial p_k}{\partial \bar{x}^i} \frac{\partial x^k}{\partial \bar{x}^j} - \frac{\partial p_k}{\partial \bar{x}^j} \frac{\partial x^k}{\partial \bar{x}^i} = 0.$$

Differentiating the second with respect to $\bar{p}_i$, we have

$$\{\bar{p}_i,\ \bar{p}_j\} = 0.$$

If we differentiate the first with respect to $\bar{p}_j$ and the second with respect to $\bar{x}^i$ and subtract the resulting equations, we have

$$\{\bar{p}_j,\ \bar{x}^i\} = \delta_j^i,$$

where

(61.11) $\delta = 1$ or 0,   as $i = j$ or $i \neq j$.

Consider $2n$ independent functions u^α, for $\alpha = 1,\ \cdots,\ 2n$, of x^i and p_i.   Since

$$\frac{\partial x^i}{\partial u^\alpha} \frac{\partial u^\alpha}{\partial x^j} = \delta_j^i, \qquad \frac{\partial p_i}{\partial u^\alpha} \frac{\partial u^\alpha}{\partial p_j} = \delta_i^j, \qquad \frac{\partial x^i}{\partial u^\alpha} \frac{\partial u^\alpha}{\partial p_j} = \frac{\partial p_j}{\partial u^\alpha} \frac{\partial u^\alpha}{\partial x^i} = 0,$$

---

* *Lie-Engel*, 1888, 1, vol. 2, p. 139.

we have

$$\sum_{\alpha}^{1,\,\cdots,\,2n} (u^\alpha, u^\beta) \{u^\alpha, u^\gamma\} = \delta^\beta_\gamma,$$

where (u^α, u^β) is defined by (59.17). Applying this identity to the above case we have

$$(\bar{x}^i, \bar{x}^i) = 0, \qquad (\bar{p}_j, \bar{x}^i) = \delta^i_j, \qquad (\bar{p}_i, \bar{p}_i) = 0.$$

Hence we have the theorem:*

[61.4] *For a homogeneous contact transformation the functions φ^i and ψ_i satisfy the identities*

$$(61.12) \qquad \{\varphi^i, \varphi^j\} = \{\psi_i, \psi_j\} = 0, \qquad \{\psi_j, \varphi^i\} = \delta^i_j,$$

and

$$(61.13) \qquad (\varphi^i, \varphi^j) = (\psi_i, \psi_j) = 0, \qquad (\psi_j, \varphi^i) = \delta^i_j.$$

If u and v are functions of x^i and p_i, the symbol (u, v) defined by

$$(61.14) \qquad (u, v) = \frac{\partial u}{\partial p_i}\frac{\partial v}{\partial x^i} - \frac{\partial u}{\partial x^i}\frac{\partial v}{\partial p_i}$$

is called the *parenthesis of Poisson*. For any three functions u, v, w we have by direct calculation that the following equation is an identity

$$(61.15) \qquad ((u, v), w) + ((v, w), u) + ((w, u), v) = 0.$$

It is called the *Jacobi identity*.

If we subject the x's and p's to a homogeneous contact transformation and denote by $\bar{u}$ and $\bar{v}$ the transforms of u and v, that is, $\bar{u}(\varphi, \psi) = u(x, p)$, we have in consequence of (61.13)

$$(\bar{u}, \bar{v}) = \frac{\partial \bar{u}}{\partial \bar{p}_i}\frac{\partial \bar{v}}{\partial \bar{x}^i} - \frac{\partial \bar{u}}{\partial \bar{x}^i}\frac{\partial \bar{v}}{\partial \bar{p}_i} = (u, v).$$

Hence we have:

[61.5] *The parenthesis of Poisson of any two functions is an invariant under a homogeneous contact transformation.*

When two functions u and v are such that (u, v) is identically zero, the two functions are said to be in *involution*. From (61.13) we have:

---

* *Lie-Engel*, 1888, 1, vol. 2, p. 137; also, *Whittaker*, 1927, 5, p. 300.

[61.6] *The functions φ^i of a homogeneous contact transformation are in involution and likewise the functions ψ_i.*

62. Infinitesimal homogeneous contact transformations.

As we wish to consider continuous groups of homogeneous contact transformations, we begin with a study of infinitesimal homogeneous contact transformations. Such a transformation is defined by equations of the form

$$(62.1) \qquad \bar{x}^i = x^i + \xi^i \delta t, \qquad \bar{p}_i = p_i + \eta_i \delta t,$$

where the ξ's and η's are functions of the x's and p's such that these equations must satisfy the conditions (59.10). This gives the conditions

$$(62.2) \qquad \eta_i + p_j \frac{\partial \xi^j}{\partial x^i} = 0, \qquad p_j \frac{\partial \xi^j}{\partial p_i} = 0.$$

If we put

$$(62.3) \qquad C = p_i \xi^i,$$

we have by differentiation and as a consequence of (62.2)

$$(62.4) \qquad \frac{\partial C}{\partial x^i} = p_j \frac{\partial \xi^j}{\partial x^i} = -\eta_i, \qquad \frac{\partial C}{\partial p_i} = \xi^i + p_j \frac{\partial \xi^j}{\partial p_i} = \xi^i.$$

From the second set of these equations and (62.3) we have

$$(62.5) \qquad C = p_i \frac{\partial C}{\partial p_i},$$

that is, C is homogeneous of first degree in the p's; this follows at once from (62.3), since (62.1) defines a homogeneous contact transformation. Differentiating (62.5) with respect to x^i and p_i, we obtain respectively

$$(62.6) \qquad \frac{\partial C}{\partial x^i} - p_j \frac{\partial^2 C}{\partial p_j \partial x^i} = 0, \qquad p_j \frac{\partial^2 C}{\partial p_i \partial p_j} = 0,$$

which are equivalent to (62.2) because of (62.4).

Conversely, if C is any function of the x's and p's homogeneous of first degree in the p's, and we take ξ^i and η_i as defined by (62.4), the conditions (62.2) are satisfied, because (62.6) are a consequence of (62.5). Accordingly we have:*

---

* *Lie-Engel*, 1888, 1, vol. 2, p. 263.

[62.1] *Every infinitesimal homogeneous contact transformation is defined by equations of the form*

$$(62.7) \qquad \bar{x}^i = x^i + \frac{\partial C}{\partial p_i}\delta t, \qquad \bar{p}_i = p_i - \frac{\partial C}{\partial x^i}\delta t,$$

where C is homogeneous of first degree in the p's; moreover, any such function C determines an infinitesimal homogeneous contact transformation.

The function C is called the *characteristic function* of the transformation.

If we form the differentials of (62.7), we obtain

$$(62.8) \qquad \begin{aligned} d\bar{x}^i &= dx^i + \left(\frac{\partial^2 C}{\partial p^i \partial x^j}dx^j + \frac{\partial^2 C}{\partial p_i \partial p_k}dp_k\right)\delta t, \\ d\bar{p}_i &= dp_i - \left(\frac{\partial^2 C}{\partial x^i \partial x^j}dx^j + \frac{\partial^2 C}{\partial x^i \partial p_k}dp_k\right)\delta t. \end{aligned}$$

Equations (62.7) and (62.8) define the extended infinitesimal transformation of the elements x^i, p_i, dx^i and dp_i. For this extended transformation $p_i dx^i$ is invariant in consequence of (62.6). Hence in accordance with the general theory of continuous groups the quantity $p_i dx^i$ is invariant under the finite group G_1 generated by the extended infinitesimal transformation, and the finite equations* of G_1 are given by the integrals of the equations

$$(62.9) \qquad \frac{dx^i}{dt} = \frac{\partial C}{\partial p_i}, \qquad \frac{dp_i}{dt} = -\frac{\partial C}{\partial x^i},$$

say

$$(62.10) \qquad \bar{x}^i = \varphi^i(x;p;t), \qquad \bar{p}_i = \psi_i(x;p;t),$$

and their differentials. Consequently (62.10) define a one-parameter group of contact transformations.

From the foregoing considerations we have:

[62.2] *The most general one-parameter group of homogeneous contact transformations is given by the solution of equations (62.9) in which C is an arbitrary analytic function of the x's and p's which is homogeneous of degree one in the latter.*

---

* Cf. *Lie-Engel*, 1888, 1, vol. 2, p. 212.

Any integral of (62.9) satisfies the condition $C = $ const. This means that in the $2n$-dimensional space of coordinates x^i and p_i each integral curve, or *trajectory*, lies in one of the hypersurfaces $C = h$. When in the right-hand members of (62.10), the quantities x^i and p_i are given particular values, these equations define such an integral curve, and the hypersurface is determined by the value of C for the particular point. Since C is homogeneous of degree one in the p's, φ^i are of degree zero and ψ_i of degree one. Hence if we replace p_i by p_i/h, we get a curve for which φ^i are unaltered and $\bar{p}_i = \psi_i/h$, and these satisfy the condition

$$(62.11) \qquad\qquad C = 1.$$

From the standpoint of contact transformations this change is of no significance, so that except for the case $C = 0$ we may in all generality take (62.11) as holding.

When real initial values of the p's can be chosen so that $C = 0$, this equation holds along the trajectory; such a trajectory we call *singular*. In this case, as follows from (62.2) and $p_i\xi^i = 0$, the ξ's are determined only to within a factor. Consequently, if C is of such a form that $\dfrac{\partial C}{\partial p_i}$ and $\dfrac{\partial C}{\partial x^i}$ involve a factor which makes them vanish or become infinite when $C = 0$, we remove this factor. Consider, for example, the case when C is the square-root of a homogeneous quadratic form in the p's, say

$$(62.12) \qquad\qquad C = \sqrt{g^{ij}p_ip_j},$$

in which the g's are functions of the x's. For the non-singular trajectories we choose the p's so that $C = 1$, in which case equations (62.9) become

$$(62.13) \qquad \frac{dx^i}{dt} = g^{ij}p_j, \qquad \frac{dp_i}{dt} = -\frac{1}{2}\frac{\partial g^{jk}}{\partial x^i}p_jp_k.$$

From (62.9) and (62.3) we have that, when $C = 0$, $p_i dx^i = 0$. This condition is satisfied by the first set of (62.13) and accordingly we take (62.13) as the equations for the singular trajectories also, when there are such trajectories. We observe also that $g^{ij}p_ip_j = const.$ is a first integral of these equations.

If we transform equations (62.9) by means of a general homogeneous contact transformation (59.8), written $x'^i = \varphi^i$, $p_i' = \psi_i$ we have

$$\frac{dx'^i}{dt} = \frac{\partial \varphi^i}{\partial x^j} \frac{\partial C}{\partial p_j} - \frac{\partial \varphi^i}{\partial p_j} \frac{\partial C}{\partial x^j}, \qquad \frac{dp'_i}{dt} = \frac{\partial \psi_i}{\partial x^j} \frac{\partial C}{\partial p_j} - \frac{\partial \psi_i}{\partial p_j} \frac{\partial C}{\partial x^j}.$$

If C' is the transform of C, we have

$$\frac{\partial \varphi^i}{\partial x^j} \frac{\partial C}{\partial p_j} - \frac{\partial \varphi^i}{\partial p_j} \frac{\partial C}{\partial x^j} = \left(\frac{\partial \varphi^i}{\partial x^j} \frac{\partial \varphi^k}{\partial p_j} - \frac{\partial \varphi^i}{\partial p_j} \frac{\partial \varphi^k}{\partial x^j} \right) \frac{\partial C'}{\partial x'^k}$$
$$+ \left(\frac{\partial \varphi^i}{\partial x^j} \frac{\partial \psi_k}{\partial p_j} - \frac{\partial \varphi^i}{\partial p_j} \frac{\partial \psi_k}{\partial x^j} \right) \frac{\partial C'}{\partial p'_k}$$

In consequence of (61.13) this reduces to $\dfrac{\partial C'}{\partial p'_i}$. In like manner the

right-hand member of the second set of equations is equal to $-\dfrac{\partial C'}{\partial x'^i}$.

Hence we have:

[62.3] *A group G_1 of homogeneous contact transformations is transformed into another group G_1 by any homogeneous contact transformation and the equations of the new group are integrals of the equations*

$$\frac{dx'^i}{dt} = \frac{\partial C'}{\partial p'_i}, \qquad \frac{dp'_i}{dt} = -\frac{\partial C'}{\partial x'^i},$$

where C' is the transform of the characteristic function of the given group.

From (62.1) and (62.4) it follows that the symbol Af of an infinitesimal homogeneous contact transformation expressed in terms of the characteristic function C is given by

$$(62.14) \qquad Af = \xi^i \frac{\partial f}{\partial x^i} + \eta_i \frac{\partial f}{\partial p_i} = (C, f).$$

Exercises

1. The homogeneous contact transformation determined (§61) by the equation $x^i \bar{x}^i + 1 = 0$ has the equations

$$\bar{x}^i = -\frac{p_i}{p_j x^j}, \qquad \bar{p}_i = x^i p_j x^j.$$

2. If $F(x; p)$ and $G(x; p)$ are transformed by a homogeneous contact transformation into $\bar{F}(\bar{x}; \bar{p})$ and $\bar{G}(\bar{x}; \bar{p})$, then $p_i \dfrac{\partial F}{\partial p_i} = \bar{p}_i \dfrac{\partial \bar{F}}{\partial \bar{p}_i}$ and $(F, G) = (\bar{F}, \bar{G})$.

3. If u is a solution of the equation $(F, f) = 0$ and v is any function such that (F, v) is a function of F, then (u, v) is a solution.

4. A necessary and sufficient condition that the two one-parameter groups of homogeneous contact transformations determined by functions C_1 and C_2 be commutative is that $(C_1, C_2) = 0$.

5. If $r(\leqslant n)$ independent (constant coefficients) infinitesimal homogeneous contact transformations with the symbols (C_a, f) for $a = 1, \cdots, r$ are such that $(C_a, C_b) = 0$, there exists a homogeneous contact transformation in variables $\bar{x}^i$ and $\bar{p}^i$ in terms of which the symbols are $(\overline{C}_a, f) = \dfrac{\partial f}{\partial \bar{x}^a}$.

Lie-Engel, 1888, 1, vol. 2, p. 267.

6. A necessary and sufficient condition that r independent (constant coefficients) characteristic functions $C_1, \ldots, C_r$ determine a group G_r of homogeneous contact transformations is that

$$(C_a, C_b) = c_{ab}^e\, C_e,$$

where the c's are constants satisfying (7.3) and (7.4).

Lie-Engel, 1888, vol. 2, p. 300.

7. Show that the functions

$$C_1 = p_1 + p_2, \quad C_2 = x^1 p_1 + x^2 p_2, \quad C_3 = (x^1)^2 p_1 + (x^2)^2 p_2$$

determine a G_3 of homogeneous contact transformations.

8. Show that the functions

$$C_1 = p_1, \quad C_2 = x^1 p_1 + \tfrac{1}{2} x^2 p_2, \quad C_3 = (x^1)^2 p_1 + x^1 x^2 p_2$$

determine a G_3 of homogeneous contact transformation with the same constants of structure as the G_3 of Ex. 7, and that it is transformed into the latter by the transformation

$$\bar{x}^1 = \frac{x^1 \sqrt{p_1} + i x^2 \sqrt{p_2}}{\sqrt{p_1} + i \sqrt{p_2}}, \quad \bar{p}_1 = p_1 + p_2,$$

$$\bar{x}^2 = \sqrt{x^1 - x^2}\, \frac{\sqrt[4]{p_1 p_2}}{\sqrt{p_1} + i \sqrt{p_2}}, \quad \bar{p}_2 = 2i \sqrt{x^1 - x^2}\, \sqrt[4]{p_1 p_2}\, (\sqrt{p_1} + i \sqrt{p_2}).$$

Lie-Engel, 1888, 1, vol. 2, p. 312.

63. Non-homogeneous contact transformations.

In his development of the theory of contact transformations Lie considers first transformations of the form

$$(63.1) \qquad \begin{aligned} Z &= \varphi(z, x^1, \cdots, x^n; p_1, \cdots, p_n), \\ X^i &= \varphi^i(z, x; p), \quad P_i = \psi_i(z, x; p), \end{aligned}$$

such that the equation

$$(63.2) \qquad\qquad dZ - P_i dX^i = 0$$

is satisfied whenever

$$(63.3) \qquad\qquad dz - p_i dx^i = 0.$$

Consequently the analytical problem reduces to the determination of functions φ, φ^i, ψ_i and also a function ρ of z, the x's and the p's, such that

$$(63.4) \qquad dZ - P_i dX^i = \rho(dz - p_i dx^i)$$

for independent values of the differentials dz, dx^i and dp_i. When, in particular, we take $p_i = \dfrac{\partial f}{\partial x^i}$, where f is a function of the x's, the differentials in (63.3) correspond to displacements in the hypersurface $z = f(x^1, \cdots, x^n)$ and the elements are tangential to it. When these values are substituted in the first $n + 1$ of equations (63.1), we have on the elimination of the x's, $Z = F(X^1, \cdots, X^n)$ and the P_i determine tangential elements at each point of this hypersurface in consequence of (63.2). We call a transformation (63.1) *non-homogeneous* as distinguished from the homogeneous transformations discussed in §§59–62.

For an n dimensional space of coordinates x^i, if we let x^1 play the exceptional role of z in the above equations, the analytical problem consists in determining transformations

$$(63.5) \qquad \begin{aligned} \bar{x}^i &= \varphi^i(x^1, \cdots, x^n; p_2, \cdots, p_n), \\ \bar{p}_i &= \psi_i(x^1, \cdots, x^n; p_2, \cdots, p_n), \end{aligned}$$

where the φ's and ψ's are such that the relation

$$(63.6) \qquad \bar{p}_i d\bar{x}^i = -dx^1 + p_\alpha dx^\alpha \qquad (\alpha = 2, \cdots, n)$$

holds for independent values of the differentials dx^i and dp_α. The resulting conditions are

$$(63.7) \quad \psi_i \frac{\partial \varphi^j}{\partial x^1} = -1, \quad \psi_i \frac{\partial \varphi^j}{\partial x^\alpha} = p_\alpha, \quad \psi_i \frac{\partial \varphi^j}{\partial p_\alpha} = 0 \quad (\alpha = 2, \cdots, n).$$

Comparing these equations with (59.10), we see that if we take any homogeneous transformation and in its equations put $p_1 = -1$, we obtain a non-homogeneous transformation.

Suppose conversely that we have a set of functions satisfying (63.7), and in the φ's we put

$$(63.8) \qquad p_\alpha = -\frac{p'_\alpha}{p'_1},$$

and denote the resulting functions by φ'^i. Since

$$(63.9) \qquad \frac{\partial \varphi'^i}{\partial p_1'} = \frac{\partial \varphi^i}{\partial p_\alpha} \frac{p_\alpha'^1}{p_1'^2} = -\frac{\partial \varphi^i}{\partial p_\alpha} \frac{p_\alpha}{p_1'}, \qquad \frac{\partial \varphi'^i}{\partial p_\alpha'} = -\frac{\partial \varphi^i}{\partial p_\alpha} \frac{1}{p_1'},$$

$$(\alpha = 2, \cdots, n),$$

where α is summed in the first set of equations, we have from (63.7), (63.8) and (63.9)

$$\psi_i' \frac{\partial \varphi'^i}{\partial x^i} = p_i', \qquad \psi_i' \frac{\partial \varphi'^i}{\partial p_i'} = 0,$$

where

$$(63.10) \qquad \psi_i' = -p_1' \psi_i \left(x^1, \cdots, x^n; -\frac{p_2'}{p_1'}, \cdots, -\frac{p_n'}{p_1'} \right).$$

Consequently the functions φ'^i and ψ_i' define a homogeneous contact transformation. In particular, we observe that, if the non-homogeneous transformation is derived from a homogeneous one as indicated in the preceding paragraph, the homogeneous transformation arising from the above process is the given one. It should be observed that a homogeneous transformation obtained in this manner does not apply to elements for which $p_1' = 0$, unless the ψ's are homogeneous of degree one in the p's. From the foregoing considerations it follows that the results of §60 apply to non-homogeneous transformations.

When we substitute the expressions (63.9) in equations of the form (59.17), we obtain

$$(63.11) \quad [\varphi^k, \varphi^l] \equiv \frac{\partial \varphi^k}{\partial p_\alpha} \left(\frac{\partial \varphi^l}{\partial x^\alpha} + p_\alpha \frac{\partial \varphi^l}{\partial x^1} \right) - \frac{\partial \varphi^l}{\partial p_\alpha} \left(\frac{\partial \varphi^k}{\partial x^\alpha} + p_\alpha \frac{\partial \varphi^k}{\partial x^1} \right) = 0.$$

Conversely, suppose we have a set of functions φ^i of $x^1, \ldots, x^n$; $p_2, \ldots, p_n$ satisfying these equations and such that the jacobian of the φ's with respect to the x's is of rank n. The set of functions ψ_i uniquely determined by the first n of equations (63.7) satisfy the other equations (63.7). In fact, if equations (63.11) be multiplied by ψ_k and k be summed, we obtain

$$(63.12) \qquad \psi_k \frac{\partial \varphi^k}{\partial p_\alpha} \left(\frac{\partial \varphi^l}{\partial x^\alpha} + p_\alpha \frac{\partial \varphi^l}{\partial x^1} \right) = 0.$$

These equations hold for $l = 1, \cdots, n$, since (63.11) are identically satisfied when $k = l$. Solving the first n of equations (63.7) for ψ_k, we have

$$\left|\frac{\partial \varphi^i}{\partial x^j}\right| \cdot \psi_k = (-1)^k \begin{vmatrix} 1 & \dfrac{\partial \varphi^1}{\partial x^1} & \cdots & \dfrac{\partial \varphi^{k-1}}{\partial x^1} & \dfrac{\partial \varphi^{k+1}}{\partial x^1} & \cdots & \dfrac{\partial \varphi^n}{\partial x^1} \\ -p_2 & \dfrac{\partial \varphi^1}{\partial x^2} & & \cdots & & & \dfrac{\partial \varphi^n}{\partial x^2} \\ & & \cdots & & & & \\ & & \cdots & & & & \\ -p_n & \dfrac{\partial \varphi^1}{\partial x^n} & & \cdots & & & \dfrac{\partial \varphi^n}{\partial x^n} \end{vmatrix}.$$

The determinant on the right is equal to the determinant of the quantities $\left(\dfrac{\partial \varphi^l}{\partial x^\alpha} + p_\alpha \dfrac{\partial \varphi^l}{\partial x^1}\right)$ for $\alpha = 2, \cdots, n; l = 1, \cdots, k-1,$ $k+1, \cdots, n.$* Since one at least of the ψ's is not zero, one of these determinants is not zero and consequently it follows from (63.12) that $\psi_k \dfrac{\partial \varphi^k}{\partial p_\alpha} = 0.$ Hence we have:

[63.1] *A necessary and sufficient condition that a set of functions φ^i of $x^1, \ldots, x^n; p_2, \ldots, p_n$ determine a non-homogeneous contact transformation, in which x^1 plays the exceptional role, is that the jacobian of the φ's with respect to the x's be of rank n and that $[\varphi^k, \varphi^l] = 0$; then the functions ψ_i of the transformation are uniquely determined.*†

From (63.10) and (63.8) we have

$$\frac{\partial \psi_j'}{\partial x^i} = -p_1 \frac{\partial \psi_j}{\partial x^i}, \qquad \frac{\partial \psi_k'}{\partial p_1'} = -\psi_k + \frac{\partial \psi_k}{\partial p_\alpha} p_{\alpha'}, \qquad \frac{\partial \psi_k'}{\partial p_\alpha'} = \frac{\partial \psi_k}{\partial p_\alpha}.$$

Hence from (61.13) in the φ''s and ψ''s we have in the notation (63.11)

(63.13)　　$[\psi_j, \psi_k] = \psi_j \dfrac{\partial \psi_k}{\partial x^1} - \psi_k \dfrac{\partial \psi_j}{\partial x^1}, \qquad [\psi_k, \varphi^j] = \delta_k^j + \psi_k \dfrac{\partial \varphi^j}{\partial x^1}.$

Equations (63.11) and (63.13) follow directly also, if we solve the equations (59.14) and $p_i \dfrac{\partial \psi_j}{\partial p_i} = \psi_j$ for $\dfrac{\partial \varphi^j}{\partial p_1}$ and $\dfrac{\partial \psi_j}{\partial p_1}$ and put $p_1 = -1,$ giving

---

* *Fine*, 1905, 1, p. 505.
† *Lie-Engel*, 1888, 1, vol. 2, p. 124.

$$\frac{\partial \varphi^j}{\partial p_1} = \left(p_\alpha \frac{\partial \varphi^j}{\partial p_\alpha} \right)_{p_1 = -1}, \qquad \frac{\partial \psi_j}{\partial p_1} = \left(p_\alpha \frac{\partial \psi_j}{\partial p_\alpha} - \psi_j \right)_{p_1 = -1},$$

and substitute in (61.13).

These results apply to the transformation in the form (63.5). In order to obtain them for (63.1) and (63.4), we let i in (63.5) take the values $0, 1, \ldots, n$, and put $\varphi^0 = Z$, $x^0 = z$, $p_0 = -1$, $\rho = -1/\psi_0$ and $P_\alpha = -\psi_\alpha/\psi_0$ for $\alpha = 1, \cdots, n$. In equations (63.11) and (63.13) now k and l take the values $0, 1, \ldots, n$. Making use of these equations, we obtain from (61.13) for a contact transformation (63.1) the following:

$$[Z, X^\alpha] = [X^\alpha, X^\beta] = [P_\alpha, P_\beta] = 0, \qquad [P_\alpha, X^\beta] = 0 \quad (\alpha \neq \beta),$$

$$(63.14) \qquad [P_\alpha, X^\alpha] = \rho, \qquad [P_\alpha, Z] = \rho P_\alpha \quad (\alpha = 1, \cdots, n).$$

In these equations the parentheses are defined by (63.11) with α taking the values $1, \ldots, n$.[*]

If the rank of the jacobian matrix $\left\| \dfrac{\partial \varphi^i}{\partial p_\alpha} \right\|$ for equations (63.5) is $n - r$, on the elimination of the p's from the first n of these equations we obtain equations

$$F_\sigma(\bar{x}^1, \cdots, \bar{x}^n; x^1, \cdots, x^n) = 0 \quad (\sigma = 1, \cdots, r).$$

Making use of (63.6), we have analogously to (59.13)

$$\bar{p}_i = u^\sigma \frac{\partial F_\sigma}{\partial \bar{x}^i}, \qquad 1 = u^\sigma \frac{\partial F_\sigma}{\partial x^1}, \qquad p_\alpha = -u^\sigma \frac{\partial F_\sigma}{\partial x^\alpha}.$$

Conversely, if we have a set of equations $F_\sigma = 0$, it can be shown, as in §61, that these determine a non-homogeneous contact transformation, provided the determinant (61.3) is of rank $n + r$ for arbitrary values of the parameters u^σ. Then the further determination of the functions φ^i of the transformation requires the solution for the $\bar{x}$'s of the equations $F_\sigma = 0$ and (61.4) in which $p_1 = -1$. In so doing it is understood that the variables are numbered in such a way that the determinant $\left| \dfrac{\partial F_\sigma}{\partial x^\tau} \right|$ for $\sigma, \tau = 1, \cdots, r$ is of rank r.[†]

---

[*] *Lie-Engel*, 1888, 1, vol. 2, p. 122.

[†] Cf. *Lie-Engel*, 1888, 1, vol. 2, p. 155.

For an infinitesimal transformation (62.1) in which $p_1 = -1$, the equations (63.7) become

$$(63.15) \quad \eta_i - \frac{\partial \xi^1}{\partial x^i} + p_\alpha \frac{\partial \xi^\alpha}{\partial x^i} = 0, \quad \frac{\partial \xi^1}{\partial p_\alpha} - p_\beta \frac{\partial \xi^\beta}{\partial p_\alpha} = 0$$

$$(\alpha, \beta = 2, \cdots, n).$$

If we put

$$W = p_\alpha \xi^\alpha - \xi^1,$$

these equations reduce to

$$\eta_i = -\frac{\partial W}{\partial x^i}, \quad \xi^\alpha = \frac{\partial W}{\partial p_\alpha}, \quad \xi^1 = p_\alpha \frac{\partial W}{\partial p_\alpha} - W.$$

Conversely, if W is any function of $x^1, \ldots, x^n$ and $p_2, \ldots, p_n$, these expressions satisfy (63.15). Hence we have:

[63.2] *The most general infinitesimal non-homogeneous contact transformation is defined by equations of the form*

$$(63.16) \quad \delta x^1 = \left(p_\alpha \frac{\partial W}{\partial p_\alpha} - W \right) \delta t, \quad \delta x^\alpha = \frac{\partial W}{\partial p_\alpha} \delta t, \quad \delta p_i = -\frac{\partial W}{\partial x^i} \delta t,$$

where W is an arbitrary function of $x^1, \ldots, x^n$; $p_2, \ldots, p_n$.[*]

64. Restricted non-homogeneous contact transformations. We consider now the particular case when the function φ^1 in equations (63.5) is of the form $x^1 + \varphi^1(x^2, \cdots, x^n; p_2, \cdots, p_n)$ and the functions φ^α for $\alpha = 2, \cdots, n$ do not involve x^1. Then equations (63.7) reduce to

$$(64.1) \quad \psi_1 = -1, \quad \psi_\beta \frac{\partial \varphi^\beta}{\partial x^\alpha} = p_\alpha + \frac{\partial \varphi^1}{\partial x^\alpha}, \quad \psi_\beta \frac{\partial \varphi^\beta}{\partial p_\alpha} = \frac{\partial \varphi^1}{\partial p_\alpha}.$$

Consequently the functions ψ_α do not involve x^1. Moreover equations (63.11) reduce in this case to

$$(64.2) \quad (\varphi^\beta, \varphi^\gamma) \equiv \frac{\partial \varphi^\beta}{\partial p_\alpha} \frac{\partial \varphi^\gamma}{\partial x^\alpha} - \frac{\partial \varphi^\beta}{\partial x^\alpha} \frac{\partial \varphi^\gamma}{\partial p_\alpha} = 0$$

and

$$(64.3) \quad (\varphi^1, \varphi^\beta) = p_\alpha \frac{\partial \varphi^\beta}{\partial p_\alpha} \quad (\alpha, \beta, \gamma = 2, \cdots, n).$$

---

[*] Cf. Ex. 3, p. 281.

In consequence of theorem [63.1] we have:

[64.1] *If φ^α are $n - 1$ independent functions of $x^2, \ldots, x^n$; $p_2, \ldots, p_n$ satisfying (64.2) and φ^1 is any function of these quantities satisfying (64.3), the equations*

$$(64.4) \qquad \bar{x}^1 = x^1 + \varphi^1, \quad \bar{x}^\alpha = \varphi^\alpha, \quad \bar{p}_\alpha = \psi_\alpha \quad (\alpha = 2, \cdots, n),$$

where the ψ's are uniquely determined by (64.1), define a contact transformation for which

$$(64.5) \qquad d\bar{x}^1 - \bar{p}_\alpha d\bar{x}^\alpha = dx^1 - p_\alpha dx^\alpha,$$

and consequently

$$(64.6) \qquad \bar{p}_\alpha d\bar{x}^\alpha = p_\alpha dx^\alpha + d\varphi^1.^*$$

We call such transformations *restricted non-homogeneous contact transformations*

If the functions φ^α are homogeneous of degree zero in the p's, equations (64.3) reduce to $(\varphi^1, \varphi^\beta) = 0$. Looked upon as linear partial differential equations in φ^1, these equations form a complete system, as can be shown with the aid of the identity of Jacobi, namely (61.15). Consequently these equations admit $n - 1$ independent integrals. But φ^α are independent integrals in consequence of (64.2). Hence φ^1 is an arbitrary function of the φ^α, say $F(\varphi^2, \ldots, \varphi^n)$. In this case we have from (64.1)

$$\left(\psi_\beta - \frac{\partial F}{\partial \varphi^\beta}\right)\frac{\partial \varphi^\beta}{\partial x^\alpha} = p_\alpha, \qquad \left(\psi_\beta - \frac{\partial F}{\partial \varphi^\beta}\right)\frac{\partial \varphi^\beta}{\partial p_\alpha} = 0,$$

so that the transformation is essentially of the homogeneous type.

We remark also that for the restricted case equations (63.13) become

$$(64.7) \quad (\psi_\alpha, \varphi^1) = \psi_\alpha - p_\beta \frac{\partial \psi_\alpha}{\partial p_\beta}, \qquad (\psi_\beta, \varphi^\alpha) = \delta^\alpha_\beta, \qquad (\psi_\alpha, \psi_\beta) = 0.$$

From these equations it follows that theorem [61.5] holds for a restricted contact transformation.

The jacobian of φ^α with respect to the p's is of rank $n - 1 - r$, where $r \geq 0$. When the p's are eliminated from the first n of equations (64.4) we get

---

* Cf. *Lie-Engel*, 1888, 1, vol. 2, p. 129.

(64.8) $\qquad \bar{x}^1 - x^1 = F(\bar{x}^2, \cdots, \bar{x}^n; x^2, \cdots, x^n)$

and r equations

(64.9) $\qquad F_\sigma(\bar{x}^2, \cdots, \bar{x}^n; x^2, \cdots, x^n) = 0 \quad (\sigma = 1, \cdots, r).$

In this case these two sets of equations take the place of equations (59.11) for the homogeneous case, and in place of (59.13) we have, on making use of (64.5),

$$(64.10) \quad \bar{p}_\alpha = \frac{\partial F}{\partial \bar{x}^\alpha} + \rho^\sigma \frac{\partial F_\sigma}{\partial \bar{x}^\alpha}, \qquad p_\alpha = -\frac{\partial F}{\partial x^\alpha} - \rho^\sigma \frac{\partial F_\sigma}{\partial x^\alpha}.$$

Conversely, if a set of equations (64.9) and the second set of (64.10) can be solved for $\bar{x}^2, \ldots, \bar{x}^n; \rho^1, \ldots, \rho^r$ in terms of the x's and p's, when these are substituted in (64.8) and the first set of (64.10), the resulting equations define a restricted non-homogeneous transformation. The conditions to be satisfied by $F, F_1, \ldots, F_r$ can be determined in a manner similar to that followed in §61.*

For an infinitesimal transformation of the restricted type the function W does not involve x^1, and consequently we have:

[64.2] *The most general infinitesimal restricted non-homogeneous transformation is defined by equations of the form*

$$(64.11) \quad \delta x^1 = \left(p_\alpha \frac{\partial W}{\partial p_\alpha} - W \right) \delta t, \quad \delta x^\alpha = \frac{\partial W}{\partial p_\alpha} \delta t, \quad \delta p_\alpha = -\frac{\partial W}{\partial x^\alpha} \delta t,$$

where the characteristic function W is an arbitrary function of $x^2, \ldots, x^n; p_2, \ldots, p_n.$

In this case we have in place of (64.6)

$$\bar{p}_\alpha d\bar{x}^\alpha = p_\alpha dx^\alpha + d\left(p_\alpha \frac{\partial W}{\partial p_\alpha} - W \right) \delta t,$$

to within terms of higher order in δt. Hence when W is homogeneous of degree one in the p's, the transformation is homogeneous (cf. §62).

From the results of §62 it follows that each infinitesimal transformation of this type generates a continuous group G_1 of restricted transformations, whose finite equations are the integrals of

---

* *Lie-Engel*, 1888, 1, vol. 2, p. 156.

(64.12) $\quad \dfrac{dx^1}{dt} = p_\alpha \dfrac{\partial W}{\partial p_\alpha} - W, \qquad \dfrac{dx^\alpha}{dt} = \dfrac{\partial W}{\partial p_\alpha}, \qquad \dfrac{dp_\alpha}{dt} = -\dfrac{\partial W}{\partial x^\alpha}.$

From (64.6) it follows that the product of two restricted transformations is one of the same kind. Consequently the product of a restricted transformation and the transformations of a group G_1 of restricted transformations yields a new G_1 of restricted transformations, whose equations therefore are of the form

$$\frac{d\bar{x}^1}{dt} = \bar{p}_\alpha \frac{\partial \overline{W}}{\partial \bar{p}_\alpha} - \overline{W}, \qquad \frac{d\bar{x}^\alpha}{dt} = \frac{\partial \overline{W}}{\partial \bar{p}_\alpha}, \qquad \frac{d\bar{p}_\alpha}{dt} = -\frac{\partial \overline{W}}{\partial \bar{x}^\alpha}.$$

Because of (64.2), (64.3) and (64.7) we have $\overline{W} = W$, that is, $\overline{W}$ is the transform of W by means of the given restricted transformation.

Equations (64.12) with the exception of the first are the general form of the Hamiltonian equations for a conservative holonomic dynamical system for which the Hamiltonian function does not involve the time t. Consequently, if we apply to these equations any transformation of the form (64.4) where the φ's satisfy (64.2) and φ^1 equations (64.3), we have a new Hamiltonian system* and the Hamiltonian function is the transform of the original Hamiltonian function.

If we adjoin the first equation (64.12) to the Hamiltonian equations of a conservative holonomic system for which W does not involve the time, we obtain

$$\frac{dx^1}{dt} = p_\alpha \frac{dx^\alpha}{dt} - W,$$

the right-hand member of which is the Lagrangian function of the dynamical system.

65. Homogeneous contact transformations of maximum rank.
From the second set of equations (62.6) it follows that the rank of the hessian of C ($\neq 0$) with respect to p's, that is, $\left| \dfrac{\partial^2 C}{\partial p_i \partial p_j} \right|$, is $n - 1$ at most. If we put

(65.1) $$H = \frac{1}{2}C^2,$$

we have

---

* Cf. *Jacobi*, 1837, 1, p. 67.

$$(65.2) \qquad \frac{\partial H}{\partial p_i} = C \frac{\partial C}{\partial p_i}, \qquad \frac{\partial^2 H}{\partial p_i \partial p_j} = C \frac{\partial^2 C}{\partial p_i \partial p_j} + \frac{\partial C}{\partial p_i} \frac{\partial C}{\partial p_j}.$$

From the second set of these equations it follows that the hessian of H with respect to the p's is equal to*

$$(65.3) \qquad C^{n-1} \begin{vmatrix} C & -\dfrac{\partial C}{\partial p_1} & \cdots & -\dfrac{\partial C}{\partial p_n} \\ \dfrac{\partial C}{\partial p_1} & \dfrac{\partial^2 C}{\partial p_1^2} & \cdots & \dfrac{\partial^2 C}{\partial p_1 \partial p_n} \\ \cdot & \cdot & \cdots & \cdot \\ \cdot & \cdot & \cdots & \cdot \\ \dfrac{\partial C}{\partial p_n} & \dfrac{\partial^2 C}{\partial p_1 \partial p_n} & \cdots & \dfrac{\partial^2 C}{\partial p_n^2} \end{vmatrix}.$$

If we multiply the last n rows of this determinant by $p_1, \ldots, p_n$ respectively and add the resulting terms of all but the last row to the corresponding terms of the last row, the terms of the latter become $C, 0, \ldots, 0$ in consequence of (62.5) and (62.6). Consequently a necessary condition that the hessian of H with respect to the p's be of rank n is that the determinant formed from (65.3) by omitting the first column and last row be of rank n. If the columns of this determinant be multiplied by $p_1, \ldots, p_n$ respectively and the terms of the first $n - 1$ columns be added to the last, the terms of the latter become $C, 0, \ldots, 0$. Hence if this determinant is to be of rank n it is necessary that the determinant $\left| \dfrac{\partial^2 C}{\partial p_\alpha \partial p_\beta} \right|$ for $\alpha, \beta = 1, \cdots, n - 1$ be of rank $n - 1$.

If we apply the above process to (65.3) using any one of the last n rows as we used the last, we find that it is necessary that every minor of order $n - 1$ of the hessian of C with respect to the p's be of rank $n - 1$. But this is a consequence of one minor being of rank $n - 1$. For, if $\left| \dfrac{\partial^2 C}{\partial p_\alpha \partial p_\beta} \right|$ for $\alpha, \beta = 1, \cdots, n - 1$ is of rank $n - 1$, it follows from the equations

$$p_i \frac{\partial^2 C}{\partial p_i \partial p_\alpha} = 0 \qquad (i = 1, \cdots, n; \alpha = 1, \cdots, n - 1)$$

---

* Cf. *Fine*, 1905, 1, p. 505.

that each of the determinants of order $n - 1$ of the matrix $\left\|\dfrac{\partial^2 C}{\partial p_i \partial p_\alpha}\right\|$ is of rank $n - 1$. When then we apply the same reasoning to the equations $p_i \dfrac{\partial^2 C}{\partial p_i \partial p_\sigma} = 0$ for σ equal to n and any $n - 2$ of the numbers $1, \cdots, n - 1$, we are led to the conclusion that, if the hessian of C with respect to the p's is of rank $n - 1$, every minor of order $n - 1$ is of rank $n - 1$. Accordingly we have:

[65.1] *When and only when the rank of the hessian of C with respect to the p's is $n - 1$, the rank of the hessian of H with respect to the p's is n.*

Without loss of generality we take (§62)

$$(65.4) \qquad\qquad C = 1.$$

Then equations (62.9) may be written because of (65.1) in the form

$$(65.5) \qquad\qquad \frac{dx^i}{dt} = \frac{\partial H}{\partial p_i}, \qquad \frac{dp_i}{dt} = -\frac{\partial H}{\partial x^i}.$$

If we assume that C satisfies the condition of the preceding theorem, the hessian of H with respect to the p's is of rank n and consequently the first set of equations (65.5) can be solved for the p's as functions of the x's and $\dot{x}$'s, where $\dot{x}^i = \dfrac{dx^i}{dt}$. Since H is homogeneous of degree two in the p's, it follows that the solutions p_i are homogeneous of first degree in the $\dot{x}$'s. When these expressions for the p's are substituted in H, we get a function $\mathcal{H}$ of the x's and $\dot{x}$'s, which is homogeneous of degree two in the $\dot{x}$'s. Thus we have

$$(65.6) \qquad\qquad \mathcal{H}(x; \dot{x}) = H(x; p).$$

Differentiating $\mathcal{H}$ with respect to $\dot{x}^i$, we have in consequence of (65.5)

$$(65.7) \qquad\qquad \frac{\partial \mathcal{H}}{\partial \dot{x}^i} = \frac{\partial H}{\partial p_j} \frac{\partial p_j}{\partial \dot{x}^i} = \dot{x}^j \frac{\partial p_j}{\partial \dot{x}^i}.$$

Hence by Euler's theorem

$$(65.8) \qquad\qquad 2\mathcal{H} = \dot{x}^i \frac{\partial \mathcal{H}}{\partial \dot{x}^i} = \dot{x}^i \dot{x}^j \frac{\partial p_j}{\partial \dot{x}^i} = \dot{x}^i p_j.$$

From this equation and (65.7) we have

$$\frac{\partial \mathcal{3C}}{\partial \dot{x}^i} = \frac{1}{2}\left(p_i + \dot{x}^j \frac{\partial p_j}{\partial \dot{x}^i}\right) = \frac{1}{2}p_i + \frac{1}{2}\frac{\partial \mathcal{3C}}{\partial \dot{x}^i},$$

and consequently

(65.9) $$p_i = \frac{\partial \mathcal{3C}}{\partial \dot{x}^i}.$$

From (65.6), (65.5) and (65.8) we have

$$\frac{\partial \mathcal{3C}}{\partial x^i} = \frac{\partial H}{\partial x^i} + \frac{\partial H}{\partial p_j}\frac{\partial p_j}{\partial x^i} = \frac{\partial H}{\partial x^i} + \dot{x}^j \frac{\partial p_j}{\partial x^i} = \frac{\partial H}{\partial x^i} + 2\frac{\partial \mathcal{3C}}{\partial x^i}.$$

Consequently

(65.10) $$\frac{\partial \mathcal{3C}}{\partial x^i} + \frac{\partial H}{\partial x^i} = 0.$$

If we denote by $\mathcal{C}$ the function of the x's and $\dot{x}$'s when the expressions for p_i from the solution of (65.5), that is, (65.9), are substituted in C, we have

(65.11) $$\mathcal{3C} = \tfrac{1}{2}\mathcal{C}^2, \qquad \mathcal{C} = 1,$$

the latter being a consequence of (65.4). From (65.9) we have

(65.12) $$p_i = \frac{\partial \mathcal{C}}{\partial \dot{x}^i}$$

and from (65.10)

(65.13) $$\frac{\partial \mathcal{C}}{\partial x^i} + \frac{\partial C}{\partial x^i} = 0.$$

When now the expressions (65.12) are substituted in the second set of equations (62.9), we have in consequence of (65.13)

(65.14) $$\frac{d}{dt}\left(\frac{\partial \mathcal{C}}{\partial \dot{x}^i}\right) - \frac{\partial \mathcal{C}}{\partial x^i} = 0.$$

Thus in the quantities x^i and $\dot{x}^i$ equations (65.12) and (65.14) replace (62.9).

Conversely, if we have any function $\mathcal{C}$ of the x's and $\dot{x}$'s, homogeneous of degree one in the $\dot{x}$'s, such that

(65.15) $$\mathcal{C}(x; \dot{x}) = 1,$$

and the rank of the hessian of $\mathcal{C}$ with respect to the $\dot{x}$'s is $n - 1$, equations (65.9) can be solved for the $\dot{x}$'s as homogeneous functions of degree one in the p's, and by repeating the above processes we obtain the first set of equations (62.9), and the second set follow from (65.12) and (65.14).

If we take a particular point $P(x^i)$ of our space, the first set of equations (62.7) defines a near-by point, which varies for each choice of the p's subject to the condition (65.4). The equation of the locus of this point is obtained on the elimination of the p's from the first set of (62.7) and (65.4). If the hessian of C with respect to the p's is of rank $n - 1$, this elimination can be accomplished by the process which led to (65.11). Consequently, if y^i are coordinates with P as origin so that $y^i = \bar{x}^i - x^i$, we have from (62.7) and (62.9) $y^i = \dot{x}^i \delta t$ and hence the equation of the locus is

$$(65.16) \qquad\qquad \mathbb{C}(x; y) = \delta t,$$

since $\mathbb{C}$ is homogeneous of the first degree in the $\dot{x}$'s. Also from this property of $\mathbb{C}$ and from (65.12) it follows that

$$(65.17) \qquad\qquad p_i = \frac{\partial \mathbb{C}(x; y)}{\partial y^i}.$$

Consequently p_i are the components of the covariant normal to the hypersurface (65.16), which we call the *elementary hypersurface* of the transformation for the point P.

If X^i are current coördinates, the tangential hyperplane at a point is defined by

$$p_i(X^i - y^i) = 0,$$

or

$$(65.18) \qquad\qquad p_i X^i = w,$$

where in consequence of (62.7) and (62.5)

$$(65.19) \qquad\qquad w = p_i \frac{\partial C}{\partial p_i} \delta t = C \delta t.$$

Consequently (65.18) with w given by (65.19) is the tangential equation of the elementary hypersurface.*

66. Geometrical properties of continuous groups of maximum rank. Waves. Consider the finite equations

$$(66.1) \qquad \bar{x}^i = \varphi^i(x; p; t), \qquad \bar{p}_i = \psi_i(x; p; t)$$

---

* Cf. *Vessiot*, 1906, 1, p. 262; also, *Levi-Civita* and *Amaldi*, 1927, 6, pp. 452, 462.

of a group G_1 of homogeneous contact transformations for which the hessian of the characteristic function C with respect to the p's is of rank $n - 1$. When the functions φ^i are written as power series in t, as solutions of (62.9), it is seen that the rank of the jacobian of the φ's with respect to the p's is $n - 1$. Consequently when the p's are eliminated from the first set of equations (66.1), we get an equation

$$(66.2) \qquad\qquad F(\bar{x}; x; t) = 0.$$

This equation for given values of the x's defines the hypersurface whose points are the transforms of the point $P(x)$ for the given value of t; when $t = 0$, the hypersurface reduces to the point P, as follows from φ^i expressed as power series in t. For other values of t equation (66.2) defines a family of hypersurfaces which may be interpreted as waves emanating from P, t being the time which has elapsed since the emanation.

When in the first set of equations (66.1) we give the x's and p's fixed values, these equations define a curve, the locus of corresponding points on the waves, and the second set of equations (66.1) defines the covariant normal to the waves at these points. We call these curves the *trajectories of the wave-motion*. When the initial values satisfy $C = 0$, we call the trajectory *singular*. The non-singular trajectories are integrals of equations (65.14). Along a singular trajectory we have $\bar{p}_i d\bar{x}^i = 0$ (§62). Consequently such a trajectory is tangential to a hypersurface (66.2) at its point of meeting with it. Substituting in (66.2) for the $\bar{x}$'s such a solution, we have on differentiating the resulting identity with respect to t

$$\frac{\partial F}{\partial \bar{x}^i} \frac{d\bar{x}^i}{dt} + \frac{\partial F}{\partial t} = 0.$$

Hence the singular trajectories meet (66.2) in points of the hypersurfaces

$$\frac{\partial F}{\partial t} = 0,$$

that is, in the points of tangency of (66.2) with its envelope considering t as a parameter.

Excluding the points on the envelope of the hypersurfaces (66.2), we can solve this equation for t, thus

$$(66.3) \qquad\qquad f(\bar{x}; x) - t = 0;$$

then $f(x; x) = 0$. Since the transformations form a group, f must be of such a form that

$$f\left(\bar{x} + \frac{\partial C}{\partial \bar{p}_i}\delta t; x\right) - (t + \delta t) = 0,$$

from which it follows that

(66.4) $$\frac{\partial f}{\partial \bar{x}^i}\frac{\partial C}{\partial \bar{p}_i} = 1,$$

where

(66.5) $$\bar{p}_i = \rho\frac{\partial f}{\partial \bar{x}^i}.$$

Since C is homogeneous of degree one in the p's and $C = 1$, we have $\rho = 1$ and consequently f is a solution of the equation

(66.6) $$C\left(\bar{x}; \frac{\partial f}{\partial \bar{x}}\right) = 1.$$

Moreover, it is a complete integral, since it involves n parameters x^i subject to the condition $f(x; x) = 0$.

We proceed to the converse problem of deriving the trajectories from a complete integral of equation (66.6), say $\varphi(\bar{x}; a^1, \ldots, a^{n-1})$, where none of the a's is additive. Because of this requirement the jacobian of the functions φ and $\frac{\partial \varphi}{\partial a^\alpha}$, where $\alpha = 1, \cdots, n - 1$, with respect to the $\bar{x}$'s is not zero. Hence the equations

(66.7) $$\varphi(\bar{x}; a) + a^n - t = 0,$$

(66.8) $$\frac{\partial \varphi}{\partial a^\alpha} = b_\alpha,$$

where a^n and the b's are arbitrary constants, admit a solution

(66.9) $$\bar{x}^i = \varphi_1^i(a^1, \cdots, a^n; b_1, \cdots, b_{n-1}, t).$$

We denote by ψ_i the functions of the a's, b's, t and h which are obtained by the substitution of φ^i for $\bar{x}^i$ in the right-hand members of the equations

(66.10) $$\bar{p}_i = h\frac{\partial \varphi}{\partial \bar{x}^i},$$

where h is an arbitrary constant. We shall show that the φ_1's and ψ's constitute a solution of equations (62.9). In fact, if we substitute (66.9) in (66.7) and (66.8) and differentiate the resulting identities with respect to t, we obtain

$$(66.11) \qquad \frac{\partial \varphi}{\partial \bar{x}^i} \frac{d\bar{x}^i}{dt} - 1 = 0, \qquad \frac{\partial^2 \varphi}{\partial a^\alpha \partial \bar{x}^i} \frac{d\bar{x}^i}{dt} = 0.$$

Also if we substitute the function φ in (66.6) and differentiate the resulting identity with respect to the a's and $\bar{x}$'s, we get in consequence of (66.10)

$$(66.12) \qquad \frac{\partial C}{\partial \bar{p}_i} \frac{\partial^2 \varphi}{\partial \bar{x}^i \partial a^\alpha} = 0, \qquad \frac{\partial C}{\partial \bar{x}^i} + h \frac{\partial C}{\partial \bar{p}_j} \frac{\partial^2 \varphi}{\partial \bar{x}^j \partial \bar{x}^i} = 0.$$

From the first set of these equations, (66.11) and (66.4) we obtain

$$\frac{\partial \varphi}{\partial \bar{x}^i}\left(\frac{d\bar{x}^i}{dt} - \frac{\partial C}{\partial \bar{p}_i}\right) = 0, \qquad \frac{\partial^2 \varphi}{\partial \bar{x}^i \partial a^\alpha}\left(\frac{d\bar{x}^i}{dt} - \frac{\partial C}{\partial \bar{p}^i}\right) = 0,$$

which are equivalent to the first set of equations (62.9), since the determinant of these equations is a non-vanishing jacobian. From this result, (66.10) and the second set of equations (66.12) we have

$$\frac{d\bar{p}_i}{dt} = h\frac{\partial^2 \varphi}{\partial \bar{x}^i \partial \bar{x}^j} \frac{d\bar{x}^j}{dt} = h\frac{\partial^2 \varphi}{\partial \bar{x}^i \partial \bar{x}^j} \frac{\partial C}{\partial \bar{p}_j} = -\frac{\partial C}{\partial \bar{x}^i},$$

as was to be proved.

If we denote by x^i and p_i the values of $\bar{x}^i$ and $\bar{p}_i$ when $t = 0$, we have from (66.7) and (66.8)

$$(66.13) \quad a^n = -\varphi(x; a), \qquad b_\alpha = \frac{\partial}{\partial a^\alpha}\varphi(x; a^1, \cdots, a^{n-1}),$$

and from (66.10)

$$(66.14) \qquad p_i\frac{\partial}{\partial x^j}\varphi(x;\ a) - p_j\frac{\partial}{\partial x^i}\varphi(x;\ a) = 0.$$

There are $n - 1$ independent equations (66.14); they define the $n - 1$ a's as functions of the x's and p's, homogeneous of degree zero in the p's. Then a^n and the b's are given by (66.13) as functions of these quantities and from (66.9) we have the first set of (66.1). For $t = 0$ we have from (66.6) and (66.10)

$$C\left(x; \frac{\partial \varphi}{\partial x}\right) = \frac{1}{h}C(x;\ p) = 1.$$

Consequently the second set of (66.1) is given by substituting the expressions for the $\bar{x}$'s, a's and b's in terms of the x's and p's in

$$(66.10') \qquad \bar{p}_i = C(x; p) \frac{\partial \varphi}{\partial \bar{x}^i},$$

which is homogeneous of degree one in the p's.

From (66.7) and (66.13) we have

$$(66.15) \qquad \varphi(\bar{x}; a) - \varphi(x; a) - t = 0.$$

For given values of the p's, the $\bar{p}$'s given by (66.10') are the components of the covariant normal to the hypersurface (66.15) for a given value of t. From (66.5) these are the components of the covariant normal to the hypersurface (66.3). Hence the latter is the envelope of the hypersurfaces (66.15) for the same value of t, on treating the p's as parameters. Since the a's are independent functions of the ratios p_α/p_n, the envelope of the hypersurfaces considering these ratios as parameters is the same as the envelope considering the a's as parameters. If then we eliminate the a's from (66.15) and

$$\frac{\partial \varphi(\bar{x}; a)}{\partial a^\alpha} - \frac{\partial \varphi(x; a)}{\partial a^\alpha} = 0,$$

we obtain the equation (66.3).

Consider any hypersurface

$$(66.16) \qquad F(x^1, \cdots, x^n) = 0,$$

F being irreducible. From the results of §§60, 61 it follows that the envelope of the waves emanating from each point of the hypersurface after a time t is defined by

$$(66.17) \qquad \bar{x}^i = \varphi^i\left(x; \frac{\partial F}{\partial x}; t\right),$$

in which the x's are in the relation (66.16), and that

$$(66.18) \qquad \bar{p}_i = \psi_i\left(x; \frac{\partial F}{\partial x}; t\right)$$

are the components of the covariant normal at each point. As t varies the envelopes constitute a series of wave-fronts whose character is determined by the *hypersurface of origin* (66.16).

Eliminating the x's from (66.16) and (66.17), we get as the equation of the wave-fronts

$$(66.19) \qquad \Phi(\bar{x}^1, \cdots, \bar{x}^n; t) = 0.$$

Excluding the points on the envelope of these hypersurfaces as in the case of (66.2), we may replace (66.19) by

$$(66.20) \qquad \psi(\bar{x}^1, \cdots, \bar{x}^n) - t = 0.$$

It follows from the considerations applied to (66.3) that ψ is a solution of equation (66.6). It is the general integral determined by the relation (66.16) between the parameters x^i in the complete integral $f(\bar{x}; x)$, since (66.20) is the envelope of the hypersurfaces (66.3) for the relation (66.16). Thus the waves (66.3) and the wave-fronts (66.20) are related in accordance with Huygen's principle.

When the coördinates x^i of a point P on the hypersurface (66.16) are substituted in (66.17), the resulting equations define a curve, the locus of the points of contact with the wave-fronts of the waves emanating from P as t varies. The congruence of curves so defined are the trajectories of motion of the wave-fronts; evidently they are integral curves of equations (65.14).

The hypersurface (66.20) for a particular value of t is the envelope of the hypersurfaces (66.7) in which the a's are functions of the x's determined by the $n - 1$ equations

$$(66.21) \qquad \frac{\partial F}{\partial x^i} \frac{\partial \varphi(x; a)}{\partial x^j} - \frac{\partial F}{\partial x^j} \frac{\partial \varphi(x; a)}{\partial x^i} = 0,$$

the x's being subject to the relation $F = 0$. If the x's are eliminated from (66.21), the first of (66.13) and $F = 0$, we get a relation

$$\psi(a^1, \cdots, a^{n-1}) = a^n.$$

Consequently (66.20) is the envelope of (66.7) for the a's in this relation. Conversely, if we have any such relation, the envelope of

$$\varphi(x; a) + \psi(a) = 0$$

is the hypersurface of origin (66.16) of the waves (66.15).

In order to make a similar discussion of the singular trajectories of the group G_1, defined by a characteristic function C, we consider the differential equation

$$(66.22) \qquad C\left(\bar{x}; \frac{\partial \varphi}{\partial \bar{x}}\right) = 0.$$

Suppose we have an integral φ of this equation involving $n - 2$

arbitrary constants $a^1, \ldots, a^{n-2}$, none of which is additive, and consider the $n - 1$ equations

$$(66.23) \quad \varphi(\bar{x}; a) = b, \qquad \frac{\partial \varphi(\bar{x}; a)}{\partial a^\sigma} = b_\sigma \quad (\sigma = 1, \cdots, n - 2),$$

where the b's are arbitrary constants. For each set of values of the a's and b's these equations define a curve along which the differentials satisfy the conditions

$$\frac{\partial \varphi}{\partial \bar{x}^i} d\bar{x}^i = 0, \qquad \frac{\partial^2 \varphi}{\partial a^\sigma \partial \bar{x}^i} d\bar{x}^i = 0.$$

If we define functions $\bar{p}_i$ by (66.10), it follows from (66.22) that

$$\frac{\partial \varphi}{\partial \bar{x}^i} \frac{\partial C}{\partial \bar{p}_i} = 0, \qquad \frac{\partial^2 \varphi}{\partial a^\sigma \partial \bar{x}^i} \frac{\partial C}{\partial \bar{p}_i} = 0.$$

Since the matrix of $\dfrac{\partial \varphi}{\partial \bar{x}^i}$ and $\dfrac{\partial^2 \varphi}{\partial a^\sigma \partial \bar{x}^i}$ is of rank $n - 1$, we have from the above sets of equations

$$\frac{d\bar{x}^i}{\dfrac{\partial C}{\partial \bar{p}_i}} = \cdots = \frac{d\bar{x}^n}{\dfrac{\partial C}{\partial \bar{p}_n}} = \rho,$$

where ρ is the factor of proportionality. From (66.10) and the second set of (66.12) we have

$$d\bar{p}_i = h \frac{\partial^2 \varphi}{\partial \bar{x}^i \partial \bar{x}^j} d\bar{x}^j = -\rho \frac{\partial C}{\partial \bar{x}^i}.$$

Consequently the curves (66.23) are integrals of (62.9).

In order to obtain the curves through a point $P(x)$, we replace (66.23) by

$$(66.24) \quad \varphi(\bar{x}; a) - \varphi(x; a) = 0, \qquad \frac{\partial \varphi(\bar{x}; a)}{\partial a^\sigma} - \frac{\partial \varphi(x; a)}{\partial a^\sigma} = 0.$$

Since the p's in this case must satisfy the condition $C(x; p) = 0$, there are $n - 2$ of equations (66.14) for the determination of the a's as functions of the x's and ratios of the p's. When these are substituted in (66.24), we have the equations of the singular trajectories, and the $\bar{p}$'s are given by (66.10) in which h is given by

$$p_i = h \frac{\partial \varphi}{\partial x^i}.$$

67. Application to geodesics of a Riemannian space. Consider a Riemannian space whose metric is defined by

$$(67.1) \qquad ds^2 = g_{ij}dx^idx^j.$$

In consequence of (47.10) and (47.11) the equations (47.22) of the non-minimal geodesics of the space are expressible in the form

$$(67.2) \qquad g_{ij}\frac{d^2x^j}{ds^2} + [jk, i]\frac{dx^j}{ds}\frac{dx^k}{ds} = 0.$$

Consider now the group G_1 of contact transformations for which the characteristic function is

$$(67.3) \qquad C = \sqrt{g^{ij}p_ip_j}.$$

In this case we have equations (62.13). When the trajectories are non-singular, it follows from (62.13) and (67.1) that $t = s$. Hence we have

$$(67.4) \qquad \frac{dx^i}{ds} = g^{ij}p_j, \qquad \frac{dp_i}{ds} = -\frac{1}{2}\frac{\partial g^{jk}}{\partial x^i}p_jp_k.$$

In consequence of (47.10) and the first set of (67.4) we have

$$(67.5) \qquad p_i = g_{ij}\frac{dx^j}{ds}.$$

Substituting this expression in the second set of (67.4), we obtain equations (67.2).

If the form (67.1) is not definite, there are real values of the p's such that
$$(67.6) \qquad g^{ij}p_ip_j = 0.$$

As in the preceding case we obtain from (62.13) equations (67.2) with s replaced by t, and from the first set of (62.13) and (67.6) it follows that $g_{ij}\dfrac{dx^i}{dt}\dfrac{dx^j}{dt} = 0$, that is, the curves are minimal geodesics. Hence we have:

[67.1] *For the group G_1 with the characteristic function (67.3) the non-singular trajectories are the non-minimal geodesics of the Riemannian space with the fundamental form (67.1) and $t = s$; and the singular trajectories are the minimal geodesics of the space.*

For the case of non-minimal geodesics equation (66.6) is

$$(67.7) \qquad g^{ij}\frac{\partial\varphi}{\partial x^i}\frac{\partial\varphi}{\partial x^j} = 1.$$

From the results of §66 it follows that, if $\varphi(x; a^1, \ldots, a^{n-1})$ is a complete integral of (67.7), the corresponding equations (66.8) define $n - 1$ hypersurfaces which intersect in the non-minimal geodesics. Moreover, from (66.11) and (67.4) it follows that for particular values of the a's these geodesics are the orthogonal trajectories of the hypersurfaces $\varphi = $ const.* Also if (67.1) is not definite and $\varphi(x; a^1, \ldots, a^{n-2})$ is an integral of

$$g^{ij}\frac{\partial\varphi}{\partial x^i}\frac{\partial\varphi}{\partial x^j} = 0,$$

the second set of equations (66.23) define the minimal geodesics.

When $\varphi(\bar{x}; a^1, \ldots, a^{n-1})$ is an integral of (67.7) and the a's are eliminated from the equations

$$(67.8) \quad \varphi(\bar{x}; a) - \varphi(x; a) - s = 0, \qquad \frac{\partial\varphi(\bar{x}; a)}{\partial a^\alpha} - \frac{\partial\varphi(x; a)}{\partial a^\alpha} = 0$$
$$(\alpha = 1, \cdots, n-1),$$

in which the x's have particular values, the resulting equation

$$(67.9) \qquad f(\bar{x}; x) - s = 0$$

defines the hypersurfaces orthogonal to the geodesics through the point $P(x)$. Each of these hypersurfaces is the locus of points at the same distance s from P, measured along the geodesics. Consequently we call them *geodesic hyperspheres* for the space. The function f is a solution of (67.7) and the geodesics are the integrals of the equations

$$\frac{d\bar{x}^1}{g^{1j}\dfrac{\partial f}{\partial \bar{x}^j}} = \cdots = \frac{d\bar{x}^n}{g^{nj}\dfrac{\partial f}{\partial \bar{x}^j}}.$$

If the integrals of these equations are

$$\psi^\alpha(\bar{x}; x) = b^\alpha, \qquad (\alpha = 1, \cdots, n-1),$$

and we put

$$y^\alpha = \psi^\alpha, \qquad y^n = f(\bar{x}; x),$$

_____

* 1926, 3, p. 58.

and denote by $a_{ij}dy^i dy^j$ the transform of (67.1), we have

$$a^{\alpha n} = g^{ij} \frac{\partial \psi^\alpha}{\partial \bar{x}^i} \frac{\partial f}{\partial \bar{x}^j} = 0, \qquad a^{nn} = g^{ij} \frac{\partial f}{\partial \bar{x}^i} \frac{\partial f}{\partial \bar{x}^j} = 1.$$

Consequently the form (67.1) becomes

$$ds^2 = [(dy^n)^2 + a_{\alpha\beta} dy^\alpha dy^\beta],$$

the hypersurfaces y^n = const. being the geodesic hyperspheres with $P(x)$ as center.

Suppose that the x's and p's, as defined by (67.5), are subjected to any homogeneous contact transformation (59.8) of maximum rank and that $\bar{C}$ is the transform of C. Then $\bar{C}$ is homogeneous of degree one in the $\bar{p}$'s. From the results of §62 it follows that equations (67.4) are transformed into

$$(67.10) \qquad \frac{d\bar{x}^i}{ds} = \frac{\partial \bar{C}}{\partial \bar{p}_i}, \qquad \frac{d\bar{p}_i}{ds} = -\frac{\partial \bar{C}}{\partial \bar{x}^i}.$$

If we define functions $\bar{g}^{ij}$ by

$$(67.11) \qquad \bar{g}^{ij} = \frac{1}{2} \frac{\partial^2 (\bar{C}^2)}{\partial \bar{p}_i \partial \bar{p}_j},$$

then $\bar{g}^{ij}$ is homogeneous of degree zero in the $\bar{p}$'s and it follows from the second set of equations (62.6) that

$$(67.12) \qquad \bar{g}^{ij} \bar{p}_i \bar{p}_j = \bar{C}^2.$$

When the $\bar{p}$'s are such that $\bar{C} \neq 0$ then $\bar{C} = 1$ and the above equations may be written

$$(67.13) \qquad \frac{d\bar{x}^i}{ds} = \bar{g}^{ij} \bar{p}_j, \qquad \frac{d\bar{p}_i}{ds} = -\frac{1}{2} \frac{\partial \bar{g}^{jk}}{\partial \bar{x}^i} \bar{p}_j \bar{p}_k,$$

since we have in consequence of (67.11) that

$$\frac{\partial \bar{g}^{jk}}{\partial \bar{p}_i} \bar{p}_j \bar{p}_k = \frac{1}{2} \frac{\partial^3 \bar{C}^2}{\partial \bar{p}_i \partial \bar{p}_j \partial \bar{p}_k} \bar{p}_j \bar{p}_k = 0.$$

The first of (67.10) become in consequence of (65.1)

$$\dot{\bar{x}}^i \equiv \frac{d\bar{x}^i}{ds} = \frac{\partial \bar{H}}{\partial \bar{p}_i}.$$

As in §65 these equations can be solved for $\bar{p}_i$ as functions of the

$\bar{x}$'s and $\dot{\bar{x}}$'s, linear homogeneous in the latter; we denote by $\overline{\mathcal{3C}}$ and $\mathcal{C}$ the functions resulting from the substitution in $\bar{H}$ and $\bar{C}$ of these expressions for $\bar{p}_i$. Since $\overline{\mathcal{C}}^2$ is homogeneous of second degree in the $\dot{\bar{x}}$'s, we have from (65.9) and Euler's theorem

$$(67.14) \qquad \bar{p}_i = \frac{\partial \overline{\mathcal{3C}}}{\partial \dot{\bar{x}}^i} = \frac{1}{2} \frac{\partial \overline{\mathcal{C}}^2}{\partial \dot{\bar{x}}^i} = \frac{1}{2} \frac{\partial^2 \overline{\mathcal{C}}^2}{\partial \dot{\bar{x}}^i \partial \dot{\bar{x}}^j} \dot{\bar{x}}^j = \bar{g}_{ij} \dot{\bar{x}}^j,$$

where $\bar{g}_{ij}$ are defined by

$$\bar{g}_{ij} \equiv \frac{1}{2} \frac{\partial^2 \overline{\mathcal{C}}^2}{\partial \dot{\bar{x}}^i \partial \dot{\bar{x}}^j}.$$

The $\bar{g}_{ij}$ as thus defined are functions of $\bar{x}^i$ and $\dot{\bar{x}}^i$, homogeneous of degree zero in the latter. From (67.13) and (67.14) it follows that $\bar{g}^{ij} \bar{g}_{jk} = \delta^i_k$. From this result, (67.13) and (67.12) we have $\overline{\mathcal{C}}^2 = \bar{g}_{ij} \dot{\bar{x}}^i \dot{\bar{x}}^j$. In consequence of (65.13) the second set of (67.10) become

$$\bar{g}_{ij} \frac{d^2 \bar{x}^j}{ds^2} + [jk, i] \frac{d\bar{x}^j}{ds} \frac{d\bar{x}^k}{ds} = 0,$$

where $[jk, i]$ are defined by (47.11) in terms of $\bar{g}_{ij}$, since

$$\frac{\partial \bar{g}_{ij}}{\partial \dot{\bar{x}}^k} \dot{\bar{x}}^j = \frac{\partial \bar{g}_{ik}}{\partial \dot{\bar{x}}^j} \dot{\bar{x}}^j = 0.$$

Thus the trajectories defined by (67.10) are the geodesics of the space with the generalized Riemannian metric $d\bar{s}^2 = \bar{g}_{ij} d\bar{x}^i d\bar{x}^j$ into which the given Riemannian space has been transformed.*

68. Application to dynamics. Consider a conservative dynamical system for which neither the constraints nor the potential energy V involve the time t. Suppose that there are n independent variables x^i; then the kinetic energy is given by

$$T = \frac{1}{2} g_{ij} \dot{x}^i \dot{x}^j,$$

where the g's are functions of the x's, and involve the masses of the system, and the Lagrangian function is

---

* Spaces with a generalized Riemannian metric have been studied by *Finsler*, 1918, 2, and by *Berwald*, 1925, 6, vol. 34, p. 213.

$$L = \frac{1}{2}g_{ij}\dot{x}^i\dot{x}^j - V.$$

Substituting in the equations of Lagrange

$$\frac{d}{dt}\left(\frac{\partial L}{\partial \dot{x}^i}\right) - \frac{\partial L}{\partial x^i} = 0,$$

we find that the resulting equations may be written

(68.1) $$g_{ij}\ddot{x}^i + [jk, i]\dot{x}^j\dot{x}^k + \frac{\partial V}{\partial x^i} = 0,$$

in consequence of (47.11). These equations admit the first integral

(68.2) $$\frac{1}{2}g_{ij}\dot{x}^i\dot{x}^j + V = E,$$

where E is the energy constant.

If we put

(68.3) $$p_i = g_{ij}\dot{x}^j, \qquad \dot{x}^i = g^{ij}p_j,$$

we have from equation (68.2)

$$g^{ij}p_ip_j = 2(E - V).$$

Hence if we take for the characteristic function of a group G_1 of homogeneous transformations the quantity*

(68.4) $$C = \sqrt{\frac{1}{2(E - V)}g^{ij}p_ip_j} = 1,$$

the equations of the transformation are

(68.5) $$\frac{dx^i}{ds} = \frac{1}{2(E - V)}g^{ij}p_j, \qquad \frac{dp_i}{ds} = -\frac{1}{4}\frac{\partial}{\partial x^i}\left(\frac{1}{E - V}g^{jk}\right)p_jp_k,$$

where the parameter s is given by (cf. §67)

(68.6) $$ds^2 = 2(E - V)g_{ij}dx^idx^j.$$

The Hamiltonian function of the dynamical system is given by

---

* This problem for the motion of a particle was discussed by *Lie*, 1889, 2, pp. 145–156; and the more general problem by *Vessiot*, 1906, 1, pp. 266–268.

$$H = p_i \dot{x}^i - L = \frac{1}{2} g^{ij} p_i p_j + V,$$

and consequently the Hamiltonian equations are

$$\frac{dx^i}{dt} = g^{ij} p_j, \qquad \frac{dp_i}{dt} = -\left(\frac{1}{2} \frac{\partial g^{jk}}{\partial x^i} p_j p_k + \frac{\partial V}{\partial x^i} \right).$$

These equations are equivalent to (68.5) with the relation

(68.7) $$\frac{ds}{dt} = 2(E - V)$$

holding along any trajectory. Thus we have the known result that the trajectories of energy E of the given system are in one-to-one correspondence with the geodesics of the space with the fundamental form (68.6).

From the results of §67 it follows that, if φ is any solution of the equation

(68.8) $$g^{ij} \frac{\partial \varphi}{\partial x^i} \frac{\partial \varphi}{\partial x^j} = 2(E - V),$$

the orthogonal trajectories of the hypersurfaces $\varphi = $ const. are geodesics. If $\varphi(x, a^1, \ldots, a^{n-1})$ is a complete integral of this equation, on eliminating the a's from the corresponding equations (67.8), the resulting equation (67.9) defines geodesic hyperspheres with $P(x)$ as center.

Consider the case of a single particle of mass m and write (68.2) in the form

(68.9) $$m \left(\frac{ds_0}{dt} \right)^2 = 2(E - V),$$

where ds_0 is the element of length of the space of the particle. Then from (68.6) we have

(68.10) $$ds^2 = 2m(E - V) ds_0^2,$$

and consequently

$$s = \int \sqrt{2m(E - V)} \, ds_0 = \int mv \, ds_0.$$

Hence s is the action and the dynamical trajectories are the extremals of the integral of action.

From the view-point of optics in accordance with Fermat's principle the trajectories are the same as the paths of light through an isotropic non-homogeneous medium of refractive index equal to $\sqrt{2m(E - V)}$. In accordance with this principle $s = \kappa\tau$, where τ is the time and κ a factor of proportionality. Then the velocity of light is

$$(68.11) \qquad u = \frac{ds_0}{d\tau} = \frac{\kappa}{\sqrt{2m(E - V)}}.$$

Comparing this with (68.9) we have

$$(68.12) \qquad u = \frac{\kappa}{mv}.$$

From (68.10) it follows that the spaces with metric (68.6) and ds_0 are conformal. Consequently in the latter metric the trajectories through a point $P(x)$ are orthogonal to the hypersurfaces (67.9). If we take $\kappa = E$, the equation of these hypersurfaces is

$$(68.13) \qquad f(\bar{x}; x) - E\tau = 0.$$

From (68.7) it follows that along any trajectory the relation between the time t of the particle and τ of the light are given by

$$(68.14) \qquad \frac{d\tau}{dt} = \frac{2(E - V)}{E}$$

which is in conformity with (68.12).

Exercises

1. Show that

$$\bar{x} = x - \frac{a}{\sqrt{1 + p^2}}, \qquad \bar{y} = y + \frac{ap}{\sqrt{1 + p^2}}, \qquad \bar{p} = p$$

define a G_1 of restricted contact transformations, each of which sends a point into a circle, and that the characteristic function is $\sqrt{1 + p^2}$.

2. Find the contact transformation in 3-space determined by

$$\bar{z} + z + x\bar{x} + y\bar{y} = 0,$$

and show that it transforms a point into its polar plane with respect to the paraboloid $x^2 + y^2 + 2z = 0$.

3. If equation (59.4) with $p_1 = -1$ is written in the form

$$d\bar{x}^1 - P_\alpha d\bar{x}^\alpha = \rho(dx^1 - p_\alpha dx^\alpha) \qquad (\alpha = 2, \cdots, n),$$

equations (63.16) are in this case

$$\delta x^1 = \left(p_\alpha \frac{\partial W}{\partial p_\alpha} - W\right)\delta t, \qquad \delta x^\alpha = \frac{\partial W}{\partial p_\alpha}\delta t, \qquad \delta p_\alpha = -\left(\frac{\partial W}{\partial x^\alpha} + p_\alpha \frac{\partial W}{\partial x^1}\right)\delta t,$$

and the symbol of the G_1 generated by this infinitesimal transformation is
$$[W, f] - W\frac{\partial f}{\partial x^1}.$$

Lie-Engel, 1888, 1, vol. 2, p. 255.

4. If an infinitesimal non-homogeneous contact transformation with the characteristic function W undergoes a general contact transformation (63.1), the characteristic function of the resulting transformation is equal to ρW.

Lie-Engel, 1888, 1, vol. 2, p. 277.

5. If $A_1 f$ and $A_2 f$ are the symbols of infinitesimal non-homogeneous contact transformations with the characteristic functions W_1 and W_2, then $(A_1, A_2)f$ is the symbol of the transformation with the characteristic function $[W_1, W_2]$.

6. The symbol of the G_1 of restricted non-homogeneous contact transformations with the characteristic function $W(x; p)$ is

$$(W, f) + \left(p_\alpha \frac{\partial W}{\partial p_\alpha} - W\right)\frac{\partial f}{\partial x^1}.$$

7. If $A_1 f$ and $A_2 f$ are the symbols of restricted contact transformations with the characteristic functions W_1 and W_2, then $(A_1, A_2)f$ is the symbol of the restricted transformation with the characteristic function (W_1, W_2).

8. A necessary and sufficient condition that a non-homogeneous contact transformation leave $dx^1 - p_\alpha dx^\alpha$ $(\alpha = 2, \cdots, n)$ invariant is that it be a restricted transformation.

9. If $n - 1$ independent functions φ^α of $x^2, \ldots, x^n, p_2, \ldots, p_n$ satisfy $(\varphi^\alpha, \varphi^\beta) = 0$, a solution φ_1 of the system (64.3) is given by a quadrature.

Lie-Engel, 1888, vol. 2, pp. 127, 128.

10. Show that (61.12) in which $i, j = 2, \cdots, n$ hold for a restricted transformation (64.4); and that the former are equivalent to the requirement that $dp_\alpha \delta x^\alpha - \delta p_\alpha dx^\alpha$ be invariant under such a transformation, where dp_α, dx^α and δp_α, δx^α are arbitrary.

11. Show that the numerical value of the jacobian of a restricted transformation is one, and consequently the volume of a region in space of coordinates $x^1, \ldots, x^n, p_1, \ldots, p_n$ is invariant under such a transformation.

12. Show that the integral
$$J_1 = \int dp_\alpha dx^\alpha \qquad\qquad (\alpha = 2, \cdots, n)$$

over any 2-dimensional region of the space of coordinates x^i and p_i, called the x, p space, is an invariant under a restricted transformation.

Born, 1925, 8, p. 40.

69. Function groups.

We have obtained in §64 the conditions which the functions φ^i and ψ^i must satisfy in the case of a restricted non-homogeneous contact transformation. The inverse problem of finding sets of functions satisfying these conditions raises questions concerning sets of functions in the $2n$ quantities x^i and p_i.[*]

Suppose that we have s such independent functions say $F_1, \ldots, F_s$, and we form the parentheses (F_α, F_β). If any of the resulting functions is not a function of the given ones, we adjoin it to the original set and thus get a set of s_1 independent functions. We proceed in this manner with the new set and get one with s_2 independent functions and so on. Since there are at most $2n$ independent functions of x^i and p_i, we obtain finally $r(\leq 2n)$ independent functions of which all the others are functions, where

$$s \leq s_1 \leq s_2 \cdots r \leq 2n.$$

If Φ_1 and Φ_2 are any two functions of these r functions, it follows from the identity

(69.1) $$(\Phi_1, \Phi_2) = \frac{\partial \Phi_1}{\partial F_\alpha} \frac{\partial \Phi_2}{\partial F_\beta} (F_\alpha, F_\beta)$$

that (Φ_1, Φ_2) is a function of the set. Hence starting with a set of functions, we get by the above process a set each of which is a function of r independent ones and such that (F_α, F_β) for any two functions of the set is a function of these r independent ones. In consequence of (69.1) any function of functions of the set may be added to it and it will continue to possess the above property. Such a set of functions is said to constitute a *function group* of *rank r*, r being the rank of the jacobian of the functions of the group, and the r independent functions are a *basis* of the group. In consequence of (69.1) we have that any r independent functions of the group are a basis, and as we change these functions we change the basis but not the group.

If a sub-set of functions of a function group form a function group, we say that the latter is a *sub-group* of the given group. Suppose that two groups of ranks r and s have p functions in common; in this case it must be possible to choose the bases F_α and G_β so that $F_\sigma = G_\sigma (\sigma = 1, \cdots, p)$. Then we must have

$$(F_\sigma, G_\tau) = \varphi_{\sigma\tau}(F_1, \cdots, F_r) = \psi_{\sigma\tau}(G_1, \cdots, G_s)(\sigma, \tau = 1, \cdots, p).$$

But $\varphi_{\sigma\tau}$ and $\psi_{\sigma\tau}$ must be the same functions of $F_1, \ldots, F_p$, otherwise there would be relations between $F_{p+1}, \ldots, F_r$ and $G_{p+1}, \ldots, G_s$. Hence we have:

---

[*] Cf. *Lie-Engel*, 1888, 1, vol. 2, pp. 178–210. Also *Vivanti*, 1904, 1, pp. 255–264.

[69.1] *If two function groups have functions in common, the latter form a sub-group of each group.*

Consider the system of differential equations

$$(69.2) \qquad (F_\alpha, f) = 0,$$

where the F's determine a basis of a function group. From the Jacobi identity (61.15), we have

$$(69.3) \qquad (F_\alpha, (F_\beta, f)) - (F_\beta(F_\alpha, f)) = ((F_\alpha, F_\beta), f).$$

Since (F_α, F_β) is a function of the F's, it follows from (69.1) that the right-hand member of (69.3) vanishes because of (69.2) and consequently equations (69.2) form a complete system. Hence they admit $2n - r$ independent solutions, $f_\sigma (\sigma = 1, \cdots, 2n - r)$. From the Jacobi identity for F_α, f_σ and f_r, it follows that (f_σ, f_r) is a solution of (69.2) and consequently (f_σ, f_r) is a function of the f's. Accordingly the f's are the basis of a function group of rank $2n - r$. Each f_σ is in involution (§61), or *commutative*, with all the F's. Hence we have:

[69.2] *A function group of rank r determines another function group, which is of rank $2n - r$; its functions are in involution with those of the given group.*

Two such groups are said to be *reciprocal*.

If any function of a group is in involution with all the members of the group it is said to be *singular*. From the definition of sub-group it follows that the independent singular functions of a group are the basis of a sub-group of the given group. Since a singular function is a member of the reciprocal group, we have in consequence of theorem [69.1]:

[69.3] *If a function group has a sub-group of singular functions, it is a sub-group of the reciprocal group also.*

If all of the functions of a group are singular, the group is said to be *commutative*. In consequence of theorems [69.2] and [69.3] we have that in this case $r \leq n$.

In order to determine the number of independent singular functions of a group, we remark that for a function Φ of the basis functions $F_1, \ldots, F_r$ we have

$$(69.4) \quad (F_\alpha, \Phi) = \frac{\partial F_\alpha}{\partial p_i} \frac{\partial \Phi}{\partial F_\beta} \frac{\partial F_\beta}{\partial x^i} - \frac{\partial F_\alpha}{\partial x^i} \frac{\partial \Phi}{\partial F_\beta} \frac{\partial F_\beta}{\partial p_i} = (F_\alpha, F_\beta)\frac{\partial \Phi}{\partial F_\beta},$$

and consequently, if Φ is to be a singular function, it must be a solution of the system of r equations

$$(69.5) \qquad \varphi_\alpha \Phi \equiv \varphi_{\alpha\beta} \frac{\partial \Phi}{\partial F_\beta} = 0,$$

where $\varphi_{\alpha\beta}$ are functions of the F's defined by

$$(69.6) \qquad\qquad \varphi_{\alpha\beta} = (F_\alpha, F_\beta).$$

The commutator of the system (69.5) is equal, in consequence of (69.4), to

$$(F_\alpha, (F_\beta, \Phi)) - (F_\beta, (F_\alpha, \Phi)) = ((F_\alpha, F_\beta), \Phi)$$

$$= (\varphi_{\alpha\beta}, F_\gamma)\frac{\partial \Phi}{\partial F_\gamma} = -\frac{\partial \Phi}{\partial F_\gamma}(F_\gamma, F_\delta)\frac{\partial \varphi_{\alpha\beta}}{\partial F_\delta} = \frac{\partial \varphi_{\alpha\beta}}{\partial F_\delta}\varphi_\delta\Phi.$$

Hence equations (69.5) form a complete system and we have:

[69.4]   *If the rank of the matrix $\|(F_\alpha, F_\beta)\|$ is $r - m$, there are m independent singular functions in the group, and they determine a subgroup of rank m.*

Since $\|(F_\alpha, F_\beta)\|$ is skew-symmetric, its rank is less than r when r is odd, and consequently a function group of odd rank has at least one singular function.

In §64 it was remarked that the parenthesis (F_α, F_β) is unaltered by a restricted contact transformation, and consequently the rank and number of independent singular functions are invariant under a restricted non-homogeneous transformation.

We shall show that a basis of a non-commutative function group can be chosen so that (F_α, F_β) have the values 1 or 0. Since the group is not commutative, there is one function which is not singular, say F_1. If $\Phi(F_1, \ldots, F_r)$ is a solution of

$$\varphi_{1\alpha}\frac{\partial \Phi}{\partial F_\alpha} = -1,$$

where $\varphi_{1\alpha}$ are defined by (69.6) and are not all zero since F_1 is not singular, it is not a function of F_1 alone since $\varphi_{11} = 0$. Denoting the independent functions F_1 and Φ by φ^1 and ψ_1 respectively, we have $(\psi_1, \varphi^1) = 1$. We take φ^1, ψ_1 and $r - 2$ other independent functions as the basis of the group and consider the two equations

$$(\varphi^1, \Phi) = \frac{\partial \Phi}{\partial \psi_1} + (\varphi^1, F_\sigma)\frac{\partial \Phi}{\partial F_\sigma} = 0,$$

(69.7)

$$(\psi_1, \Phi) = -\frac{\partial \Phi}{\partial \varphi^1} + (\psi_1, F_\sigma)\frac{\partial \Phi}{\partial F_\sigma} = 0 \quad (\sigma = 3, \cdots, r).$$

These equations are obviously independent, and since

$$(\varphi^1, (\psi_1, \Phi)) - (\psi_1, (\varphi^1, \Phi)) = ((\varphi^1, \psi_1), \Phi) = (1, \Phi) = 0,$$

they form a complete system. We denote by $\bar{F}_3, \ldots, \bar{F}_r$ any $r - 2$ independent solutions, functions of $\varphi^1, \psi_1, F_3, \ldots, F_r$. From the Jacobi identity it follows that $(\bar{F}_\sigma, \bar{F}_\tau)$ is a solution of (69.7), and consequently $\bar{F}_\sigma$ for $\sigma = 3, \cdots, r$ is a sub-group of the given group. If it is commutative, we have

$$(\psi_1, \varphi^1) = 1, \qquad (\varphi^1, \bar{F}_\sigma) = (\psi_1, \bar{F}_\sigma) = (\bar{F}_\sigma, \bar{F}_\tau) = 0$$
$$(\sigma, \tau = 3, \cdots, r).$$

If the sub-group is not commutative, we proceed in the same way with the sub-group as above, and continue the process until we use up all the functions or arrive at a commutative sub-group. Hence we have:[*]

[69.5] *For a non-commutative function group of rank r a basis exists, say* $\varphi^1, \ldots, \varphi^{m+q}, \psi_1, \cdots, \psi_m (2m + q = r, q < r)$ *such that*

(69.8) $\quad (\varphi^\lambda, \varphi^\mu) = 0, \qquad (\psi_\alpha, \psi_\beta) = 0, \qquad (\psi_\alpha, \varphi^\lambda) = \delta_\alpha^\lambda$
$$\begin{pmatrix} \lambda, \mu = 1, \cdots, m + q; \\ \alpha, \beta = 1, \cdots, m \end{pmatrix}.$$

If the group is commutative, we have only the functions φ^λ for $\lambda = 1, \cdots, r$ and these are the original functions F^λ. When the basis is chosen so that equations (69.8) are satisfied, the function group is said to be in *canonical form*.

When the basis of a non-commutative function group is such that (69.8) holds with $q > 0$, the functions $\varphi^{m+1}, \ldots, \varphi^{m+q}$ are functions of the reciprocal group. Furthermore, if any one of these functions, say φ^{m+1}, is omitted from the functions of the given group, the remaining ones determine a sub-group of the latter. Then φ^{m+1} is a function of the reciprocal group of this sub-group and this reciprocal group must contain a function G such that $(\varphi^{m+1}, G) \neq 0$; otherwise

---

[*] *Lie-Engel*, 1888, 1, vol. 2, pp. 199, 200.

the reciprocal group of the sub-group would be the reciprocal group of the given group. Hence the basis of the reciprocal group of the sub-group can be chosen so that there is a function ψ_{m+1} such that $(\psi_{m+1}, \varphi^{m+1}) = 1$ and ψ_{m+1} is independent of the functions of the original group. If we adjoin ψ_{m+1} to the functions of the given group, we have a group of rank $r + 1$ for which (69.8) hold when α and β take the values $1, \cdots, m + 1$, and the given group is a sub-group of it. Continuing this process q times we get a group of order $r + q(=2(r - m))$ of which the original group is a sub-group and for which (69.8) holds when λ, μ, α, β take the values $1, \cdots, r - m$. Because of the last set of equations (69.8) in this case it follows that this group has no singular functions. Its reciprocal group is of the same kind and its basis may be chosen so that for it we have equations (69.8) holding when λ, μ, α, β take the values $r - m + 1, \cdots, n$. The functions of these two reciprocal groups satisfy the conditions

$$(69.9) \qquad (\varphi^i, \varphi^j) = 0, \qquad (\psi_i, \psi_j) = 0, \qquad (\psi_j, \varphi^i) = \delta_j^i,$$
$$(i, j = 1, \cdots, n)$$

and are the functions of a group of rank $2n$ of which the original group is a sub-group. If $q = 0$ in (69.8), we adjoin the reciprocal group and get (69.9). Hence we have:

[69.6] *A non-commutative function group of rank r is a sub-group of a group of rank $2n$ whose basis can be chosen so that the functions of the basis satisfy* (69.9) *identically.*

In consequence of theorem [64.1] and equations (64.7) we have that these functions determine a restricted non-homogeneous contact transformation.

Suppose that we have two function groups of the same rank and with the same number of singular functions, and denote their respective bases by $F_\alpha(x, p)$ and $F'_\alpha(x', p')$. Because of theorem [69.6] we have that for the former a basis can be chosen so that we have a group of rank $2n$ for which the functions of the basis satisfy the equations (69.9) identically. In like manner we obtain a similar group of rank $2n$, whose basis functions $\varphi'^i(x', p'), \psi'_i(x', p')$ satisfy these conditions. In accordance with the results of §64 functions $\varphi^0(x, p)$ and $\varphi'^0(x', p')$ can be chosen so that the equations

$$\bar{x}^0 = x^0 + \varphi^0(x, p), \qquad \bar{x}^i = \varphi^i(x, p), \qquad \bar{p}_i = \psi_i(x, p);$$
$$\bar{x}'^0 = x'^0 + \varphi'^0(x', p'), \qquad \bar{x}'^i = \varphi'^i(x', p'), \qquad \bar{p}'_i = \psi'_i(x', p')$$

define two restricted non-homogeneous contact transformations. Then the equations

$$x'^0 + \varphi'^0 = x^0 + \varphi^0, \qquad \varphi'^i = \varphi^i, \qquad \psi'_i = \psi_i$$

define a contact transformation which transforms the one group into the other. Hence because of the remarks in the paragraph following theorem [69.4] we have:*

[69.7]   *A necessary and sufficient condition that two function groups be equivalent under a restricted non-homogeneous contact transformation is that they have the same rank and the same number of independent singular functions.*

70. Homogeneous function groups.   Let $H_1, \ldots, H_r$ be the basis of a function group of rank r and assume that H_α is homogeneous of degree m_α in the p's for $\alpha = 1, \cdots, r$; in this case the group is said to be *homogeneous*.   If F is any function of the H's, it follows from Euler's theorem that

$$p_i \frac{\partial F}{\partial p_i} = p_i \frac{\partial F}{\partial H_\alpha} \frac{\partial H_\alpha}{\partial p_i} = \sum_\alpha m_\alpha H_\alpha \frac{\partial F}{\partial H_\alpha}.$$

Since the right-hand member is a function of the H's, the quantity $p_i \dfrac{\partial F}{\partial p_i}$ is a function of the group, or zero.   We shall show conversely that, if this condition is satisfied by every F, all the functions of the group are homogeneous.   By hypothesis

$$(70.1) \qquad\qquad p_i \frac{\partial H_\alpha}{\partial p_i} = \Omega_\alpha(H_1, \cdots, H_r) \qquad (\alpha = 1, \cdots, r).$$

If $\Omega_\alpha = 0$ for every α, then it follows that all the functions of the group are homogeneous of degree zero.   Since then (H_α, H_β) is of degree -1, it cannot be a function of the H's and consequently the group is commutative in this case.

If all of the Ω's are not zero, it follows from the identity

$$p_i \frac{\partial F}{\partial p_i} = p_i \frac{\partial F}{\partial H_\alpha} \frac{\partial H_\alpha}{\partial p_i} = \Omega_\alpha \frac{\partial F}{\partial H_\alpha}$$

---

* *Lie-Engel*, 1888, 1, vol. 2, p. 204.

that, if $\bar{H}_1, \ldots, \bar{H}_{r-1}$, are $r - 1$ independent solutions of $\Omega_\alpha \dfrac{\partial F}{\partial H_\alpha} = 0$, they are homogeneous of degree zero. If we denote by $\bar{H}_r$ a solution of $\Omega_\alpha \dfrac{\partial F}{\partial H_\alpha} = F$, it belongs to the group and is homogeneous of degree one. Consequently we have:*

[70.1] *A necessary and sufficient condition that a function group be homogeneous is that for every function F of the group the quantity $p_i \dfrac{\partial F}{\partial p_i}$ is a function of the group, or zero;*

and also:

[70.2] *For a homogeneous function group there exists a basis all of whose functions but one are of degree zero in the p's and the remaining one is of first degree, or all are of degree zero; in the latter case the group is commutative.*

We shall prove the theorem:†

[70.3] *The reciprocal group of a homogeneous function group is homogeneous.*

For, if F is a function of the reciprocal group we have

$$(70.2) \qquad \frac{\partial H_\alpha}{\partial p_i}\frac{\partial F}{\partial x^i} - \frac{\partial H_\alpha}{\partial x^i}\frac{\partial F}{\partial p_i} = 0 \qquad (\alpha = 1, \cdots, r).$$

Operating on these equations with $p_j \dfrac{\partial}{\partial p_j}$ and making use of (70.2) and the identity

$$p_j\frac{\partial^2 H_\alpha}{\partial p_j \partial p_i} = (m_\alpha - 1)\frac{\partial H_\alpha}{\partial p_i},$$

we find that $p_j \dfrac{\partial F}{\partial p_j}$ is a solution of (70.2) and consequently a function of the reciprocal group. Since this holds for every F, we have, as shown above, that the reciprocal group is homogeneous.

Theorem [69.5] evidently applies to non-commutative homogeneous groups but the functions φ^λ and ψ_α can be given more

---

* *Lie-Engel*, 1888, 1, vol. 2, p. 215.

† *Lie-Engel*, 1888, 1, vol. 2, p. 217.

specialized form. If in accordance with theorem [70.2] we take as basis $N_1, \ldots, N_{r-1}, H$, such that H is of degree one and the N's of degree zero, one of the latter, say N_1, is not singular; otherwise the group would be commutative. We have

$$(H, N_\alpha) = \theta_\alpha(N), \quad H(N_\alpha, N_\beta) = \theta_{\alpha\beta}(N) \quad (\alpha, \beta = 1, \cdots, r-1),$$

since each member of these equations is homogeneous of degree zero. The function $\psi = H\Omega(N)$ satisfies the condition $(\psi, N_1) = 1$, if Ω is a solution of the equation

$$H\theta_{\alpha 1}\frac{\partial \Omega(N)}{\partial N_\alpha} + \theta_1\Omega(N) = 1.$$

Since all of the functions $\theta_{\alpha 1}$, θ_1 are not zero, because N_1 is not singular, there exists such a function ψ_1, which is of degree one, non-singular and may be used to replace H in the basis. The function $\varphi^1(N)$ satisfies the condition $(\psi_1, \varphi^1) = 1$, if φ^1 is a solution of

(70.3) $$\frac{\partial \varphi^1}{\partial N_\alpha}(\psi_1, N_\alpha) = 1.$$

Since not every (ψ_1, N_α) is zero, there exists such a function. Then as in §69 we may take as basis $\varphi^1, \psi_1, \bar{H}_3, \ldots, \bar{H}_r$ such that $(\varphi^1, \bar{H}_\sigma) = (\psi_1, \bar{H}_\sigma) = 0$ for $\sigma = 3, \cdots, r$ and $\bar{H}_\sigma$ is the basis of a sub-group. The basis of the latter may be chosen in accordance with theorem [70.2]. If this sub-group is not commutative, we may proceed as above and continue until we arrive at a commutative sub-group. Thus we arrive at the basis

$$\psi_1, \cdots, \psi_m, \quad \varphi^1, \cdots, \varphi^m, \quad H_1, \cdots, H_q \quad (2m + q = r),$$

in which the ψ's are of degree one, the φ's of degree zero, and the H's are the basis of a commutative sub-group of the given group chosen in accordance with theorem [70.2]. If they are of degree zero, we denote them by $\varphi^{m+1}, \ldots, \varphi^{m+q}$. If H_1 is of degree one and the others of degree zero, we take as basis $H_1, H_1H_2, \ldots, H_1H_q$ and denote them by $\psi_{m+1}, \ldots, \psi_{m+q}$, all of degree one. Hence we have:

[70.4] *For a non-commutative homogeneous group the basis functions can be chosen so that they satisfy the conditions*

(70.4) $$(\varphi^\lambda, \varphi^\mu) = (\psi_\alpha, \psi_\beta) = 0, \quad (\psi_\alpha, \varphi^\lambda) = \delta_\alpha^\lambda,$$

where $\lambda, \mu = 1, \cdots, m + q;\ \alpha, \beta = 1, \cdots, m,$ *or* $\lambda, \mu = 1, \cdots, m;$ $\alpha, \beta = 1, \cdots, m + q,$ *where* $2m + q = r,$ *the* φ*'s being of degree zero and the* ψ*'s of degree one.*

If the group is commutative and all the functions are of degree zero, we have $(\varphi^\alpha, \varphi^\beta) = 0,\ (\alpha, \beta = 1, \cdots, r)$; if all are not of degree zero, they can be chosen to be of degree one and we have $(\psi_\alpha, \psi_\beta) = 0.$ Theorem [70.4] defines the *canonical form* of the basis.

We return to the case of the non-commutative group whose basis $\psi_1, \ldots, \psi_m, \varphi^1, \ldots, \varphi^{m+q}$ satisfies (70.4). If we take the sub-group obtained by omitting φ^{m+1}, as in §69, and consider the reciprocal group of this sub-group, φ^{m+1} is a non-singular function of it and a function ψ_{m+1} of this reciprocal group exists which is of the first degree and such that $(\psi_{m+1}, \varphi^{m+1}) = 1$; it is determined as ψ_1 was with the aid of (70.3). Moreover ψ_{m+1} is not a function of the original group, and if it is added to the basis of this group, we obtain a group of rank $r + 1$ satisfying (70.4) for

$$\alpha, \beta = 1, \cdots, m + 1$$

and having the given group as a sub-group. Continuing this process q times we obtain a group of rank $2(m + q)$, whose basis satisfies (70.4) for $\lambda, \mu, \alpha, \beta = 1, \cdots, m + q$, containing the given group as a sub-group and having no singular functions. If we add to this group its reciprocal group with basis in canonical form we have a group of rank $2n$, whose basis satisfies the identities

$$(70.5) \quad (\varphi^i, \varphi^j) = (\psi_i, \psi_j) = 0, \qquad (\psi_i, \varphi^j) = \delta_i^j \qquad (i, j = 1, \cdots, n).$$

If we proceed with a group having the basis (70.4) with $\lambda, \mu = 1, \cdots, m;\ \alpha, \beta = 1, \cdots, m + q$, we arrive at the same result. Hence we have:*

[70.5] *A non-commutative homogeneous function group of rank* r *is a sub-group of a homogeneous group of rank* $2n$, *whose basis can be chosen so that equations (70.5) are satisfied identically, the* φ*'s being of degree zero and the* ψ*'s of degree one.*

If a homogeneous group of rank r is subjected to a homogeneous contact transformation, the rank is unchanged and so also is the number of independent singular functions and of independent

---

* *Lie-Engel*, 1888, 1, vol. 2, p. 225.

singular functions of degree zero as follows from theorem [61.5]. Suppose we have two homogeneous groups of the same rank and with the same number of independent singular functions and the same number of independent singular functions of degree zero, and that the bases of corresponding groups of rank $2n$ are taken in the form (70.5) and similar equations in $\varphi'^i(x', p')$ and $\psi'_i(x', p')$. The equations $\varphi'^i = \varphi^i$ and $\psi'_i = \psi_i$ determine a homogeneous contact transformation transforming either group into the other. Hence we have:*

[70.6] *A necessary and sufficient condition that two homogeneous function groups be equivalent under a homogeneous contact transformation is that they have the same rank, the same number of singular functions and the same number of singular functions of zero degree.*

Exercises

1. Show that, if ξ_a^i are the vectors of a group G_r of point transformations, the functions $\xi_a^i p_i$ constitute a homogeneous function group; determine the number of independent singular functions in the group.

2. A necessary and sufficient condition that a function group have m independent singular functions is that there be m independent functional relations between the functions of the group and of its reciprocal group.

3. If $F_\alpha(x; p)$ and $\bar{F}_\alpha(\bar{x}; \bar{p})$ are the bases of two function groups, there exists a restricted non-homogeneous transformation which sends $F_1, \ldots, F_r$ into $\bar{F}_1, \ldots, \bar{F}_r$ respectively, when and only when $(\bar{F}_\alpha, \bar{F}_\beta)$ is the same function of the $\bar{F}$'s as (F_α, F_β) is of the F's.

Lie-Engel, 1888, 1, vol. 2, p. 209.

4. Show that the constants s, s_1, $\ldots$ for a set of functions (cf. §69) are invariant under a restricted non-homogeneous contact transformation; these and the number of independent singular functions of the set are called its *numerical invariants*.

5. A necessary and sufficient condition that functions $F_1(x; p), \ldots, F_r(x; p)$ be equivalent respectively to $\bar{F}_1(\bar{x}; \bar{p}), \ldots, \bar{F}_r(\bar{x}; \bar{p})$ under a restricted non-homogeneous contact transformation is that corresponding numerical invariants (Ex. 4) be equal, and that all dependent functions and parentheses of Poisson of the second set be expressible in terms of members of the set in the same form as for the respective functions of the first set.

6. If $F_\alpha(x; p)$ constitute the basis of a function group of rank r, the functions $\varphi_{\alpha\beta}$ defined by (69.6) satisfy identically the equations

$$\varphi_{\alpha\beta} + \varphi_{\beta\alpha} = 0$$

$$\varphi_{\alpha\beta}\frac{\partial \varphi_{\gamma\delta}}{\partial F_\alpha} + \varphi_{\alpha\gamma}\frac{\partial \varphi_{\delta\beta}}{\partial F_\alpha} + \varphi_{\alpha\delta}\frac{\partial \varphi_{\beta\gamma}}{\partial F_\alpha} = 0 \quad (\alpha, \beta, \gamma, \delta = 1, \cdots, r),$$

the second set being a consequence of (61.15).

---

* *Lie-Engel*, 1888, 1, vol. 2, p. 226.

7. If $\varphi_{\alpha\beta}$ are given functions of r quantities F_α such that the equations of Ex. 6 are satisfied, there exists a number n such that the functions F_α are expressible in $2n$ variables $x^1, \ldots, x^n; p_1, \ldots, p_n$, they constitute a function group and satisfy (69.6); their determination requires the solution of a system of linear partial differential equations of the first order.

Lie-Engel, 1888, 1, vol. 2, pp. 235–242.

8. Show that the functions $\varphi_{\alpha\beta} = c_\alpha{}^\gamma F_\gamma$, where the c's are constants satisfying (7.3) and (7.4), meet the conditions of Exs. 6, 7 and thus define a function group.

9. Show by means of the results of §§29, 39, that, if $C_1, \ldots, C_r$ are the characteristic functions of a G_r of homogeneous contact transformations and

$$\bar{x}^i = \varphi^i(x; p; a), \qquad \bar{p}_i = \psi_i(x; p; a)$$

are the finite equations of the group, then

$$C_a(\bar{x}; \bar{p}) = w_a^b(a) C_b(x; p) \qquad (a, b = 1, \cdots, r);$$

show also that these equations determine a linear homogeneous group Γ_r with the C's as variables and the a's as parameters, and that the symbols of the group are $c_{ab}^e C_e \dfrac{\partial f}{\partial C_b}$.

Lie-Engel, 1888, 1, vol. 2, p. 333.

10. The number of singular functions of a homogeneous function group is determined by the rank of the matrix $\left\|(H^\alpha, H^\beta)\right\|$; the singular functions of degree zero must satisfy

$$(H^\alpha, H^\beta)\frac{\partial f}{\partial H^\beta} = 0, \qquad H^\beta\frac{\partial f}{\partial H^\beta} = 0.$$

BIBLIOGRAPHY

1837. 1. *Jacobi, C. G. J.:* Note sur l'intégration des equations différentielles de la dynamique. Comptes Rendus, vol. 5, pp. 61–67.

1854. 1. *Riemann, B.:* Ueber die Hypothesen, welche der Geometrie zu Grunde liegen. Gesammelte Werke, 1876, pp. 254–269. Also an edition edited by H. Weyl, Springer, Berlin, 1919.

1885. 1. *Lie, S.:* Allgemeine Untersuchungen über Differentialgleichungen, die eine continuirliche, endliche Gruppe gestatten. Math. Annalen, vol. 25, pp. 71–151.

1886. 1. *Lie, S.:* Untersuchungen über Transformationsgruppen. Archiv för Math. og Naturv., vol. 10, pp. 74–128, 353–413.

1888. 1. *Lie, S.* and *Engel, F.:* Theorie der Transformationsgruppen, vols. 1 and 2. Teubner, Leipzig. Reprinted in 1930.

   2. *Maurer, L.:* Ueber allgemeinere Invariantensysteme. Bayer. Akad. d. Wiss., Ber., vol. 18, pp. 103–150.

   3. *Killing, W.:* Die Zusammensetzung der stetigen endlichen Transformationsgruppen. Math. Annalen, vol. 31, pp. 252–290.

1889. 1. *Killing, W.:* Die Zusammensetzung der stetigen endlichen Transformationsgruppen, II. Math. Annalen, vol. 33, pp. 1–48.

   2. *Lie, S.:* Die infinitesimalen Berührungstransformationen der Mechanik. Sächs. Akad. d. Wiss., Ber., vol. 41, pp. 145–156.

   3. *Lie, S.:* Ueber irreducible Berührungstransformationsgruppen. Sächs. Akad. d. Wiss., Ber., vol. 41, pp. 320–327.

1891. 1. *Schur, F.:* Zur Theorie der endlichen Transformationsgruppen. Math. Annalen, vol. 38, pp. 263–286.

   2. *Engel, F.:* Die kanonische Form der Parametergruppen. Sächs. Akad. d. Wiss., Ber., vol. 43, pp. 308–315.

   3. *Lie, S.* and *Scheffers, G.:* Vorlesungen über Differentialgleichungen mit bekannten infinitesimalen Transformationen. Teubner, Leipzig.

   4. *Umlauf, K. A.:* Ueber die Zusammensetzung der endlichen continuierlichen Transformationsgruppen, insbesondre der Gruppen vom Range Null. Inaug. Diss., Leipzig.

1892. 1. *Killing, W.:* Ueber die Grundlagen der Geometrie. Journal für die reine und angew. Math. (Crelle), vol. 109, pp. 121–186.

1893. 1. *Lie, S.* and *Scheffers, G.:* Vorlesungen über continuierliche Gruppen mit geometrischen und anderen Anwendungen. Teubner, Leipzig.

   2. *Lie, S.* and *Engel, F.:* Theorie der Transformationsgruppen, vol. 3. Teubner, Leipzig. Reprinted in 1930.

1894. 1. *Cartan, E.:* Sur la structure des groupes de transformations finis et continus. Thèse, Nony, Paris.

1897. 1. *Bianchi, L.:* Sugli spazii a tre dimensioni che ammettono un gruppo continuo di movimenti. Soc. Italiana delle Scienze, Mem. di mat., ser. 3, vol. 11, pp. 267–352.

2. *Frobenius, G.:* Über die Darstellung der endlichen Gruppen durch lineare substitutionen, I.   Preus. Akad., Sitz. ber., pp. 994–1015.

1899. 1. *Frobenius, G.:* Über die Darstellung der endlichen Gruppen durch lineare Substitutionen, II.   Preus. Akad., Sitz. ber., pp. 482–500.

1902. 1. *Bianchi, L.:* Lezioni di geometria differenziale.   Second edition, vol. 1.   Spoerri, Pisa.

1903. 1. *Fubini, G.:* Sugli spazii che ammettono un gruppo continuo di movimenti.   Annali di Mat., ser. 3, vol. 8, pp. 39–81.

1904. 1. *Vivanti, G.:* Leçons élémentaires sur la théorie des groupes de transformations.   Gauthier-Villars, Paris.

2. *Fubini, G.:* Sugli spazii a quattro dimensioni che ammettono un gruppo continuo di movimenti.   Annali di Mat., ser. 3, vol. 9, pp. 33–90.

1905. 1. *Fine, H. B.:* A college algebra.   Ginn and Company, Boston.

2. *Dickson, L. E.:* Definitions of a group and a field of independent postulates.   Amer. Math. Soc., Trans., vol. 6, pp. 198–204.

3. *Huntington, E. V.:* Note on the definitions of abstract groups and of fields by sets of independent postulates.   Amer. Math. Soc., Trans., vol. 6, pp. 181–197.

4. *Moore, E. H.:* On a definition of abstract groups.   Amer. Math. Soc., Trans., vol. 6, pp. 179–180.

5. *Schur, I.:* Neue Begründung der Theorie der Gruppencharactere.   Preus. Akad., Sitz. ber., pp. 406–432.

1906. 1. *Vessiot, E.:* Sur l'interprétation mécanique des transformations de contact infinitésimals.   Soc. Math. de France, Bull., vol. 34, pp. 230–269.

1907. 1. *Bôcher, M.:* Introduction to higher algebra.   Macmillan, New York.

1909. 1. *Eisenhart, L. P.:* A treatise on the differential geometry of curves and surfaces.   Ginn, Boston.

1911. 1. *Cohen, A.:* An introduction to the Lie theory of one-parameter groups.   Heath, New York.

2. *Burnside, W.:* The theory of groups.   2nd edition.   Cambridge University Press.

1917. 1. *Levi-Civita, T.:* Nozione di parallelismo in una varietà qualunque e consequente specificazione geometrica della curvatura Riemanniana.   Circ. mat. di Palermo, Rendic., vol. 42, pp. 173–205.

1918. 1. *Bianchi, L.:* Lezioni sulla teoria dei gruppi continui finiti di trasformazioni.   Spoerri, Pisa.

2. *Finsler, P.:* Ueber Kurven und Flächen in allgemeinen Raümen.   Inaug. Diss., Göttingen.

1924. 1. *Dickson, L. E.:* Differential equations from the group standpoint, Annals of Math., ser. 2, vol. 25, pp. 287–378.

2. *Cartan, E.:* Les récentes généralizations de la notion d'espace.   Bull. des Sci. Math., vol. 48, pp. 294–320.

1925. 1. *Eisenhart, L. P.:* Linear connections of a space which are determined by simply transitive groups.   Nat. Acad. Sci., Proc., vol. 11, pp. 246–250.

2. *Schreier, O.:* Abstrakte kontinuierliche Gruppen.   Abhandlungen aus dem Math. Seminar der Hamburgischen Universität, vol. 4, pp. 15–32.

3. *Cartan, E.:* La géométrie des espaces de Riemann. Mémorial de Sciences Mathématiques, fasc. 9. Gauthier-Villars, Paris.

4. *Weyl, H.:* Theorie der Darstellung kontinuierlichen halbeinfacher Gruppen durch lineare Transformationen, I. Math. Zeits., vol. 23, pp. 271–309.

5. *Eiesland, J.:* The group of motions of an Einstein space. Amer. Math. Soc., Trans., vol. 27, pp. 213–245.

6. *Berwald, L.:* Über Parallelübertragung in Raümen mit allgemeiner Massbestimmung. Deut. Math. Verein., Jahresber., vol. 34, pp. 213–220.

7. *Bortolotti, E.:* Parallelismo assoluto e vincolato negli S_3 a curvatura costante. R. Ist. Veneto di sci., Atti, vol. 84, pp. 821–858.

8. *Born, M.:* Vorlesungen über Atommechanik, vol. 1. Springer, Berlin.

1926. 1. *Cartan, E.* and *Schouten, J. A.:* On the geometry of the group-manifold of simple and semi-simple groups. Akad. van Wetens., Amsterdam, Proc., vol. 29, pp. 803–815.

2. *Weyl, H.:* Theorie der Darstellung kontinuierlichen halbeinfacher Gruppen durch lineare Transformationen, II. Math. Zeits., vol. 24, pp. 328–395.

3. *Eisenhart, L. P.:* Riemannian geometry. Princeton University Press.

1927. 1. *Eisenhart, L. P.:* Non-Riemannian geometry. Colloquium Publications of the Amer. Math. Soc., vol. 8, New York.

2. *Cartan, E.:* La géométrie des groupes de transformations, Journal de Mathématiques, ser. 9, vol. 6, pp. 1–119.

3. *Eisenhart, L. P.* and *Knebelman, M. S.:* Displacements in a geometry of paths which carry paths into paths. Nat. Acad. Sci., Proc., vol. 13, pp. 38–42.

4. *Fine, H. B.:* The calculus. Macmillan, New York.

5. *Whittaker, E. T.:* Analytical dynamics. Cambridge University Press.

6. *Levi-Civita, T.* and *Amaldi, U.:* Lezioni di meccanica razionale, vol. 2, part 2, pp. 452–469. Zanichelli, Bologna.

1928. 1. *Franklin, P.:* The canonical form of a one parameter group. Annals of Math., ser. 2, vol. 29, pp. 113–122.

1929. 1. *Schouten, J. A.:* Zur Geometrie der kontinuierlichen Transformationsgruppen, Math. Annalen, vol. 102, pp. 244–272.

2. *Einstein, A.:* Zur einheitlichen Feldtheorie. Preus. Akad., Sitz. ber., pp. 2–8.

3. *Michal, A.:* Scalar extensions of an orthogonal ennuple of vectors. Amer. Math. Monthly, vol. 37, pp. 529–533.

4. *Eisenhart, L. P.:* Contact transformations. Annals of Math., ser. 2, vol. 30, pp. 211–249.

1930. 1. *Cartan, E.:* La théorie des groupes finis et continus et l'analysis situs. Mémorial des Sciences Mathématiques, fasc. 42. Gauthier-Villars, Paris.

2. *Mattioli, G. D.:* Sulla determinazione delle varietà riemanniane che ammettono gruppi semplicemente transitivi di movimenti. Accad. naz. dei lincei, Rendic., ser. 6, vol. 11, pp. 369–371.

3. *Knebelman, M. S.:* On groups of motions in related spaces.   Amer. Jour. of Math., vol. 52, pp. 280–282.

4. *Thomas, T. Y.:* On the unified field theory, I.   Nat. Acad. Sci., Proc., vol. 16, pp. 761–776.

1931. 1. *Weyl, H.:* The theory of groups and quantum mechanics.   Translated by H. P. Robertson.   Methuen and Co., London.

2. *Wigner, E.:* Gruppentheorie und ihre Anwendungen auf die Quantenmechanik der Atomspektren.   Vieweg, Braunschweig.

1932. 1. *Engel, F.* and *Faber, K.:* Die Liesche Theorie der partiellen Differentialgleichungen erster Ordnung.   Teubner, Leipzig.

2. *Eisenhart, L. P.:* Intransitive groups of motions.   Nat. Acad. Sci., Proc., vol. 18, pp. 193–202.

3. *Robertson, H. P.:* Groups of motions in spaces admitting absolute parallelism.   Annals of Math., ser. 2, vol. 33, pp. 496–521.

4. *Eisenhart, L. P.:* Equivalent continuous groups.   Annals of Math., ser. 2, vol. 33, pp. 665–670.

INDEX

297